# Key Concepts in Geography

# Key Concepts in Geography

Edited by

Sarah L. Holloway, Stephen P. Rice and Gill Valentine

SAGE Publications
London • Thousand Oaks • New Delhi

Editorial arrangement © Sarah L. Holloway, Stephen P. Rice and Gill Valentine 2003
Chapter 1 © Mike Heffernan 2003
Chapter 2 © Keith Richards 2003
Chapter 3 © Ron Johnston 2003
Chapter 4 © Alison Blunt 2003
Chapter 5 © Nigel Thrift 2003
Chapter 6 © Martin Kent 2003
Chapter 7 © John B. Thornes 2003
Chapter 8 © Peter J. Taylor 2003
Chapter 9 © Noel Castree 2003
Chapter 10 © Ken Gregory 2003
Chapter 11 © Tim Burt 2003
Chapter 12 © Andrew Herod 2003
Chapter 13 © Cindi Katz
Chapter 14 © Barbara A. Kennedy 2003
Chapter 15 © Nick Spedding 2003
Chapter 16 © Ian G. Simmons 2003
Chapter 17 © Karen M. Morin 2003

First published 2003

SAGE Publications Ltd
6 Bonhill Street
London EC2A 4PU

SAGE Publications Inc.
2455 Teller Road
Thousand Oaks, California 91320

SAGE Publications India Pvt Ltd
B-42, Panchsheel Enclave
Post Box 4109
New Delhi 100 017

**British Library Cataloguing in Publication data**

A catalogue record for this book is available from the British Library

ISBN 0 7619 7388 5
ISBN 0 7619 7389 3 (pbk)

**Library of Congress Control Number 2002112358**

Typeset by Photoprint, Torquay, Devon
Printed in Great Britain by The Cromwell Press Ltd, Trowbridge, Wiltshire

# Contents

# Notes on Contributors

**Alison Blunt** is Lecturer in Geography at Queen Mary, University of London. Her research interests include geographies of home and identity, cultures of imperial travel and domesticity, and feminist and postcolonial geographies. She is author of *Travel, Gender, and Imperialism: Mary Kingsley and West Africa* (Guilford, 1994), co-author (with Jane Wills) of *Dissident Geographies: an Introduction to Radical Ideas and Practice* (Prentice Hall, 2000), co-editor (with Gillian Rose) of *Writing Women and Space: Colonial and Postcolonial Geographies* (Guilford, 1994), and co-editor (with Cheryl McEwan) of *Postcolonial Geographies* (Continuum, 2002). She is currently writing a book on Anglo-Indian women and the spatial politics of home in India, Britain and Australia.

**Tim Burt** is Professor of Geography and Master of Hatfield College at the University of Durham. He has degrees from Cambridge, Carleton (Ottawa) and Bristol and previously worked at Huddersfield Polytechnic and Oxford University before moving to Durham in 1996. His main research interests are in hydrology and geomorphology.

**Noel Castree** is a Reader (Associate Professor) in Human Geography at the University of Manchester. His interests are in the political economy of environmental change, with a specific focus on Marxian theories. He is co-editor (with Bruce Braun) of *Remaking Reality: Nature at the Millennium* (Routledge, 1998) and *Social Nature* (Blackwell, 2001. He is currently researching how economic and cultural value are constructed in the 'new' human genetics.

**Ken Gregory** was Warden of Goldsmiths College, University of London, from 1992 to 1998 and is now Emeritus Professor, University of London, and Visiting Professor, University of Southampton. His research interests are in river-channel change and management, palaeohydrology and the development of physical geography.

**Mike Heffernan** is Professor of Historical Geography and Head of the School of Geography at the University of Nottingham. He is interested in the historical, cultural and political geographies of Europe and

North America from the eighteenth to the twentieth centuries and the history of geographical thought in the same period.

**Andrew Herod** is Professor of Geography at the University of Georgia. He has written widely on issues related to globalization and the geography of labour unionism. He is author of *Labor Geographies: Workers and the Landscapes of Capitalism* (Guilford Press, 2001), editor of *Organizing the Landscape: Geographical Perspectives on Labor Unionism* (University of Minnesota Press, 1998) and co-editor (with Gearóid Ó Tuathail and Susan Roberts) of *An Unruly World? Globalization, Governance and Geography* (Routledge, 1998). His most recent book, *Geographies of Power: Placing Scale* (Blackwell, 2002), is co-edited with Melissa W. Wright.

**Ron Johnston** is a professor in the School of Geographical Sciences at the University of Bristol, having previously worked at Monash University and the Universities of Canterbury, Sheffield and Essex. His main research interests are in electoral and urban social geography, and in the history of geography. He is author of *Geography and Geographers: Anglo-American Human Geography since 1945* (5th edn, 1997) and co-editor of *The Dictionary of Human Geography* (4th edn, Blackwell, 2000). In 1999 he was elected a Fellow of the British Academy and awarded the Prix Vautrin Lud.

**Cindi Katz** teaches at the Graduate School of the City University of New York, where she is Chair of the Environmental Psychology Program. She is co-editor (with Janice Monk) of *Full Circles: Geographies of Women over the Life Course* (Routledge) and also of the forthcoming *Disintegrating Developments: Global Economic Restructuring and the Struggle over Social Reproduction*.

**Barbara A. Kennedy** is a fluvial geomorphologist who studied at Cambridge University and the University of British Columbia. She is particularly interested in the development of ideas about landforms. Since 1979 she has been a lecturer at Oxford University.

**Martin Kent** is Professor of Biogeography at the University of Plymouth. He has published widely in the fields of plant ecology and biogeography. His book on *Vegetation Description and Analysis: a Practical Approach* (Wiley, 1992), with Dr Paddy Coker, has reached a wide audience and is being revised at the present time. Current research interests include spatial analysis of plant community boundaries, machair vegetation in the Outer Hebrides, vegetation burial, Culm grassland and poor fen vegetation in Devon and Cornwall and problems of water supply in relation to tourism development in Mallorca.

**Karen M. Morin** is Associate Professor of Geography at Bucknell University in Pennsylvania. Much of her work to date has focused on relationships among nineteenth-century American and British imperialisms, travel writing and gender relations. She is currently working on a number of projects that link postcolonialism to the historical geography of the USA.

**Keith Richards** is a graduate of the Department of Geography at the University of Cambridge, where he is now Professor of Geography, and has been Head of Department and Director of the Scott Polar Research Institute. He is a fluvial geomorphologist whose interests include river-channel forms and processes in a wide range of environments; modelling of fluvial systems; river management, river and floodplain restoration; and interrelationships between hydrological and ecological processes in floodplain environments. He also has interests in hydrological processes, sediment production and transfer processes in drainage basins, glacial hydrology and fluvial processes in proglacial environments; and in the philosophy and methodology of geography and the environmental sciences. He is a former Secretary and Chairman of the British Geomorphological Research Group.

**Ian G. Simmons** retired in 2001 from the Department of Geography at the University of Durham, where he happily remembers being first-year tutor to one of the editors of this volume. His scholarly work has focused on the small-scale effects of prehistoric cultures on moorland evolution in England and on the large-scale changes wrought by humanity on the whole earth in the last 10 000 years, together with our ideas about both.

**Nick Spedding** completed his PhD at the University of Edinburgh and currently holds a lectureship in the Department of Geography and Environment at the University of Aberdeen. His teaching specialisms include glacial hydrology and geomorphology, landscape development and the history, philosophy and methodology of physical geography.

**Peter J. Taylor** is Professor of Geography at Loughborough University. Among his recent books are *The Way the Modern World Works: World Hegemony to World Impasse* (Wiley, 1996), *Modernities: A Geohistorical Interpretation* (Polity and University of Minnesota Press, 1999) and *Political Geography: World-Economy, Nation-State and Locality* (4th edn, Prentice Hall, 2000). His current researches focus upon relations between world cities, and he is Associate Director of the Metropolitan Institute at Virginia Tech. He is founder of the Globalization and World Cities (GaWC) Study Group and Network which operates out of Loughborough University and Virginia Tech.

**John B. Thornes** is Research Chair in Physical Geography at King's College London. He was previously head of geography departments at King's College and Royal Holloway College, University of London, and at Bristol University. He is an honorary fellow of Queen Mary College, London University. He was formerly President of the Institute of British Geographers, Vice-President of the Royal Geographical and Chairman of the British Geomorphological Research Group. His research is mainly on semi-arid geomorphology and recently on the invasion of grasslands by woody vegetation in South Africa.

**Nigel Thrift** is a geographer known for his work on international finance, on time, and on nonrepresentational theory. He has been Head of the School of Geographical Sciences at the University of Bristol and Chair of the University's Research Assessment Panel, which is concerned with maximizing the outcome of the RAE. He is currently Chair of the University Research Committee. His most recent work has been concerned with the likely social and cultural effects of mobile telecommunications. In particular, he is interested in new kinds of content that will produce new kinds of geographies.

# List of Figures and Tables

**FIGURES**

**TABLES**

# Preface

Defining the core of geography is harder than one might expect. Sociologists have society, biologists living things, economists the economy and physicists matter and energy. But what is at the very core of geography? What are its key concepts? For the general public, the answer is very often maps and perhaps encyclopaedic knowledge of other people and places, from the world's longest rivers to the names of capitals in far-away countries. While such knowledge will serve you well in a pub quiz and might earn you a considerable amount of money on *Who Wants to be a Millionaire?*, it is less likely to provide the resources that you need to answer a university examination question well. Geography at university is about much more than maps, facts and figures, although these are, of course, also necessary.

When reading geography at university, then, many students are obliged to explain to their family and friends that no, they do not spend hours learning the height of the world's greatest mountains or colouring around the coastline in blue. They are learning about the earth, about humans' relationships with the earth, and about people's relationships with one another – all of which we know vary across time and space. Such a definition of geography does not lead to one clearly identifiable concept which lies at the heart of the discipline, in the same way that the past becomes the focus of history or elementary and compound substances the core of chemistry. Rather than having one central organizing concept, geography has many. As students of geography – and in this we include those of us paid to work in the discipline as well as the undergraduates we teach – we can take a step back from this kind of definition of geography and identify a number of concepts which lie at the centre of our discipline. Key among these are space, place, landscape, environment, system, scale and time.

The aim of this book is to help you to understand the use (and abuse) of these concepts within the discipline of geography. Naming them as core to our discipline might lead the unsuspecting reader to assume they are all well defined and theorized concepts. On some occasions this is indeed the case. Castree (Chapter 9), for example, provides an insightful review of the multiple ways in which place has been conceptualized in human geography. In doing so, he reveals place to be a much theorized although, importantly, also contested concept. On other occasions, however, these concepts are implicitly assumed but not defined, or only sporadically considered (see, for example, Kent, Chapter 6, on space in physical geography). When we asked potential authors to write these chapters some ran a metaphorical mile, but others replied that they would enjoy the challenge of thinking about something that is ever present but implicit in their work, essential but unsaid.

In naming this book *Key Concepts in Geography* and giving a list of key concepts as we did previously, you should not let us dupe you into thinking that our choice of concepts is either self-evident or unproblematic. Most geographers would agree that the concepts we've included are important to geography, though individuals might choose to question the relevance of some to their own work, and could perhaps suggest others they would like to have seen included. Some of these alternatives have partly become subsumed in the analysis of others included in the book, but the fact remains that our choice of concepts here is inevitably partial, reflecting the contested nature of the discipline in which we all work and study (see Chapters 1–4).

If further evidence were needed of the temporary and unstable nature of these concepts, the degree of overlap and, in some cases, contradiction between chapters provides this. Thinking about this in more detail, we can see that the end of one concept and the beginning of another is not neatly defined. In human geography for example, Taylor (Chapter 8) argues cogently that you cannot think about time without thinking about space, while in defining space Thrift (Chapter 5) also talks about place and images, the respective topics of the chapters by Castree (Chapter 9) and Morin (Chapter 17). Moreover, these authors do not always have the same take on the concepts in question, illustrating that the concepts are not stable and bounded but, like geography itself, open, temporary and mobile. Rather than being problematic, this is what makes geography a dynamic and fascinating discipline.

Lying as they do at the core of geography, these concepts are an important aspect of all undergraduate degree programmes. Here we have chosen to bring both sides of the discipline together and discuss human and physical geographers' understandings of key concepts in one volume. At one level this is a practical decision, as it combines all the approaches an undergraduate studying human and physical geography would need together in one volume. However, it is also more than this. One of the claims geographers can often be heard to make about our discipline is that it, uniquely, bridges the divide between the physical and social sciences (and to a lesser extent the humanities), thus bringing us unrivalled insights into human–environment interactions. In turn we often impose on our undergraduates modules in both human and physical geography with the aim of providing you with a rounded geographical education. However, relatively few academic geographers combine an interest in physical and human geography in their own research. Thus there is scope for both students and teachers to engage with the 'other' side of the discipline.

This volume attempts to bring the two sides together in thinking about our key concepts and, in doing so, to produce a text which will be of use to all students of geography, whether first-year undergraduates or established academics. The keen-eyed reader will observe, however, that within the volume human and physical geography remain separated as we have devoted two chapters to each concept, one for either side of the discipline. This lack of integration within the book reflects the different treatments the concepts have been given in human and physical geography. While some concepts are explicitly theorized in both sides of the discipline (e.g. time and scale), others are much more explicit in one and implicit in the other (e.g. space and systems) as concepts go in and out of academic fashion.

What is striking in the chapters is the degree to which we draw upon and reinforce other disciplines. In order to put our treatment of these concepts in context, the first section of the book ('Geographical traditions: emergence and divergence') traces the origins of geography and the ways in which it has interacted with the physical science, social science and humanities traditions at different points in time. These chapters illustrate the degree of two-way traffic in ideas between geography and an exceedingly wide range of disciplines from chemistry to literary criticism. Some of the chapters in the subsequent section ('key concepts') of the book then show how these interdisciplinary links have influenced geographers' understandings of their own key concepts – for example, as developments in ecology are reawakening interest in space on physical geography and ideas in cultural studies are helping us rethink the concept of landscape in cultural geography.

The apparent vitality of interdisciplinary linkages, alongside the evident intradisciplinary tensions, might lead some to declare the end of geography. Such a declaration, however, would be premature. A more thorough reading of this volume will show that there are indeed common threads in, and cross-linkages between, human and physical geography, as the two sides of the discipline exchange and rework ideas. For example, see Richards (Chapter 2) on the debunking of science as positivism and the true nature of scientific endeavour as a multifaceted process that is relevant in investigations of nature, environment *and* society, or see Chapters 15 and 17 on the enduring influence of Carl Sauer's seminal work on landscape. That these similarities are sometimes less evident than our differences merely reinforces the importance of the discipline. Geography does indeed occupy a unique position at the intersection of a variety of different traditions, and this often uncomfortable position brings us unique insights because of, rather than despite, our differences.

This passion for geography inspires this volume. In it we aim to provide an accessible introduction to the ways geography has been shaped by (and shapes) the physical science, social science and humanities traditions, and to the concepts which lie at the heart of the discipline. The chapters are written by experts in their field: the different subject-matter they tackle, as well as their own particular style of communicating, are evident in the ways they have chosen to write the chapters. Some authors have chosen to write an historical review of the way a diverse range of geographers have thought about a particular concept or tradition, ending with the diversity of thought evident today. Others have written pieces with a more contemporary focus, analysing the competing strands evident in current writing and identifying new ways forward. Some pieces appear neutral, others are more overtly polemical.

You should see these chapters as a way into geography's key concepts. Do not under any circumstance expect to begin reading a chapter over your cornflakes and to become an expert on that particular concept by lunchtime. While we do aim to provide you with an accessible introduction to these issues, the material is intellectually demanding and will require effort on your part. As the average age of the authors (45) suggests understanding these chapters is the beginning of a process and not the end. For this reason we have included a section on further reading at the end of each chapter. These have been annotated to guide your travels through the literature, allowing you to follow up themes and counter-themes which have sparked your interest. It is in this way, and only in this way, that you will truly read for your degree in geography.

*Sarah L. Holloway, Stephen P. Rice and Gill Valentine*

# Acknowledgements

We would like to thank Robert Rojek at Sage for commissioning this book and for his editorial support. We are eternally grateful to Mark Szegner for redrawing so many of the figures in this book.

The following gave their permission to reproduce material: Figures 7.3 and 7.4, Island Press; Figure 7.5, Oxford University Press; Figure 14.1, The American Meteorological Society (AMS); Figure 15.1 (Sugden, 1970), Blackwell Publishing; Figure 17.2, Ministry of Defence.

Every attempt has been made to obtain permission to reproduce copyright material. If any proper acknowledgment has not been made we would invite copyright holders to inform us of the oversight.

# Geographical Traditions: Emergence and Divergence

# Histories of Geography

## Mike Heffernan

### Definition

There is no single history of 'geography', only a bewildering variety of
different, often competing versions of the past. One such
interpretation charts the transition from early-modern navigation to
Enlightenment exploration to the 'new' geography of the late
nineteenth century and the regional geography of the interwar period.
This contextualist account – like all other histories of geography –
reflects the partialities of its author.

### INTRODUCTION

The deceptively simple word 'geography' embraces a deeply contested
intellectual project of great antiquity and extraordinary complexity.
There is no single, unified discipline of geography today and it is
difficult to discern such a thing in the past. Accordingly, there is no
single history of 'geography', only a bewildering variety of different,
often competing versions of the past. Physical geographers under-
standably perceive themselves to be working in a very different
historical tradition from human geographers, while the many perspectives employed on either side of this crude binary division also
have their own peculiar historical trajectories (see, as examples,
Glacken, 1967; Chorley et al., 1964; 1973; Beckinsale and Chorley,
1991; Livingstone, 1992).

Until recently, the history of geography was written in narrow,
uncritical terms and was usually invoked to legitimize the activities
and perspectives of different geographical constituencies in the present. The discipline's past was presented in an intellectual vacuum,
sealed off from external economic, social, political or cultural forces.
More recently, however, the history of geography has been presented

in a less introspective, self-serving and teleological fashion. Drawing on skills, techniques and ideas from the history of science, a number of scholars (some based in geography departments, others in departments of history or the history of science) have revealed a great deal about various kinds of geography in different historical and national contexts. We now have a substantial body of historical research on the development of geography in universities and learned societies, in primary and secondary schools, and within the wider cultural and political arenas. This research has focused mainly on Europe and North America and extends the longer and richer vein of scholarship on the history of cartography (on the latter, see Harley, 2001). Summarizing this research is no easy task and the following represents only a crude, chronologically simplified outline account of geography's history from the sixteenth to the mid-twentieth centuries.

## FROM NAVIGATION TO EXPLORATION: THE ORIGINS OF MODERN GEOGRAPHY

The classical civilizations of the Mediterranean, Arabia, China and India provided many of the geographical and cartographical practices that European geographers would subsequently deploy (Harley and Woodward, 1987; 1992–4). That said, the origins of modern geography can be dated back to western Europe in the century after Columbus. The sixteenth century witnessed far-reaching economic, social and political upheavals, linked directly to the expansion of European power beyond the continent's previously vulnerable limits. By *c.* 1600, a new, mercantilist Atlantic trading system was firmly established linking the emerging, capitalist nation-states of western Europe with the seemingly unlimited resources of the American 'New World'. Whether this expansion proceeded from internal changes associated with the transition from feudalism to capitalism (as most historians of early-modern Europe have argued) or whether it preceded and facilitated these larger transformations (as revisionist historians insist) is a 'chicken-or-egg' question that has been extensively debated in recent years (Blaut, 1993; Diamond, 1997). All we can say for certain is that early-modern innovations in shipbuilding, naval technology and navigation progressively increased the range of European travel and trade, particularly around the new Atlantic rim, and in so doing transformed European perceptions of the wider world as well as the European self-image (Livingstone, 1992: 32–62).

Firmly rooted in the practical commercial business of long-distance trade, early-modern geography – 'the haven-finding art,' as Eva Taylor (1956) memorably called it – encompassed both the technical, mathematical skills of navigation and map-making as well as the literary and descriptive skills of those who wrote the numerous accounts of the flora, fauna, landscapes, resources and peoples of distant regions (see, for early accounts, Taylor, 1930; 1934). As Figure 1.1 suggests, based on evidence from France, geographical descriptions of the non-European world became steadily more popular through the sixteenth century – the staple fare of the expanding European libraries which were also the principal repositories for the politically important archive of maps produced by Europe's growing army of cartographers (see, for example, Konvitz, 1987; Buisseret, 1992; Brotton, 1997). Equipped with this developing body of geographical fact (liberally sprinkled though it was with speculative fiction), the larger European universities began to offer specialized courses in geography and related pursuits, including chorography, navigation and cartography (see, for example, Bowen, 1981; Cormack, 1997).

During the seventeenth century – the era of the scientific revolution – the epistemological foundations of modern science were established on a new footing and the basic structures and institutions of the modern academy were laid down in the midst of widespread religious and political conflict throughout Europe. Although it was generally viewed as a practical, navigational skill that merely facilitated scientific discovery, the inchoate science of geography was nevertheless transformed by the seismic political and religious upheavals of the 1600s (Livingstone, 1988; 1990; 1992: 63–101). Geography's place in the educational programmes of European universities began to reflect its status as an intellectual arena within which weighty moral, philosophical and religious matters could be debated (Withers and Mayhew, 2002). By the early and middle decades of the eighteenth century – the era of the European Enlightenment – the still emerging

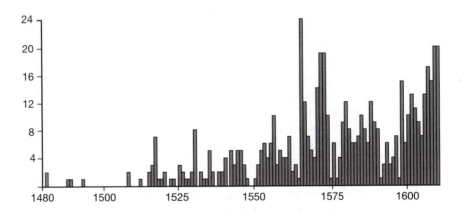

*Source*: Atkinson (1927; 1936)

**FIGURE 1.1** Books published in France concerned with geographical descriptions of the non-European world from 1481 to 1609

science of geography became further implicated in moral and philosophical debate, notably the anxious discussions about the possibilities of human development within and beyond Europe and the
relative merits of the different societies, cultures and civilizations
around the world (see, more generally, Broc, 1981; Livingstone, 1992:
102–38; Livingstone and Withers, 1999; Mayhew, 2000). At the same
time, an interest in travel as an educational activity, beneficial in and
of itself, spread from the European aristocracy into the ranks of the
newly enriched urban bourgeoisie. From this emerged the 'Grand
Tour' of the Mediterranean heartlands of the ancient world so beloved
of wealthier and lettered European men and women (Chard and
Langdon, 1996; Chard, 1999).

Partly as a result of these developments, the simple idea of geography as navigation gave way to a new formulation: geography as
exploration. This was, to be sure, a shift of emphasis rather than a
fundamental transformation but it reflected and engendered an
entirely new geographical language and rationale. While scientific
discoveries might emerge as more or less fortunate by-products of
navigation, such discoveries were seen as the planned and considered
objectives of the kind of purposeful, self-consciously scientific
exploration that developed during the eighteenth century, backed up
by new cartographic and navigational techniques and by the substantial resources of modern nation-states (see, for example, Sobel,
1996; Edney, 1997; Burnett, 2000): '[W]hat distinguishes geography as
an intellectual activity from . . . other branches of knowledge', claims
David Stoddart (1986: 29), 'is a set of attitudes, methods, techniques
and questions, all of them developed in Europe towards the end of the
eighteenth century.' Elsewhere in the same text, Stoddart (1986: 33) is
even more specific about the point of departure for the new geography
of exploration. The year 1769, when James Cook first sailed into the
Pacific, was a genuine turning point in the development of modern
geography, claims Stoddart, and not simply because Cook's journeys
opened up the Australian landmass with its unique flora and fauna to
the inquisitive European gaze. Unlike earlier generations of navigators, claims Stoddart, Cook's explorations were specifically
intended to achieve scientific objectives, to be carried out by the
illustrious international savants who accompanied him.

The idea that geography developed from navigation to exploration
through the early-modern period should not be seen as evidence of a
progressive or virtuous evolution from a speculative commercial
practice to an objective scientific pursuit. Columbus and Cook were
both sponsored by European nation-states eager to exploit the resources that might be uncovered by their voyages. Despite the rhetoric of
scientific internationalism, Cook's explorations reflected, a fortiori,
the same imperial objectives that had motivated earlier sea-faring

navigators. Neither can it be claimed that the wilder speculations of early-modern navigators and their chroniclers were more extravagant than the fantasies of later generations of explorers and their ghost-writers (see Heffernan, 2001). Geography as a practical navigational and cartographic activity was not supplanted by geography as an organized scientific pursuit based on detailed assessments of the human and environmental characteristics of different regions; rather, the same activity acquired new layers of meaning and a new scientific language through which its findings could be expressed.

The new Enlightenment geography was probably best exemplified by Alexander von Humboldt, the Prussian polymath who was born as Cook and his fellow explorers were charting their way across the Pacific. An inveterate explorer and a prolific author, von Humboldt was a complex figure: the archetypal modern, rational and inter-national scientist, his ideas were also shaped by the late eighteenth-century flowering of European romanticism and German classicism. His travels, notably in South America, were inspired by an insatiable desire to uncover and categorize the inner workings of the natural world, and his many published works, especially the multi-volume *Cosmos* which appeared in the mid-nineteenth century, sought to establish a systematic science of geography that could analyse the natural and the human worlds together and aspire to describe and explain all regions of the globe (Godlewska, 1999b; Buttimer, 2001). His only rival in this ambitious discipline-building project was his German near contemporary, Carl Ritter, a more sedentary writer of relatively humble origins whose unfinished 19-volume *Erdkunde*, also published in the mid-nineteenth century, reflected its author's Christian worldview but was inspired by the same objective of creating a generalized world geography, even though the analysis was to advance no further than Africa and Asia.

### INSTITUTIONALIZING EXPLORATION: THE GEOGRAPHICAL SOCIETIES

At this juncture, the European exploratory impetus was still dependent on the personal resources of the individuals involved. By the end of the eighteenth century, however, new institutional structures began to emerge within and beyond the agencies of the state dedicated to spon-soring exploration and geographical discovery. In 1782, Jean-Nicolas Buache was appointed geographer to the court of Louis XVI in France and, with royal approval, Buache attempted unsuccessfully to launch a geographical society to co-ordinate French exploration (Lejeune, 1993: 21–2). Stung into action by this failed initiative, a group of London scientists and businessmen, led by Sir Joseph Banks (President of the Royal Society and a veteran of Cook's Pacific expeditions) and Major

James Rennell (Chief Surveyor of the East India Company), launched the Association for Promoting the Discovery of the Interior Parts of Africa in 1788. Over the next decades, the African Association sponsored several pioneering expeditions, including those of Mungo Park, Hugh Clapperton and Alexander Gordon Laing (Heffernan, 2001).

The French Revolution and the Napoleonic wars brought a halt to the best forms of Enlightenment geographical inquiry (Godlewska, 1999a) but gave a fresh impetus to the strategically important sciences of cartography and land survey. By 1815, affluent, educated and well travelled former soldiers were to be found in virtually every major European city, and these men were the natural clientele for the first geographical societies, the building blocks of the modern discipline. The earliest such society was the Société de Géographie de Paris (SGP), which held its inaugural *séance* in July 1821. A fifth of the 217 founder members were born outside France, including von Humboldt and Conrad Malte-Brun, the Danish refugee who became the society's first Secretary-General (Fierro, 1983; Lejeune, 1993). A second, smaller geographical society was subsequently established in Berlin, the Gesellschaft für Erdkunde zu Berlin (GEB), at the instigation of the cartographer Heinrich Berghaus in April 1828 with a foundation membership of just 53, including von Humboldt and Carl Ritter, who became the society's inaugural president (Lenz, 1978).

The establishment in 1830 of the Royal Geographical Society (RGS) of London, under the patronage of William IV, marked a significant new departure. Several London societies committed to fieldwork and overseas travel already existed, including the Linnean Society for natural history (established in 1788), the Palestine Association (1804), the Geological Society (1807), the Zoological Society (1826) and the Raleigh Club (1826), the last named being a dining club whose members claimed collectively to have visited every part of the known world. The RGS was to provide a clearer London focus for those with an interest in travel and exploration. Even at its foundation, it was far larger than its existing rivals in Paris and Berlin. The 460 original fellows included John Barrow, the explorer and essayist, and Robert Brown, the pioneer student of Australian flora. Within a year, the RGS had taken over the Raleigh Club, the African Association and the Palestine Association to gain a virtual monopoly on British exploration (Brown, 1980).

The pre-eminence of the RGS as the focal point of world exploration increased over subsequent decades. By 1850, there were nearly 800 fellows (twice the number in Berlin and eight times more than Paris where the SGP membership had slumped) and, by 1870, the fellowship stood at 2400. Most fellows were amateur scholars but a number of prominent scientists also joined the society's ranks, including the young Charles Darwin who was elected after his return from the voyage of the *Beagle* in 1838. The dominant figure in the RGS during the middle years of the nineteenth century was Sir Roderick Murchison who was president on three separate occasions: 1843–5, 1851–3 and 1862–71. A talented publicist and entrepreneur, Murchison advocated geographical exploration as a precursor to British commercial and military expansion (Stafford, 1989). While other societies offered only *post hoc* awards and medals for successfully completed voyages, the RGS used its substantial resources to sponsor exploration in advance and on a large scale by providing money, setting precise objectives, lending equipment and arbitrating on the ensuing disputes. It also published general advice through its *Hints to Travellers* which began in 1854 (Driver, 2001: 49–67) and developed what was probably the largest private map collection in the world.

The success of the RGS reflected the strength of British amateur natural science, the wealth of the country's upper middle class (which provided the bulk of the fellowship) and the confidence that a large navy and overseas empire gave to prospective British explorers (Stoddart, 1986: 59–76). By concentrating on exploration and discovery, the RGS exploited a vicarious national passion for muscular 'heroism' in exotic places that was enthusiastically promoted by the British press. The explorer was the ideal masculine hero of Victorian society (the notion of a female geographer seemed almost a contradiction in terms), selflessly pitting himself against the elements and hostile 'natives' in remote regions for the greater glory of science and nation. Africa loomed especially large in the public imagination and the exploration of the 'Dark Continent', particularly the quest for the source of the Nile, provided an exciting and popular focus for the society's activities. All the major African explorers of the day – Burton, Speke, Livingstone, Stanley – were sponsored in some degree by the RGS, although their relationships with the society were not always cordial (Driver, 2001: 117–145). As the blank spaces on the African map were filled in, the RGS more than any other organization was able to bask in the reflected glory, whilst always shifting its focus to new regions. By the late nineteenth and early twentieth centuries, under the powerful influence of Sir Clements Markham and Lord Curzon, attention was directed mainly towards central Asia, the polar ice caps and the vertical challenges of high mountain ascents in the Himalayas.

The remarkable success of the RGS inspired similar surges of activity in Paris and Berlin where the geographical societies expanded rapidly after 1850 under the direction of the Marquis de Chasseloup-Laubat (Napoleon III's former Naval and Colonial Minister) and Heinrich Barth (the leading German African explorer). New geographical societies were also established elsewhere in Europe and in the new cities of North and South America (Figure 1.2).

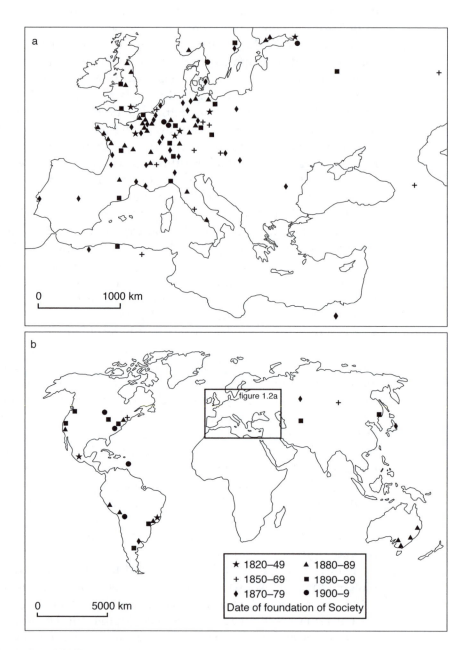

*Source*: Kolm (1909)

**FIGURE 1.2**   European and non-European geographical societies, by date of foundation

## A SCIENCE OF EMPIRE: EXPLORATION, THE 'NEW' GEOGRAPHY AND MODERN IMPERIALISM

The vision of the discipline promoted by these mid-nineteenth cen-tury geographical societies perfectly illustrates the soaring ambition of the European imperial mind (Bell et al., 1994; Godlewska and Smith, 1994; Driver, 1992; 2001). The navigational and cartographic skills of the geographer during the 'heroic' age of exploration and discovery paved the way for European military and commercial colonization of the Americas, Asia and Africa. The principal geographical 'tool' was, of course, the map. By representing the huge complexity of a physical and human landscape in a single image, geographers and cartographers provided the European imperial project with arguably its most potent device. European exploration and mapping of the coastlines of the Americas, Africa, Asia and the Pacific and the subsequent terrestrial topographic surveying of these vast continents were self-evidently an exercise in imperial authority. To map hitherto 'unknown' regions (unknown, that is, to the European), using modern techniques in tri-angulation and geodesy, was both a scientific activity dependent on trained personnel and state-of-the art equipment and also a political act of appropriation which had obvious strategic utility (Edney, 1997; Burnett, 2000; Harley, 2001).

The shift in the European balance of power following the Franco-Prussian war of 1870 gave an unexpected boost to geography. Aggress-ive colonial expansion outside Europe was identified as one way to reassert a threatened or vulnerable national power within Europe, and the later decades of the nineteenth century were characterized by a surge of colonial expansion (particularly the so-called 'Scramble for Africa') as each imperial power sought comparative advantage over its enemies, both real and imagined. This frenzied land grab emphasized the practical utility of geography and cartography. By the end of the nineteenth century, the 'high-water mark' of European imperial expansion, geography had become 'unquestionably the queen of all imperial sciences . . . inseparable from the domain of official and unofficial state knowledge' (Richards, 1993: 13; see also Said, 1978; 1993). In Germany, 19 new geographical societies had been estab-lished, including associations in the former French towns of Metz (1878) and Strasbourg (1897). In France, there were 27 societies, one in virtually every French city, and no fewer than four in French Algeria. A number of the French provincial societies were devoted to commer-cial geography and sought to encourage trading links with the French empire (Schneider, 1990). At this point, one third of the world's geo-graphers were based in France (Figure 1.3). The British were by no means immune to this late century geographical fever and the RGS remained the largest and wealthiest geographical society in the world.

A handful of provincial societies were established in the UK during the late nineteenth century, notably in Edinburgh (the Royal Scottish Geographical Society) and Manchester (both 1884) but, unlike the countries of continental Europe, the RGS retained its dominance of the British geographical movement (MacKenzie, 1994).

Backed by a new generation of civic educational reformers, a 'new' geography began to emerge in schools and universities, with Germany and France leading the way. The German university system had been significantly reformed during the nineteenth century (based in part on the ideas of Wilhelm von Humboldt, Alexander's brother, the architect of the University of Berlin) and geography already had a powerful presence in the tertiary and secondary educational programmes. The same republican politicians in Paris who championed colonial expansion as a route to national rejuvenation were also convinced that France needed to learn from Germany by completely revising its school and university system to inculcate the patriotic values that had seemed shockingly absent from the French armies of 1870. A carefully constructed geography curriculum was identified as the key to such a system. This would introduce the next generation to the beauty, richness and variety of France's *pays* while informing them of their nation's role and responsibility in the wider world. The French universities would need to train the next generation of geography teachers, and a dozen new chairs were established during the 1880s and 1890s for this purpose (Broc, 1974). Germany, eager to sustain its reputation as the leading intellectual centre of the discipline, responded with a similar educational drive.

The fiercely independent British universities initially resisted this trend, to the dismay of the geographers in the RGS. A chair of geography had been established at University College London as early as 1833 (filled by Captain Maconochie, the RGS President at the time) but this lapsed almost immediately and a full-time British university post in geography was not created until 1887 when an Oxford University readership was awarded to Halford Mackinder, partly financed by the RGS (Stoddart, 1986: 41–127). The RGS, along with the new Geographical Association (established in 1893 to promote geography in schools), worked hard to change attitudes. Sir Harry H. Johnston, the explorer, colonial administrator and prominent RGS fellow, argued that geography should become a compulsory school subject, for it was only through detailed geographical description, complete with authoritative and regularly updated topographical and thematic maps, that a region could be known, understood and therefore fully possessed by those in authority (Heffernan, 1996: 520). By dividing the world into regions and ordering the burgeoning factual information about the globe into regional segments, he insisted, geography offered one solution to the yearned-for objective of classifying and understanding the

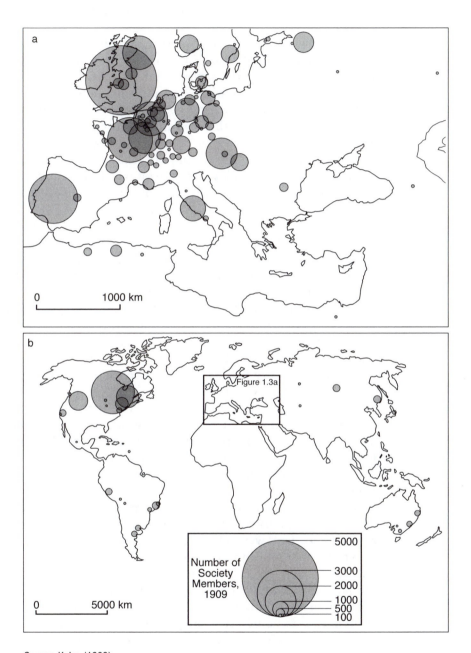

Source: Kolm (1909)

**FIGURE 1.3** Size of European and non-European geographical societies, late nineteenth century

human and environmental characteristics of the entire globe. Through geography the world could, at last, be visualized and conceptualized as a whole.

The 'new' school and university geography was no less an imperial science than its exploratory predecessor, as a cursory glance at the textbooks of the period makes clear (Hudson, 1977). The principal representatives of academic geography – notably Mackinder at Oxford and Friedrich Ratzel in Leipzig – sought not only to explain the human and natural features of the world but also to justify the existence of European empires (Heffernan, 2000a). Ratzel, in particular, was profoundly influenced by the writings of Charles Darwin and insisted (as did many so-called 'social Darwinists' within and beyond geography) that the principles of 'natural selection' applied equally to the natural, social and political realms (Stoddart, 1986: 158–79; Bassin, 1987). Nation-states, like species, struggled for space and resources, and the 'fittest' were able to impose their will on less fortunate 'races'. For many geographers of this period, including those who were fashioning a distinctively American geographical tradition, such as Ellsworth Huntington and Ellen Churchill Semple, the dominance that certain peoples exerted over others was either divinely preordained or the product of environmentally conditioned racial characteristics. Building on Enlightenment ideas about the environmentally determined nature of different peoples, a new brand of 'scientific racism' infused geographical theory in the later nineteenth and early twentieth centuries. The expansive, imperial 'races' of Europe and the European settler communities in the Americas benefited from unique climatic and environmental circumstances, it was claimed, and these advantages had created energetic, expansive civilizations. The very different climates and environments of the colonial periphery had created inferior societies and weaker civilizations in need of an ordering and benign European presence (Peet, 1985).

Such distasteful ideas reflected a prevailing orthodoxy but they also provoked spirited debate. While environmental determinism and scientific racism were often mutually reinforcing ideas, they could also contradict one another. Some racial theorists assumed that different 'races' were fixed in an unchanging 'natural' hierarchy, the contemporary manifestation of quite distinct evolutionary sequences from different points of departure (polygenesis). 'External' environmental factors could have only limited impact on this preordained racial system. This argument presupposed the need for a permanent imperial presence of intellectually and racially superior rulers in order to manage the irredeemably inferior peoples and environments of the colonial world (Livingstone, 1992: 216–60). By emphasizing the overriding significance of climatic and physical geographical factors on the process of social and economic progress and the essential unity of humanity (monogenesis), many environmental determinists tended to focus on the possibilities of human development through the judicious intervention in the natural world. If scientifically advanced European societies could overcome the worst aspects of the challenging environments of the colonial periphery by draining the pestilential marsh or irrigating the barren desert, this would not only improve colonial economic productivity but would also, in time, improve the nature of the local societies and cultures. Environmental and moral 'improvement' were thus intimately interlinked and both were dependent on the 'benign' intervention of a 'superior' external force. In time, and if coupled with a wise cultural and educational policy, a colonized people would eventually be able to take control of their own resources and manage their own affairs.

It should also be emphasized, moreover, that late nineteenth-century geography spawned some very different perspectives from any of those discussed above, notably the radical, anarchist ideas of Petr Kropotkin and Elisée Reclus, leading figures in the Russian and French anarchist circles, respectively. For Kropotkin and Reclus, the new science of geography suggested ways of developing a new harmony of human societies with the natural world, freed from the pernicious influence of class-based, nationalist politics (Blunt and Wills, 2000).

## THE NEW WORLD: GEOGRAPHY AND THE CRISIS OF THE EARLY TWENTIETH CENTURY

The onset of the twentieth century provoked a rather anxious debate about the future of the 'great powers'. Many believed that 1900 would mark a turning point in world history, the end of a 400-year period of continuous European expansion. The unexplored and unclaimed 'blank' spaces on the world map were rapidly diminishing, or so it seemed, and the sense of a 'global closure' was palpable. Different versions of this *fin de siècle* lament were rehearsed in several contexts. The German geomorphologist, Albrecht Penck used the idea of global closure in the early 1890s to justify his inspirational but sadly inconclusive scheme for a new, international 1:1 million map of the world (Heffernan, 2002). At the same time, the American historian, Frederick Jackson Turner, delivered a famous lecture at the Columbian Exposition in Chicago in 1893 (an event designed to commemorate the quatercentenary of Colombus's voyage to the Americas) which suggested the need for the newly established transcontinental USA to seek out new imperial frontiers beyond the traditional limits of the national homeland, particularly in the Pacific. And in 1904, Halford Mackinder addressed the RGS on the likely end of the 'Columbian era' of maritime, trading empires and the emergence of a twentieth-century world order dominated by cohesive land-based empires (such

as the USA), bound together by railways. Mackinder dubbed the great
Eurasian landmass – the largest expanse of territory on the planet – as
the 'geographical pivot of history' and argued that whichever power
could control the limitless resources of this huge region would dominate world affairs in the coming century. The 'closed' system Mackinder described would be extremely dangerous, he implied, because
the frontiers on the new empires would straddle the globe (for a
comparison of Turner and Mackinder, see Kearns, 1984).

The outbreak of the First World War, the first truly global conflict,
confirmed many of these fears. Although it reached its peak of savage
intensity on the Western Front, Mackinder later insisted that the war
had erupted from precisely the territorial struggle he had foreseen in
1904. Germany's pitch for global hegemony had been based on the
idea of winning what Ratzel had famously called 'living space'
(*Lebensraum*) in the east, at the expense of Russia, the region Mackinder now called 'the heartland' of the 'world island' (Mackinder,
1919). Mackinder was not asked to advise the British delegation which
negotiated the peace treaties in Paris in 1919 (to his considerable
frustration), but leading geographers from other countries were prominently involved in the redrawing of the postwar political map. The
larger geographical societies in all belligerent countries had been fully
mobilized by the intelligence services of each state (not least because
of their extensive map collections) and had generated a mass of new
geographical information and cartography for their paymasters. In the
USA, the President of the American Geographical Society, Isaiah
Bowman, was an important adviser to President Woodrow Wilson
during the peace negotiations and had previously recruited many of
America's leading geographers (including William Morris Davis and
Ellen Churchill Semple) on to the so-called House Inquiry to help
formulate US policy on postwar Europe and the wider world. Bowman
also wrote the main geographical text on the postwar order, *The New
World* (1921). Several French geographers, led by Paul Vidal de la
Blache, fulfilled a similar role as members of the Comité d'Études that
advised the French government during the war and at the peace
conferences. The RGS, for its part, was also prominently involved as a
metropolitan 'centre of calculation' for both the Naval and War Office
intelligence services (Heffernan, 2000b).

In these countries, geography emerged from the carnage of the First
World War with its reputation significantly enhanced. New geography
appointments quickly followed in the leading schools and universities, notably in Britain, where the teaching of geography still lagged
behind the continent and where university courses had previously
been taught by a single lecturer. The first British honours schools of
geography were established during the war itself (in Liverpool in 1917

and at the LSE and Aberystwyth in 1918) or immediately afterwards (at University College London and Cambridge in 1918, Manchester in 1923 and Sheffield in 1924) (Stoddart, 1986: 45–6). Although the RGS had overseen the initial appointments to geography positions in British universities, the subsequent expansion of the discipline eroded the society's control of the British geographical agenda. Anxious to develop a more rigorous, scientific geography to match the developments taking place in other countries, British university geographers established their own independent organization in 1933, the Institute of British Geographers (IBG), which only recently remerged with the older society.

By the interwar years, the 'new' geography that had arisen before 1914 had evolved into a sophisticated and popular discipline, prominent at all levels in the educational system. In the universities, a host of new subdisciplines arose, most of which continue to the present, but two wider interwar trends are worthy of special mention. The first was the conviction that geography should be an integrative, regional science. Physical and human geography should always be brought together in the analysis of specific regions, it was repeatedly argued, and the otherwise vague and undeveloped idea of the region emerged as the single most important intellectual contribution of interwar geography, particularly in Britain and France. The importance of the region can easily be explained. For the geographers who rose to prominence after 1918, the traditional nation-state was a suspect entity, the focus and the engine of the discredited nineteenth-century nationalism that had culminated with the disasters of 1914–18. The region, whether subnational or supranational, offered the prospect of radically alternative forms of government in the future. The French school of geography (dominated until his death in 1918 by Paul Vidal de la Blache and continued after the war by his many students) saw the region as the discipline's fundamental building block. Alongside the numerous regional monographs that were produced with assembly-line efficiency by French geographers, the Vidalians also proffered various schemes for devolved regional government from below (based in part on Vidal's own recommendations from 1910) and for integrated, European government from above (the most prophetic coming from Albert Demangeon). Similar ideals inspired regional geographers in Britain, including A.J. Herbertson, C.B. Fawcett, L. Dudley Stamp and H.J. Fleure, most of whom were influenced by the radical idealism of the Scottish natural scientist, planner and general polymath, Patrick Geddes (Livingstone, 1992: 260–303; Heffernan, 1998: 98–106, 128–31). Similar ideas were also influential in other national contexts, principally in Germany but also in the USA, where the school of cultural geography established by Carl Sauer at

Berkeley celebrated the idea of historical and geographical particularism and the unique qualities of diverse regions (see Chapters 15 and 17
on the importance of Sauer's work).

The second, very different, interwar trend was associated with
fascist Italy and Nazi Germany, where a new generation of academic
geographers sought to relaunch their discipline as an overtly political
science dedicated to questioning the geopolitical order established in
Paris after 1918. The Italian and German geopolitical movements
(developed by Giorgio Roletto and Karl Haushofer and associated with
the journals *Geopolitica* and *Zeitschrift für Geopolitik*) had much in
common, including a penchant for bold, black-and-white propaganda
cartography and hard-hitting, journalistic articles. Despite their
overtly nationalist stance, both movements imagined a future integrated Europe though of a very different kind than was proposed by
French and British regional geographers. The influence of Italian
geopolitical theorists on government policy was minimal, and the
impact of their German equivalents on Nazi programmes was even
smaller, despite the close relationship between Haushofer and Rudolf
Hess, one of Hitler's chief acolytes. Haushofer and his fellow academics had remarkably little to say on the central question of race and
this, more than anything else, limited their appeal to Hitler and his
Nazi ideologues (Heffernan, 1998: 131–49; Dodds and Atkinson,
2000).

The Second World War spelt the end of the geopolitical movements
of Italy and Germany (and also brought about the temporary collapse
of political geography *tout court*, and not only in these two countries).
While the interwar regional geographical tradition continued into the
post-1945 era, this too came under increasing pressure from new
developments, particularly the forms of quantitative geographical
inquiry pioneered in the USA and in Britain during the 1960s and
1970s (see Chapter 3). Although it had arisen from a practical concern
with the region as an alternative level of government and administration, the particularism of interwar regionalism, with its focus on the
uniqueness of place, sat uncomfortably beside the new idea of geography as a law-seeking, 'spatial science'. Instead of the old, more
historical form of regionalism, a new and more rigorously scientific
regional science developed strongly during the postwar years years to
play its part alongside the many other branches of geographical
research and teaching (Johnston, 1997).

## CONCLUSION

The preceding survey is a personal account and should certainly not be
read as a story of radical departures or revolutionary changes. The

rough sequence of events charted here – the transition from early-modern navigation to Enlightenment exploration to the 'new' geography of the late nineteenth century and the regional geography of the interwar period – represents a process of accretion rather than displacement; an evolution in which traditions merged, overlapped and persisted alongside later developments to create an ever more complex picture. It is impossible to distil from these stories an essential core theme that has always animated geographical inquiry, but one thing is clear: geography, whether defined as a university discipline, a school subject or a forum for wider debate, has always existed in a state of uncertainty and flux. While some have lamented this as a sign of disciplinary weakness, it might equally be argued that the absence of conceptual conformity has been one of the discipline's great strengths. If the developments of the last few decades can be taken as a guide, it would seem that this is one 'geographical tradition' that is destined to continue.

## Summary

- The deceptively simple word 'geography' embraces a deeply contested intellectual project of great antiquity and extraordinary complexity. There is no single, unified discipline of geography today and it is difficult to discern such a thing in the past.
- A rough sequence of events can be charted from the early-modern navigation, to Enlightenment exploration, to the 'new' geography of the late nineteenth century and the regional geography of the interwar period.
- It is impossible to distil from these stories an essential core theme that has always animated geographical inquiry. This could be seen either as a sign of disciplinary weakness or as a strength.

## Further reading

The literature on the history of geography is large and varied. The best starting point is David Livingstone's (1992) *The Geographical Tradition: Episodes in the History of a Contested Enterprise*, which is excellent on wider intellectual and philosophical contexts. David Stoddart's (1986) *On Geography and its History* offers a spirited defence of geography's place within the natural sciences. On the Enlightenment, Robert Mayhew's (2000) *Enlightenment Geography: The Political Languages of British Geography 1650–1850* and Anne Godlewska's (1999) *Geography Unbound: French Geographic Science from Cassini to Humboldt* have different perspectives but survey the British and French experiences very effectively. The collection edited by David Livingstone and Charles Withers (1999) *Geography and Enlightenment* contains a useful introductory essay and some strong chapters on specific topics. Anne Godlewska and Neil Smith's (1994) *Geography and Empire* is an admirable collection on the imperial theme in general and can be supplemented, for the nineteenth and early twentieth centuries, by the essays in Morag Bell et al. (1994) *Geography and Imperialism, 1820–1940*. Felix Driver's (2001) *Geography Militant: Cultures of Exploration and Empire* is a sparkling and highly imaginative study on the nineteenth century.

*Note*: Full details of the above can be found in the references list below.

## References

Atkinson, G. (1927) *La Littérature géographique française de la Renaissance: Répertoire bibliographique*. Paris: Auguste Picard.

Atkinson, G. (1936) *Supplément au Répertoire bibliographique se rapportant à la Littérature géographique française de la Renaissance*. Paris: Auguste Picard.

Bassin, M. (1987) 'Imperialism and the nation-state in Friedrich Ratzel's political geography', *Progress in Human Geography*, 11: 473–95.

Beckinsale, R. and Chorley, R. (1991) *The History of the Study of Landforms, or The Development of Geomorphology. Vol. III. Historical and Regional Geomorphology, 1890–1950*. London: Routledge.

Bell, M., Butlin, R. and Heffernan, M. (eds) (1994) *Geography and Imperialism, 1820–1940*. Manchester: Manchester University Press.

Blaut, J. (1993) *The Colonizer's Model of the World: Geographical Diffusionism and Eurocentric History*. London: Guilford.

Blunt, A. and Wills, J. (2000) *Dissident Geographies: An Introduction to Radical Ideas and Practice*. London: Prentice Hall.

Bowen, M. (1981) *Empiricism and Geographical Thought: From Francis Bacon to Alexander von Humboldt*. Cambridge: Cambridge University Press.

Broc, N. (1974) 'L'établissement de la géographie en France: diffusion, institutions, projets (1870–1890)', *Annales de Géographie*, 83: 545–68.

Broc, N. (1981) *La Géographie des Philosophes: Géographes et Voyageurs français au XVIIIe siècle*. Paris: Éditions Ophrys.

Brotton, J. (1997) *Trading Territories: Mapping the Early-Modern World*. London: Verso.

Brown, E. (ed.) (1980) *Geography Yesterday and Tomorrow*. Oxford: Oxford University Press.

Buisseret, D. (ed.) (1992) *Monarchs, Ministers and Maps: The Emergence of Cartography as a Tool of Government in Early-Modern Europe*. Chicago, IL: University of Chicago Press.

Burnett, D. (2000) *Masters of all they Surveyed: Exploration, Geography, and a British El Dorado*. Chicago, IL: University of Chicago Press.

Buttimer, A. (2001) 'Beyond Humboldtian science and Goethe's way of science: challenges of Alexander von Humboldt's geography', *Erdkunde*, 55: 105–20.

Chard, C. (1999) *Pleasure and Guilt on the Grand Tour: Travel Writing and Imaginative Geography, 1600–1830*. Manchester: Manchester University Press.

Chard, C. and Langdon, H. (eds) (1996) *Transports: Travel, Pleasure and Imaginative Geography, 1600–1830*. New Haven, CT: Yale University Press.

Chorley, R., Beckinsale, R. and Dunn, A. (1964) *The History of the Study of Landforms, or The Development of Geomorphology. Vol. I. Geomorphology before Davis*. London: Methuen.

Chorley, R., Beckinsale, R. and Dunn, A. (1973) *The History of the Study of Landforms, or The Development of Geomorphology. Vol. II. The Life and Works of William Morris Davis.* London: Methuen.

Cormack, L. (1997) *Charting an Empire: Geography and the English Universities 1580–1620.* Chicago, IL: University of Chicago Press.

Diamond, J. (1997) *Guns, Germs and Steel: The Fates of Human Societies.* London: Jonathan Cape.

Dodds, K. and Atkinson, A. (eds) (2000) *Geopolitical Traditions: A Century of Geopolitical Thought.* London: Routledge.

Driver, F. (1992) 'Geography's empire: histories of geographical knowledge', *Environment and Planning D: Society and Space*, 10: 23–40.

Driver, F. (2001) *Geography Militant: Cultures of Exploration and Empire.* Oxford: Blackwell.

Edney, M. (1997) *Mapping an Empire: The Geographical Construction of British India, 1765–1843.* Chicago, IL: University of Chicago Press.

Fierro, A. (1983) *La Société de Géographie de Paris (1826–1946).* Geneva and Paris: Librairie Groz and Librairie H. Champion.

Glacken, C. (1967) *Traces on the Rhodian Shore: Nature and Culture in Western Thought from Ancient Times to the End of the Eighteenth Century.* Berkeley and Los Angeles, CA: University of California Press.

Godlewska, A. (1999a) *Geography Unbound: French Geographic Science from Cassini to Humboldt.* Chicago, IL: University of Chicago Press.

Godlewska, A. (1999b) 'From Enlightenment vision to modern science: Humboldt's visual thinking', in D. Livingstone and C. Withers (eds) *Geography and Enlightenment.* Chicago, IL: University of Chicago Press, pp. 236–75.

Godlewska, A. and Smith, N. (eds) (1994) *Geography and Empire.* Oxford: Blackwell.

Harley, J. (ed. P. Laxton) (2001) *The New Nature of Maps: Essays in the History of Cartography.* Baltimore, MD: Johns Hopkins University Press.

Harley, J. and Woodward, D. (eds) (1987) *The History of Cartography. Vol. I. Cartography in Prehistoric, Ancient, and Medieval Europe and the Mediterranean.* Chicago, IL: University of Chicago Press.

Harley, J. and Woodward, D. (eds) (1992–4) *The History of Cartography. Vol. II. Book 1. Cartography in the Traditional Islamic and South Asian Societies. Book 2. Cartography in the Traditional East and Southeast Asian Societies.* Chicago, IL: University of Chicago Press.

Heffernan, M. (1996) 'Geography, cartography and military intelligence: the Royal Geographical Society and the First World War', *Transactions, Institute of British Geographers*, 21: 504–33.

Heffernan, M. (1998) *The Meaning of Europe: Geography and Geopolitics.* London: Arnold.

Heffernan, M. (2000a) 'Fin de siècle, fin du monde: on the origins of European geopolitics, 1890–1920', in K. Dodds and D. Atkinson (eds) *Geopolitical Traditions: A Century of Geopolitical Thought.* London: Routledge, pp. 27–51.

Heffernan, M. (2000b) 'Mars and Minerva: centres of geographical calculation in an age of total war', *Erdkunde*, 54: 320–33.

Heffernan, M. (2001) ' "A dream as frail as those of ancient Time": the in-credible geographies of Timbuctoo', *Environment and Planning D: Society and Space*, 19: 203–25.

Heffernan, M. (2002) 'The politics of the map in the early 20th century', *Cartography and Geographical Information Systems*, 29: 207–26.

Hudson, B. (1977) 'The new geography and the new imperialism', *Antipode*, 9: 12–19.

Johnston, R. (1997) *Geography and Geographers: Anglo-American Human Geography Since 1945*. London: Arnold.

Kearns, G. (1984) 'Closed space and political practice: Frederick Jackson Turner and Halford Mackinder', *Environment and Planning D: Society and Space*, 22: 23–34.

Kolm, G. (1909) 'Geographische Gessellschaften, Zeitschriften, Kongresse und Austellungen', *Geographisches Jahrbuch*, 19: 403–13.

Konvitz, J. (1987) *Cartography in France, 1660–1848: Science, Engineering and Statecraft*. Chicago, IL: University of Chicago Press.

Lejeune, D. (1993) *Les Sociétés de Géographie en France et l'Expansion coloniale au XIXe siècle*. Paris: Albin Michel.

Lenz, K. (1978) 'The Berlin Geographical Society 1828–1978', *Geographical Journal*, 144: 218–22.

Livingstone, D. (1988) 'Science, magic and religion: a contextual reassessment of geography in the sixteenth and seventeenth centuries', *History of Science*, 26: 269–94.

Livingstone, D. (1990) 'Geography, tradition and the scientific revolution: an interpretative essay', *Transactions, Institute of British Geographers*, 15: 359–73.

Livingstone, D. (1992) *The Geographical Tradition: Episodes in the History of a Contested Enterprise*. Oxford: Blackwell.

Livingstone, D. and Withers, C. (1999) *Geography and Enlightenment*. Chicago, IL: The University of Chicago Press.

MacKenzie, J. (1994) 'The provincial geographical societies in Britain, 1884–1894', in M. Bell et al. (eds) *Geography and Imperialism, 1820–1940*. Manchester: Manchester University Press, pp. 31–43.

Mackinder, H. (1919) *Democratic Ideals and Realities: A Study in the Politics of Reconstruction*. London.

Mayhew, R. (2000) *Enlightenment Geography: The Political Languages of British Geography 1650–1850*. Basingstoke: Macmillan.

Peet, R. (1985) 'The social origins of environmental determinism', *Annals of the Association of American Geographers*, 75: 309–33.

Richards, T. (1993) *The Imperial Archive: Knowledge and the Fantasy of Empire*. London: Verso.

Said, E. (1978) *Orientalism*. London: Routledge.

Said, E. (1993) *Culture and Imperialism*. London: Jonathan Cape.

Schneider, W. (1990) 'Geographical reform and municipal imperialism in France, 1870–80', in J. MacKenzie (ed.) *Imperialism and the Natural World*. Manchester: Manchester University Press, pp. 90–117.

Sobel, D. (1996) *Longitude: The True Story of a Lone Genius who Solved the Greatest Scientific Problem of his Time*. London: Fourth Estate.

Stafford, R. (1989) *Scientist of Empire: Sir Roderick Murchison, Scientific Exploration and Victorian Imperialism*. Cambridge: Cambridge University Press.

Stoddart, D. (1986) *On Geography and its History*. Oxford: Blackwell.

Taylor, E. (1930) *Tudor Geography 1485–1583*. London: Methuen.

Taylor, E. (1934) *Late Tudor and Early Stuart Geography 1583–1650*. London: Methuen.

Taylor, E. (1956) *The Haven-Finding Art: A History of Navigation from Odysseus to Captain Cook*. London: Hollis & Carter.

Withers, C. and Mayhew, R. (2002) 'Rethinking "disciplinary" history: geography in British universities, c. 1580–1887', *Transactions, Institute of British Geographers*, 27: 11–29.

# 2  Geography and the Physical Sciences Tradition

**Keith Richards**

## Definition

The physical sciences provide a role model for many disciplines, but the model is a contested one. Some of the successes of the physical sciences have been the product of lengthy gestation over hundreds of years, and the methods employed and the philosophical frameworks underpinning them have changed in response to emerging understandings. Thus there is no single tradition, except that of pluralism. Within *this* 'tradition', there is a rich source for geography of methods drawing on observation, measurement, various forms of experimentation, theory development and testing. These diverse but closely related practices are the true legacy of the sciences, and one to which geography can contribute as well as from which it may draw.

## INTRODUCTION

This chapter examines the relationship between the philosophy and method of the physical sciences and those of geography, with a view to exploring some similarities and differences. The physical sciences have been very successful in providing us with understanding of many aspects of the world, as a result of powerful procedures for revealing the nature of that world. The basic argument of this chapter is that the so-called scientific method on which the physical sciences have been based is really only a model itself, and that the balance of procedures in scientific method has changed as knowledge and understanding have developed, and may also change as a particular investigation proceeds. This is illustrated with many examples, but the Gas Laws and a research problem in eco-hydrology are emphasized as particular cases. There are many different dimensions to the scientific method and no single 'tradition'. This has implications for the conduct of debate in geography where particular assumed traditions may sometimes be misrepresented, especially in support of an anti-quantitative stance. It also suggests that there are clear possibilities for naturalism – a common methodology applicable to a wide range of kinds of scientific inquiry, including environmental and social – because the pluralist nature of the procedures of the physical scientific method have many counterparts in these other sciences.

This chapter begins by stressing the dangers of assuming the existence of a single tradition. The importance of identifying researchable questions as an early component of scientific problem-solving is then reviewed. In turn, aspects of scientific practice are considered in detail. It is emphasized that practical research procedures, adapted to present conditions of understanding, constitute a more important tradition or, rather, legacy, than the oversimplified labels attached to particular methodological contexts (e.g. positivism). The procedures of science which are reviewed include observation and measurement, experimentation and the theoretical issues of hypothesis generation and testing, identification of mechanisms behind regularities and criteria for warranting the claims of competing theories. In the conclusion the importance of the pluralism demonstrated by these procedures is emphasized, and an argument is made for naturalism and a 'rational criticism' model of the scientific method.


## CONTESTED HISTORIES

In this review of some relationships between geography and the physical sciences, it is useful to expose two challenges at the outset. The first is that any discussion of methodology is fraught with difficulty because the terms employed are so unstable. This instability can result in both confusion and artifice; the former when protagonists unwittingly adopt different meanings, the latter when a particular meaning is chosen deliberately, even politically, to support a specific position or cause. The second challenge arises because traditions are often not what they seem. They frequently turn out to be more modern (a term used deliberately) than the use of the word 'tradition' implies, to have a history in a politics of representation, and also to be unstable in content and to take different forms for different actors. For these reasons, discussing any 'tradition' in geography is liable to seem like walking across quicksand while, at the same time, it is likely to expose some of the discipline's internal tensions and divisions.

One might begin by asking what 'science' actually is – and even this turns out to be a less than straightforward question to answer. In the English language the word is commonly assumed to refer to the physical sciences (leading a former British minister of education to try to deny a national research funding agency the title 'The Social *Sciences* Research Council'). But in France *la science* and in Germany *wissenschaft* have much broader implications. Even if we consider one linguistic context and examine the history of something we have called 'science' in English, it appears to have changed markedly in its nature over time. Woolgar (1988) traces three main phases, although these could easily be subdivided and qualified. A period of 'amateur' science, in which doctors and clergymen played significant roles as natural historians, characterizes the seventeenth and eighteenth centuries, after which science increasingly became 'academic' in nature as it moved into the developing universities. Then, in the mid-twentieth century it became 'professional' as government, industry and the military dominated scientific agendas. The purpose of science has changed, its sociology has altered and the extraction of a single 'tradition' is accordingly not only difficult but also even quite inappropriate. An additional implication is that any important scientific discoveries, especially those that have unfolded over a significant period of time (such as the Gas Laws), are likely to have been affected by this changing context in which science is undertaken and practised. Finally, there is a historiography to scientific traditions which we should recognise; in geomorphology, we label whole approaches as 'Davisian' or 'Gilbertian', using historical figures as convenient symbolic representations of whole 'traditions', parts of which they might not recognize themselves were they to reappear (e.g. Sack, 1992). Such symbolic labelling also gives the impression of step changes in a history which may often be rather more gradual and with multiple origins (consider Wallace, Darwin and evolutionary theory). Thus, for all these reasons, we should begin with the sceptical view that any methodological 'tradition' to which we might refer is itself a selection, not a given; a device, not a reality.

## SCIENCE AND PROBLEM-SOLVING

Turning more formally to the question of definition, science may be defined as '. . . systematic and formulated knowledge; the pursuit of this; the principles regulating such pursuit; any branch of such knowledge' (*Oxford English Dictionary*). On this basis, whether there is a physical science tradition hardly matters; geography is itself a science (at least it is if we consider our discipline to pursue and codify knowledge in such a systematic manner). Perhaps the interesting point about the formal definition of science is that it almost immediately introduces *method* – science as an activity or entity seems to be defined less by what it is, than by how it is done.

However, before dealing with the details of scientific method there is one additional, critical characteristic to review, and this is science's sense of 'problem' or 'question'. This is itself an early aspect of the application of scientific method, although it is often not treated explicitly as such. But as Bird (1989: 2) notes, science is a problem-solving activity, and 'the scientific method starts with some kind of problem: we might go as far as to say that problem orientation is the *raison d'être* of scientific enquiry. Problem recognition is a very important part of scientific endeavour, and often a very difficult intellectual exercise.' Anyone who has tried to develop a research project will know the significance of the final phrase in this quotation. It is often easy to find a general area in which research would be interesting, even valuable, but less straightforward to turn that interest into a researchable set of questions. This is the essence of problem formulation, and a scientific tradition which has made science such a successful enterprise.

An example will illustrate this difficulty of turning a general issue into a researchable problem: the decline of floodplain woodland in many river valleys in Europe and, indeed, globally. Channelization and river-flow management have separated rivers from their flood-plains, leading to reductions of flood frequency, of rates of channel migration and of recharge of floodplain sedimentary aquifers. In turn, this has caused reduced opportunities for the regeneration of woody riparian species, a loss of habitat and a reduction in biodiversity. One initial question we may need to ask is: 'Does society value the natural functions of floodplains sufficiently to wish to reverse these trends?' In one sense this question has already been answered because society has established institutions capable of monitoring the decline, and the existence of knowledge of the trend is a partial answer. However, social research may be needed to establish the sense in which this is regarded as a general 'problem' that needs to be placed high on the agenda of the environmental sciences. At this point the problem must be converted into more specific researchable questions, the answers to which help to explain the changed relationship between rivers and floodplains and the observed loss of ecosystem health. Examples might be: 'Does the timing of high flows under a controlled flow regime no longer match the timing of seed production?'; 'Does reduced flooding and recharge prevent the establishment and growth of seedlings?' or 'Is seedling growth affected by the rate of draw-down of the floodplain water table?' Then we can develop various strategies to attempt to answer the questions we have posed; we may, for example, deploy the range of scientific methods, which include observation, measurement, experimentation and theorization. During the discussion of scientific method following, we will return to this example to assess how these approaches can be helpful, using the last of these three questions as a case study.

One way in which geography frequently parts with a modern physical science tradition is in the manner in which it poses questions, given the nature of its general subject-matter. Geography is concerned with phenomena that unfold in an unconstrained social and environmental space, across a wide range of scales. In the physical sciences, the subject-matter can often be controlled in the laboratory and, accordingly, there is a tradition of posing questions which are answerable given the methods known to be both available in such a context, and capable of providing answers. Part of the very success of science therefore lies in its characteristic tradition of asking 'well posed' (as opposed to 'ill posed') questions. An example of this can be drawn from an application of probability theory. We may ask: 'How many people are required in order to have a better than evens chance that two or more of them have the same birthday?' This is an 'ill posed question' which we cannot answer until we are provided with a number of assumptions and conditions. If we assume that each individual's birthday is equally likely to be on any of the 365 days of the year, and that one person's birthday has no effect on anybody else's, the answer is that just 23 people suffice. This is rather a small number, which seems surprising. Indeed, a TV personality once denied the answer on the grounds that he had asked an audience of 100 people whether anyone had the same birthday as him, and the answer was 'no'. In fact, however, he was then asking a different question. The answer to his new question 'How many people are required in order that there is a better than evens chance that someone has my birthday?' is actually 253 (Grimmett, pers. comm.).

It is also often true of the physical sciences that research 'programmes' exist in which series of such problems or questions are posed in a planned, systematic and even evolutionary manner. It might be argued that geography again parts from the physical sciences in this respect because it seems to have been much less able to develop this systematic approach to the indentification and planning of specific, tractable research questions. Concern about an apparently dilettante approach to the identification of problems underlies Stoddart's (1987) plea for geography to rediscover larger questions. However, perhaps this is partly a characteristic of geography as an 'open-system' science which operates in the uncontrollable natural and social world where programmatic and answerable questions are more difficult to define; and partly it may reflect another geographical tradition, that of exploration – which may need planning but must also respect serendipity. Notwithstanding such reasons, practitioners of the discipline need to confront the possibility that this may also in part be a failing, in so far as it may reflect the lack of a coherent sense of systematic problem-formulation and problem-solving.

## ASPECTS OF SCIENTIFIC PRACTICE

From the foregoing it should come as no surprise that the scientific method is itself, in some senses, also unstable (for example, if the nature of science as an activity is historically contingent). There are several reasons for this instability, both external and internal to the specific practice of science. The former relate to the social context of scientific activity; scientific agendas are driven by changing political, industrial and business agendas, with science, technology and society forming a single domain (Latour, 1987), and with changes emerging from the feedbacks among them. The internal sources of instability arise because the appropriate methodology to employ may need to change as partial answers are found to the questions raised by the scientific problem in hand. Thus, what we have come to know about a problem may change our view of what more is knowable and of the means of acquiring this additional knowledge (Richards, 1996). This implies an interdependence between ontology, which is concerned with the nature of phenomena and existence, and epistemology, which is concerned with the means of gaining knowledge about them. There is no linear connection between these but, rather, a spiral of change over time. Since the acquisition of knowledge may alter our ontological view of the world, the method we employ subsequently may have to change. That this occurs implies that the scientific method cannot be considered as a normative set of rules but only as a more or less temporary model of how things may be done. Thus, 'the' scientific method is no more than the best, or simply the most appropriate, available description of the way in which reliable scientific knowledge can be acquired at a particular time or at a particular stage in the investigation of a problem. As a model, therefore, it is subject to criticism, test and reformulation just like any other model.

Given this view of the scientific method, debates about positivism, critical rationalism and realism may seem less important than consideration of the methodological contributions these have made to the range of practices available to scientists, and therefore also to geographers. These practices are the abiding legacies of science, if not its tradition, and it is appropriate to consider their role in geographical inquiry. Box 2.1 suggests, very broadly, the kinds of contributions to scientific practice associated with particular ontological and epistemological traditions. At the beginning of any scientific investigation, it is likely that we will draw on one of the central contributions

---

**Box 2.1** *Generalized relationships between philosophies and aspects of scientific method*

| | |
|---|---|
| **Positivism** | Representing: observation, measurement (Auguste Comte, 1798–1857) |
| **Logical empiricism** | Intervening: experimentation, laws of constant conjunction (David Hume, 1711–76) |
| **Critical rationalism** | Theorizing: hypothesize, experiment, test, falsify, refine (Karl Popper, 1902–94) |
| **Realism** | Uncovering hidden structures and mechanisms (Roy Bhaskar, 1944–; Andrew Sayer) |
| **(Post)modern science** | Perception reflects the perspective of the observer; the uncertainty principle (Albert Einstein, 1879–1955; Werner Heisenberg, 1901–76); at the limits of observation and intervention; chaotic behaviour of non-linear dynamical systems; sensitivity to initial conditions |
| **Rational criticism?** | |

---

of positivism to the range of scientific practices: its focus on *observables*. This contribution reflected the ontological project of positivism, which was to separate science from metaphysics by arguing that the former should be restricted to the study of those phenomena capable of being 'represented' (Hacking, 1983) through either observation or measurement, while it was the latter that was concerned with matters of faith. This, however, does not mean that positivism is exclusively associated with quantification, an erroneous assumption that often appears in geographical discussions. When this is perpetrated in error it is, perhaps, a reflection of the degree to which the practice of 'measurement' has almost effaced the more subtle process of 'observation'. But it may also be perpetrated as a form of disciplinary politics when positivism in its considerable diversity is rendered as a straw man to support an anti-quantitative stance which both oversimplifies 'positivism' by reducing it to numerical data and ignores the potential value, at some stage of many investigations, of such data and the tools for their analysis.

## *Representing: observation and measurement*

Thus, if we return to our question about the growth of seedlings, we may begin by using our (field) observations of riparian and floodplain environments. We may have noticed that in an Alpine river in which we are interested, the hydropower dams release water in ways that result today in the timing, magnitude and duration of the spring high-water period all differing from the historical situation when flows were natural; and we may also notice that riparian trees that seeded to coincide with the *natural* annual high-water period now do so too soon. We may also observe that where there are patches of damp, fine sediment on bar surfaces, germination and initial establishment of seedlings tend to occur preferentially in these locations (Figure 2.1), indicating the relationship among the physical nature of the flood-plain substrate, hydrology and regeneration of the population of certain species. Then (and only then, after we have observed and considered) we might undertake some measurements, for example by installing instruments to monitor groundwater levels in the floodplain to confirm that its moisture status closely tracks the now controlled fluctuations of river-flow (Figure 2.2). These observations and measurements do not always answer our question; indeed, they often only

**FIGURE 2.1**   An example of the association of recruitment of seedlings (of *Populus nigra*) with patches of fine sediment in a braided reach of the River Drôme, southeast France, which has a dominantly gravel-sand sediment load and substrate
Photograph by Francine Hughes

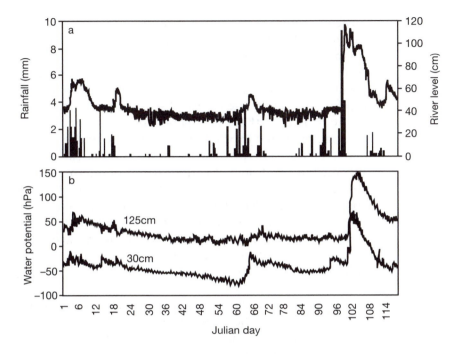

**FIGURE 2.2**  Data on rainfall and river level in the River Ouse near St Ives, Cambridgeshire, UK (a), and on water potential at two depths beneath the surface (30 and 125 cm) in the floodplain adjacent to the water-level recorder (b). At 30 cm, the potential is generally negative except during and after rainfall, especially from Julian day 100. At 125 cm, the potential is positive, indicating that the water table is generally above this tensiometer; at around Julian day 104, the potential is equivalent to a water depth of greater than 125 cm, reflecting the occurrence of flooding

suggest new ones. However, they provide us with information on patterns that begin to trigger ideas about underlying structures, mechanisms and processes about which we can theorize before undertaking new observations and measurements. Sometimes, of course, it is difficult to extract a pattern from what we see in the world because there are many pieces of conflicting evidence in a world that is unconstrained and in which some effects are cancelled by others. It is therefore difficult to theorize; but we can help ourselves to solve this problem by not only 'representing' but also by 'intervening' with experimental manipulation as discussed in the next section.

Several further issues arise when considering measurement: those of resolution, instrumentation, precision and accuracy, and dimensionality, all provide interrelated grounds for comparison of the physical sciences and geography. The reductionist tendency in science has driven it to examine phenomena of ever smaller scale, demanding increasing resolution of techniques for observation and measurement. Linked with this is the primacy of the visual sense in observation and the development of instruments to assist in 'seeing' these phenomena. Hacking (1983) provides an excellent account of the implications of this process of 'representing' the world for our appreciation of its nature. We cannot 'see' the human cell but we can have a theoretical model of its form and structure; it begins as a theoretical entity, therefore, not an immediately observable one. What can convince us that it is 'real' is that correspondence occurs between the form and structure predicted by the biological theory of the cell and that in the images of the cell revealed by either the optical or physical theory embodied in the design of various forms of microscope. Similar issues arise in the study of velocity variations in rivers. Here, the traditional propellor- or cup-based electromechanical current meter has been supplanted by electromagnetic and acoustic Doppler current meters as the subject of concern has shifted from time-averaged velocities to the structure of turbulence, requiring measurement of velocity variations with frequencies of 10–20 Hz. A problem, however, is that there are no theoretical descriptions of the structure of turbulence in particular circumstances – it has no generally definable character – against which to evaluate the evidence revealed by these instruments. Furthermore, two different instruments measuring at the same location can generate data with different temporal structures (Richards et al., 1997), resulting in a lack of confidence about the reality of those structures.

An area in which the physical sciences have a tradition which geography lacks is in their close alliance with the technology required to further scientific inquiry. This has been manifest in different ways at different times. For example, those working in the Cavendish Laboratory in Cambridge in the 1930s and early 1940s were adept at cobbling together apparatus for their experiments on subatomic particles and radioactive elements, using bits and pieces left over from other experiments and deploying practical skills of technical construction that were allied to their theoretical understanding of the scientific problems at hand. Today, the technology required to build scanning electron or scanning tunnelling microscopes is beyond a single research group or laboratory, and such equipment is provided by commercial scientific instrument makers, albeit with detailed design input from the scientists intending to use the apparatus. This use of 'off-the-shelf' instrumentation is prevalent in measurement practice in geographical research and characterizes the study of turbulence discussed above. Here, the manufacturers' decisions are less commonly influenced by the scientific users, and the result can be data collected in specific research circumstances which are contaminated with design decisions made for general construction. The most obvious of these are the physical properties of the sensor head (its shape and size) and the properties of the signal conditioning and filtering built into the electronic circuits. The best and most 'scientific' research into turbulence has been that which seeks a detailed understanding of the effects of these factors on the data collected, and seeks redesign when necessary.

In science, measurements are normally quoted with due consideration for their precision and accuracy (Figure 2.3); until recently this has been more honoured in the breach in the geographical literature. This is partly because there is rarely independent evidence of the 'true' value of a variable, and therefore difficulty in determining precision and accuracy. Consider the measurement of rainfall using

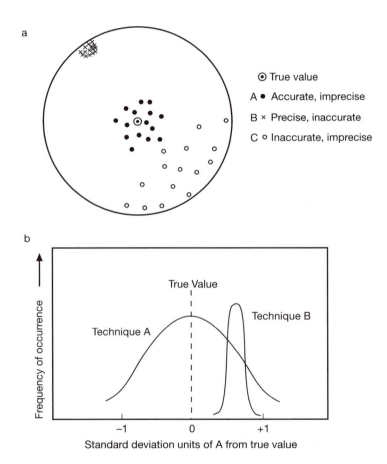

**FIGURE 2.3** The concepts of accuracy and precision are illustrated in (a) with the example of an archery target. The bull's-eye is the true value; attempts labelled A are accurate (they cluster around the true value) but imprecise (the variability is high). In (b) the concepts are illustrated using hypothetical frequency distributions; here, measurement method A gives an accurate but imprecise distribution of values, while method B gives a distribution which is precise but inaccurate

(point) raingauge data and (spatially distributed) rainfall radar; calibrating the latter with measurements from the former requires an extensive samping exercise before an approximation to a 'true' value can be defined. Nevertheless, the use of photogrammetric methods to derive digital terrain models (DTMs) has led to increasing emphasis on precision and accuracy. The DTM is also interesting because it provides three-dimensional information on the morphology of the earth's surface. Early quantitative geomorphology used one-dimensional indexes to represent the topographic form, such as the relative relief and drainage density. These provided a much needed alternative to imprecise and generalized qualitative descriptions of, for example, 'rolling landscapes'. However, the collapse of a three-dimensional entity (the landscape) into one-dimensional measurements loses a great deal of information, and these variables proved equally imprecise and not very accurate. The example of measuring flow velocity in rivers also draws attention to the change in dimensionality of much measurement in geography. Where the electromechanical current meter measured a time-averaged velocity in one direction (the main flow direction), the modern instruments that measure at high frequencies also measure in two or three orthogonal directions. This means that attention can be given to the vertical and cross-stream components of velocity and to questions of secondary circulation. It also means that the accuracy and precision of the measurements are not contaminated by folding into a single velocity component the velocities in the other two orthogonal directions. This is therefore an example of the rigour of geographical method adapting to new technology that permits a closer approximation to the traditional norms of the physical sciences.

### Intervening: experimentation

By 'intervening' (Hacking, 1983) we imply 'experimenting' – controlling, simplifying and manipulating the world so that we can see the patterns in a part of it more clearly. A typical illustration of a physical science experiment is that which reveals the pattern that was eventually codified as the Gas Law, which states:

$$pV/T = nR$$

where $p$ is pressure, $V$ is volume, $T$ is temperature, $n$ is the number of moles (the number of atoms or molecules) and $R$ is the gas constant. If a known quantity of a given gas is heated in a container of fixed volume, the pressure increases at a rate peculiar to that gas represented by its gas constant. In such an experiment, measurements of the temperatures and pressures would define a Humean 'law of constant conjunction', which is an empirical generalization deriving from the experimental measurements. This is a classic example of empirical chemistry, although its history spans two centuries, from Torricelli's invention of the mercury barometer in 1643, through the experiments of Robert Boyle and Jacques Charles, to the definition of Amadeo Avogadro's Number (which was based on a principle not accepted until 1860). The law is a very powerful one and, although it was derived under controlled laboratory conditions, it allows us to account for changes in an uncontrolled world; without it, it would be difficult to explain adiabatic air movements in the atmosphere and meteorology would be impoverished.

If we return to the example of floodplain ecology, where we have observed and measured some possible effects of altered hydrological regime on the regeneration of woody riparian plant species, we could conceive of an 'intervention' to assess our intuition here more rigorously. This might involve a field experiment; we could enclose some patches of floodplain (to exclude animal interference) and seed them under different but carefully controlled conditions. Because every site will experience slight variations in character – for example, in soils and sediments or in competitive species already existing in the seedbank in the soil – an experiment of this kind requires replication to employ statistical control of these possible extraneous influences. Such replication requires very careful experimental design, and agricultural science and ecology have done much to create the framework for this design following the pioneering work of Ronald Fisher (1890–1962) (Underwood, 1997). Although there are many variants on this theme, a key point is that experimental units (plots) must be located along the gradient of a hypothesized control variable so as to avoid bias in the event of co-variance of an unknown variable with that control. This requires randomization of replicate plots within 'blocks' defined at different points on the gradient of the control variable (Figure 2.4), and subsequent analysis of variance (ANOVA), a very powerful statistical tool. An alternative approach, however, is to seek a more physical than statistical form of experimental control by extracting a piece of the world into a laboratory (in this case, a greenhouse), and using apparatus like that shown in Figure 2.5 to grow seedlings under conditions in which soils are as standard as possible and the rates of water-table lowering are controlled. This enables the growth of seedlings to be monitored and related to the varying behaviour of the water-table in different soil types, and allows us to reinforce our conclusions about the effects of changes in hydrology and soil moisture on the regeneration performance of different species. Figure 2.6 illustrates some results of an experiment of this kind, with *Alnus incana* seedlings showing differential growth between two different substrates (a silty fine sand and a gravelly coarse sand), and with different rates of water-table lowering.

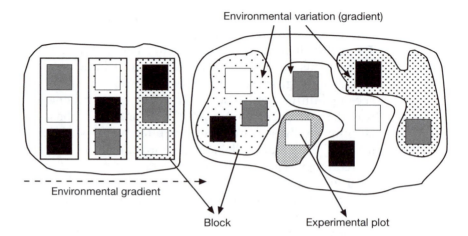

**FIGURE 2.4**  Examples of randomized plot experiments. In the left-hand case there is a relatively smooth environmental gradient and, at intervals along the gradient, blocks are defined within which random plots are chosen. In the right-hand case, the three 'blocks' are actually patches with a typical character, within which plots are randomly located

**FIGURE 2.5**  Rhizopods – soil-filled tubes connected at the base to central water reservoirs which allow control of the water table. Seedlings growing in the tubes show the effect of different water-table lowering treatments
Photograph by Francine Hughes

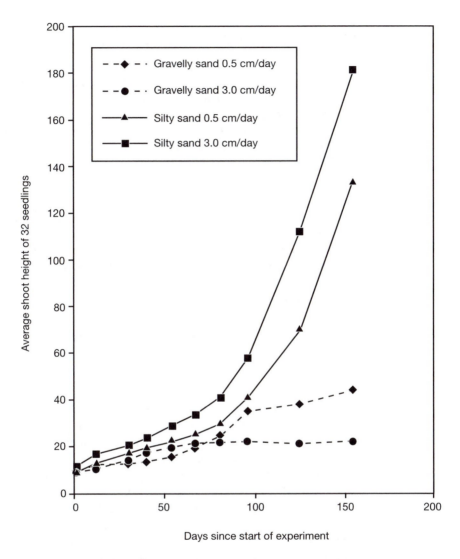

**FIGURE 2.6** Average shoot lengths in growth experiments with two soil types and two rates of water-table draw-down, conducted in the apparatus shown in Figure 2.5

Returning to the scepticism about the stability of traditions, it is worth noting that the concept of the 'experiment' is itself highly variable (as the very different statistical and physical ecological experiments described above indicate). There are many kinds and purposes of experiment, as is clearly demonstrated in Box 2.2. Harré (1981) uses Boyle's Gas Law experiments to make a 'measure of the Force of the Spring of the Air compressed and dilated' to exemplify the practice of 'finding the form of a law inductively', and identifies this as a kind of experiment which constitutes a formal aspect of scientific

---

**Box 2.2  The uses of experiment as identified by Harré (1981)**

A  As formal aspects of method:

1  To explore the characteristics of a naturally occurring process.
2  To decide between rival hypotheses.
3  To find the form of a law inductively.
4  As models to simulate an otherwise unresearchable process.
5  To exploit an accidental occurrence.
6  To provide null or negative results.

B  In the development of the content of a theory:

7  Through finding the hidden mechanism of a known effect.
8  By providing existence proofs.
9  Through the decomposition of an apparently simple phenomenon.
10  Through demonstration of underlying unity within apparent variety.

C  In the development of technique:

11  By developing accuracy and care in manipulation.
12  Demonstrating the power and versatility of apparatus.

---

method. However, some experiments are more for (im)proving a technique than discovering a property of the world and, even more striking, others are for 'finding the hidden mechanism of a known effect'. This latter kind of experiment, since it appears not to relate to observable entitities, cannot really be a positivist method; thus, it is hardly acceptable to regard a desire to conduct experiments as a sign of a positivist ontology. Experiments can include approaches as diverse as questionnaires and interviews as long as these are rigorously designed to investigate a particular problem and the issues of sampling and bias are given appropriate consideration (see Pawson, 1989). The tradition of experimental activity is therefore as diverse as science itself. However, experiments will not always *explain* why the regularities in behaviour that they reveal arise, which is why scientific inquiry into a particular problem has to adapt both its ontology and its epistemology as it uncovers information and we gain a deeper understanding of the world as a result. This is where theory (Box 2.1) enters the range of practices which constitute scientific activity, and where the critical rationalism of Popper and the realism of Harré and Bhaskar part company from positivism; science is clearly not simply about observables.

An experimental approach common in geography is one which combines extensive and intensive research, often at different stages in an investigation but not necessarily in a sequential manner – there is often a switching back and forth between these modes. Extensive research is empirical and concrete and often requires large samples to be gathered so as to demonstrate the variability of the phenomena and the degree to which specific cases studied in depth are representative. Intensive research, by contrast, is abstract and theoretical, involves small samples and seeks to uncover the underlying causal mechanisms that generate the patterns revealed by extensive investigation. One simple illustration of these two elements is the randomly sampled household questionnaire and the semi-structured interview with key respondents. Another is the systematic monitoring of solute concentrations in the discharge of water from a small catchment, combined with analysis of the mineral stability fields for interstitial water in the hillslope soils in order to explain variation in the slope drainage water chemistry. However, this is similar to the combination of empirical evidence and theoretical analysis that characterizes a classic physical science problem such as the behaviour of gases subjected to variable pressure and temperature (A1, B10 and B7 in Box 2.2); geography and the physical science tradition are thus not far apart.

The intensive phase of geographical inquiry often leads to case-study research. Such small sample investigation can nevertheless lead to generalization because the aim is to theorize about and understand the underlying mechanisms and observe how they cause observable events in one location. A case-study experiment requires recontextualization – particular mechanisms interact with the specific contextual character of the location under study to cause observable events which give us clues about the way nature works. An example is the observed event of a change in channel pattern on the South Platte River after 1850, from a braided to a meandering form, following reservoir impoundment in its headwaters. This control of flow reduced the stream power at the mean annual flood. The threshold between meandering and braided patterns occurs at a stream power of about $50$ W m$^{-2}$ but, after impoundment, the stream power in the South Platte declined to about $2.5$ W m$^{-2}$ at the mean annual flood. Thus, the post-1850 conditions were conducive to a change from dynamic braided patterns to more stable meandering. The contextualization of a general statement about channel pattern in this specific case is similar to the covering law model of the physical sciences; events are accounted for by the combination of a general statement (the covering law) and a specific statement (the initial conditions). Again, therefore, there are echoes of a physical science tradition in the methods of geography. In fact, a great deal of scientific activity requires defining the circumstances in which general statements (laws) are applicable and can be defined. This is even true of the experimental apparatus used in investigating the behaviour of gases. Place-based case-study research is to geography what the laboratory is to science. The only question is whether geographers can describe the open-system character of their field areas with both sufficient and necessary detail to account for the events generated by particular mechanisms or structures in this context, and to enable other geographers to replicate the findings of their case study.

### Theory: hypothesis generation, testing and falsification

One way in which theorizing enters the scientific process is through the control we impose on observation, measurement and experiment. Popper's critical rationalism (Box 2.1) denies inductivism and concludes that observation and its counterparts are theory laden. We do not gather observations randomly and *then* structure them according to some *post hoc* logic; the logic is mobilized at the outset because our theoretical preconceptions must determine what we choose to observe, what and how we measure, and the design of the experiments by which we collect data from which we may then generalize. Furthermore, experiments are created with a (theorized) purpose, one of which may often be to test a hypothesis, which will itself have been constructed as an outcome of theoretical consideration. However, such an experiment cannot logically be intended to *prove* or verify the hypothesis because there is always a possibility that further observations will be made which are contrary to the theory or model thus far supported by the empirical evidence. But, once a statement has been disproved, it cannot be 'un-disproved', and so experiments are designed to *falsify* a hypothesis (see Table 2.1). In the randomized plot-field experiments noted above (Figure 2.4) – the results of which are investigated using analysis of variance – this is achieved by stating a null hypothesis, such as 'there is no effect of moisture availability on the growth rate of seedlings of *Alnus incana*', and then seeking to disprove this in order to accept the logical alternative that an effect does exist.

If a null hypothesis is accepted, this suggests that the model that generated its logical alternative is wrong, which forces renewed theorization. If it is rejected, this is not the end either, because new hypotheses and experiments must be devised to provide even more stringent tests of the model. The limits of the Gas Laws were identified, for example, by experiments under temperature and pressure conditions at which the gases liquefied. As Underwood (1997: 19) notes, the cycle illustrated in Figure 2.7 is never ending:

> there is no way out of the procedure once you have started it, until you die or change research fields . . . Re-examination, novel testing, and more rigorous experimental analysis are part of the framework of

investigation. No single study – whether it rejects or retains some null
hypothesis – is sufficient to declare a problem solved.

The reference to rigour here is a reminder that whole books (including
that by Underwood) have been written on the subtleties required of
experimental design to ensure logical hypothesis testing, and the
pitfalls are numerous; unstructured predictions, lack of replication,
pseudo-replication, type I and type II errors (Table 2.1), lack of experimental power, conflation of interaction effects with error variance,
autocorrelation in temporal data, are just a selection. Consider Figure
2.6 again; these experimental data appear reasonably convincing, but
later differences in growth are determined in part by those that arise
early in the time series, possibly for abnormal reasons. There is also
an interaction between the effects of substrate and rate of water-table
lowering. What all these problems imply is that experimental tests
must be based on a clear understanding of underlying theory and the
logical implications that allow the generation of testable hypotheses.
Experimentation is never pure empiricism; it relies on theoretical
depth and careful logic. But the converse is also true; theory requires
evaluation against empirical evidence to demonstrate its value as a
mirror held up to the world. The lessons learnt from books on
experimental design in the field sciences are that such empirical
evaluation is an extremely arduous process and, in some areas of

**TABLE 2.1**    Type I and II errors and illustrations of their implications

A type I error in a test on pollution may lead to continued sampling and control,
but this means there is no harm to the environment. A type II error in a drug-
testing process could result in a harmful drug being marketed. Type II errors are
often more critical, but more attention is paid to type I errors.

| Outcome of statistical test is to | Null hypothesis ($H_0$) is (unknown to us) | |
|---|---|---|
| | True | False |
| Reject $H_0$ | Type I error: rejection of true null hypothesis (*Risk*: Falsely accept that an industrial discharge does not pollute) | Correct conclusion: false $H_0$ is rejected |
| Accept $H_0$ | Correct conclusion: true $H_0$ is accepted | Type II error: accept a false null hypothesis (*Risk*: Falsely accept that a drug has no side-effects and market it) |

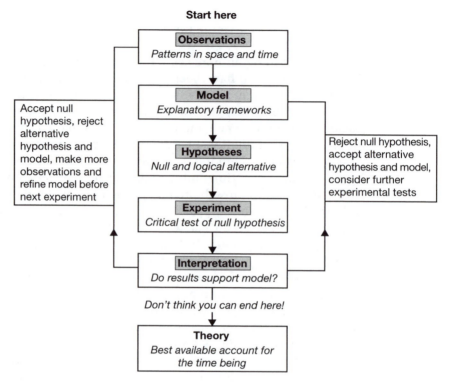

**FIGURE 2.7** The hypothesis-testing model of scientific method, illustrating its never-ending, 'cyclic' character and the role of the null hypothesis in a falsification process

geography, the tradition of applying the necessary rigour has been abandoned.

### Theory: mechanisms behind regularities

A second role for theory arises because having shown something experimentally – for example, that there is a simple relationship between temperature and pressure in a fixed volume of gas (at least, for a defined range of these controls) – the question then arises as to 'why' or 'how' nature engineers this pattern. This is a request for an explanatory account, not just a description. To answer these kinds of question, a theoretical approach is required which addresses unobservable mechanisms (see Box 2.1 for 'realism'). In the case of the Gas Laws, this theoretical explanation of the regularity is provided by the kinetic theory of gases. The pressure exerted by a gas is caused by collision of its molecules on the walls of its container; heat the gas and the speed of the molecules increases, the frequency of collision increases and the pressure rises; this is the essence of Daniel Bernoulli's suggestion in 1743 that was eventually quantified through the work of James Clerk Maxwell and Ludwig Boltzmann between 1860 and 1880. Keying 'gas law demonstration' into an Internet search engine will find several 'virtual laboratory' experiments that show this mechanism quite well, in addition to indicating the empirical regularity it generates. The regularity can be distorted very easily, however, by varying both the volume and temperature simultaneously, so the revelation of the regularity, and the theorizing that it triggers, depends on the skill in designing the experiment.

How, then, does this relate to the question of floodplain ecology? Considering Figure 2.6, a little thought reveals that seedling growth rates are probably not really *determined* by the rate of water-table lowering; this is a surrogate variable for something else. Especially when the water-table lowers very rapidly, a seedling's growth depends on its capacity to extract water from the the soil-pore spaces in the unsaturated zone above the water-table. Clearly this is likely to be more difficult when the water-table is receding rapidly, so there is at least a *correlation* between the depth of the water-table and the moisture status of the unsaturated zone above it, if not a direct causal connection. What determines the capacity of the plant to extract moisture from the soil is the inter-relationship of the rate of transpiration, the osmotic pressure at the roots and the soil moisture tension. When there is little moisture in the soil pores, the surface tension of the films of water held against the soil particles becomes very high and, even if the transpiration pull is also large, the diffusion of moisture into the roots is inhibited as cell walls fail and the plant wilts. Thus, the critical factors determining growth rates are the limiting osmotic pressure exerted through the roots, relative to the soil moisture tension. In an unsaturated soil, soil moisture tension prevents plants from accessing water when the moisture content is below the wilting point, and this depends on the soil moisture characteristic represented in the tension–moisture relationship for the soil (Figure 2.8a) and on the height above the water-table. The 'real' control variable is the soil moisture tension, for which the water-table depth is a surrogate. The tension can of course be measured using a tensiometer (Figure 2.8b), but it can also be theoretically predicted using numerical models of the saturated-unsaturated soil moisture dynamics above and below the water-table. In their simplest forms, such models are one-dimensional, finite–difference approximations of a column of soil (Figure 2.9), which numerically solve the Darcian equation for the soil-water flow velocity ($v$) at depths $h$:

$$v = -K \left( \partial\phi / \partial h \right)$$

where $K$ is the hydraulic conductivity, which is related to the soil-water tension and therefore the moisture content (as in Figure 2.8a),

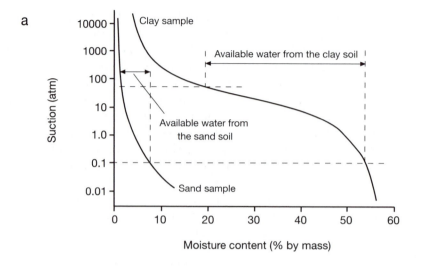

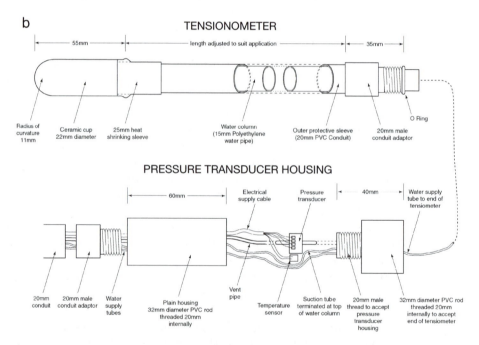

**FIGURE 2.8**   (a) Typical soil moisture characteristics curves (tension–moisture curves) for sand and clay soils, illustrating the different amounts of water available to plants, defined by the suction range across which plants can extract water from the soil-pore spaces; (b) a typical design of a tensiometer based on a pressure transducer, which measures suction when the water in the instrument is extracted through the porous tip to equilibrate with the soil moisture potential. This instrument is designed to be topped up with water from the surface and to be connected to a continuously monitoring data logger
Design by Adrian Hayes; diagram by Owen Tucker

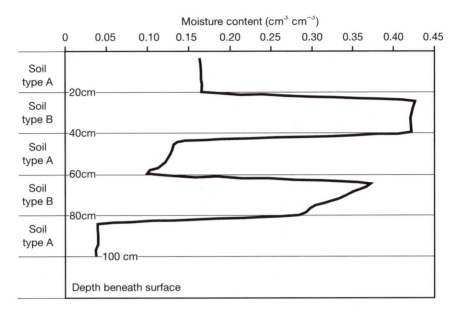

Moisture content (cm³ cm⁻³)

**FIGURE 2.9**  A modelled soil-moisture profile in a soil column with sandy and clayey horizons, with the finite–difference computational structure (25 cells each 4 cm deep) shown at the left

and ϕ is the soil-water potential (positive beneath the water-table where the water is under pressure, negative above the water-table where there is tension). The continuity equation is also required to account for the changing flux and storage of water in each cell of the model. Such theoretically based models are now very sophisticated, handling three-dimensional flow and multiple soil types. However, their use regularly reveals shortcomings in knowledge, such as that relating to the effect of large macro-pores in causing drainage at flow velocities faster than Darcy's law predicts, or that relating to the effects of plant roots in extracting water from storage at different depths. Thus theoretically based models also have an important role to play in probing the limits of existing understanding.

### Theory: demarcation criteria and the warranting of claims to truth

An important issue in scientific inquiry is that of determining which theory is closest to being a 'true' account of the aspect of the world being investigated (note: not '*the* true account'). This is why empirical investigation, albeit closely linked with theory, forms such an important part of science. In some areas of geography there is now almost a view that theory can be an end in itself, and a related view that all accounts are equally valid. However, this is fraught with danger; there must be some criteria for determining the trustworthiness of an account since, as Phillips (1992: 114) notes, 'a swindler's story is coherent and convincing'. More tellingly, Phillips notes that to apply to a theoretical argument about the world the criteria of value that we might apply to literature, poetry, art or music is to make a category mistake; these may be things of beauty, they may be evocative and they may illuminate the human condition, but they cannot be said to be 'true'. To apply the same standards to geographical theorizing and to deny claims to truthfulness is also to deny the application of theories in intervention or policy and to undermine any political project the research may have. There is also in some geographical writing a narrow interpretation of the criteria for warranting claims to truth employed by science; in the simplified model of the scientific method commonly presented in geographical literature, the criterion for selection of a preferred theory is that of predictive success. This relates to a misreading of Figure 2.7, in which hypotheses are seen to be tested by experiments, leading to a comparison of the prediction from the hypothesis with the data generated by the experiment. The misreading is in the assumption that measurement of 'success' is provided by a strong correlation between prediction and observation. As explained above, such an outcome of experimentation merely leads to another experiment until a falsification can bring about modification of the underlying theory; it is therefore empirical failure that brings successful theoretical development. This Popperian revision of the meaning of experimental success implies that warranting the truth claims of a given theory is a much more subtle and complex process in the sciences, just as it is in other branches of knowledge.

Phillips (1992) provides some valuable insights into the procedures employed to test the validity, or the proximity to truth, of hypotheses, models or theories in a discussion of the warranting of knowledge claims in qualitative social science research. This is part of a sustained argument against 'fabled threats' posed to naturalistic social science (that is, social science in which the ontology and epistemology are similar to that in the physical or natural sciences; see also Chapter 3). There are several practices that aid us in demarcating one theory as better than another in addition to the conventional one of predictive success (which, as argued above, gives only an illusion of validity). The first is a judgement about the *explanatory power* of a theory. This introduces the asymmetry between explanation and prediction; the capacity to predict does not imply an ability to explain, as any statistical forecasting model in economics or hydrology demonstrates. A predictive test of a theory that purports to provide explanation is inherently inconsistent, therefore, and something is required in addition. Of course, measuring the quality of an explanation is difficult, but two further criteria assist us in this: these are *structural corroboration* and *referential adequacy*. The first requires that the various parts of a theory give each other mutual support, and the second that the theory suggests new ideas about other, related phenomena. A glaciological example of this described by Richards (1996) involved experiments on surface melting, water balance, subglacial hydrology, hydrochemistry and basal water pressure, which all combined to provide support for a theory of the seasonal evolution of the glacier and its drainage system. The research itself evolved as new theoretical connections were identified and suggested additional experiments. However, these explanatory qualities alone cannot provide a reliable warrant for theory, and they must be supplemented by more familiar strategies: *multiplicative replication* and the *search for negative evidence*. The first involves repeated testing which generates comparable results, often with others undertaking the additional experiments. This is consistent with the practice in the physical sciences when new and unexpected results emerge and the scientific community seeks to replicate them; pathological events are discovered by this procedure, such as the spurious claims to have achieved atomic fusion at room temperature. The search for negative evidence is, of course, precisely the Popperian ideal. Thus, although this procedure for warranting the knowledge claims of social science seems to lack rigour and to be messy and imprecise, this is balanced by a spurious emphasis on predictive success in oversimplified accounts of demarcation in the physical sciences. The multiple methods of assessment required by all the sciences in their struggles to explain aspects of a world that hides its complex secrets from us have many elements in common, and the traditions of one are those of the many.

**CONCLUSION**

Thus, we have moved in this discussion, and in the associated interdisciplinary example involving hydrology and ecology, through a chain of practices from observation, through measurement, hypothesis generation and experimentation, to theoretical and physical analysis, just as physicists have done over the years in elucidating the behaviour of gases. Science requires this wide range of practices and geographical inquiry is equally reliant upon it and has, in some cases, developed its own strategies that have close parallels in the physical sciences (e.g. field-based case studies in place of laboratory investigation). The history of the Gas Laws shows that the only scientific tradition that now has meaning and value is that of plurality of method; this is what marks the practice of the physical sciences, and the flexibility, innovation and power that it bestows have led to their greatest successes. As Box 2.1 mischievously suggests, a (post)modern science now exists for which a simplified positivistic tradition is almost completely irrelevant. Particle physics and astronomy theorize about elementary particles and gravitational waves well beyond the limits of observation (but try to solve that problem); science widely grapples with non-linear dynamic processes with apparently stochastic outputs but deterministic internal structures; and physical science has had to deal with the problematic relationship between observer and observed since the quantum mechanical revolution and Schrödinger's cat. Thus, the negative interpretation of a straw positivistic science as part of an anti-quantitative purpose in some geographical writing is both anachronistic and inaccurate.

Geography and geographical research are all the better when they espouse a pluralist ontological and epistemological tradition of the kind that the physical sciences have evolved over centuries of uncovering the nature of the world. And, just as the practices and procedures have evolved as science has generated new knowledge and understanding, so, within a given research project, there is a continual movement between the various available practices as additional exploratory investigation vies with more developed theorizing about mechanisms in the established areas of research. All this implies that 'naturalism' need not be feared as an imperialist imposition of method by the physical sciences, but rather the normal set of potential procedures and practices for undertaking a logical, rigorous investigation that can be regarded as 'scientific', *sensu lato*. And, as Box 2.1 concludes, what we now require is a new tradition of rational criticism to cope with the diversity of modern scientific inquiry, whether physical, natural, environmental or social.

## Summary

- Traditions are rarely quite what they are claimed to be.
- Science has varied in its nature over time, so it is difficult to select a 'tradition'.
- Science is concerned with problems and problem-solving; an important early stage of the procedure of investigation is thus identifying researchable questions.
- 'The' scientific method is itself a model, and the procedures of scientific inquiry have evolved as knowledge and understanding have developed.
- Initially, procedures are required to 'represent' the world – through observation and measurement.
- It may then be necessary to 'intervene' by controlling, simplifying and manipulating part of the world in some form of experiment in order to indentify regularities of behaviour. This may require carefully designed field experiments or laboratory manipulation. The Gas Laws and greenhouse experiments on variations in

seedling growth rate in response to hydrological control are examples of the kinds of experiments that generate information on regularities.

- Experiments should be used to falsify theories, and this requires theorizing, hypothesis-testing and the use of a null hypothesis as a falsifiable alternative to the logical hypothesis derived from theory.

- Once regularities have been revealed they are explained through a process of theorizing about underlying mechanisms.

- A complex array of theoretical and practical procedures is required (in addition to the conventional approach based on prediction) in order to assess the relative merits of competing theories.

- All these procedures are employed in geography, either directly or in a suitably adapted form.

- Thus, geography has had at its disposal the same pluralist methodology that characterizes the physical sciences but which is, in fact, common to many areas of science (physical, natural, environmental and social).

## Further reading

Bird (1989), in *The Changing Worlds of Geography*, provides a balanced general introduction to ideas and methods in geography. An intriguing demonstration of the diversity of the sciences and their practice can be found in Harré (1981), *Great Scientific Experiments*. Phillips' (1992) volume, *The Social Scientist's Bestiary*, is an excellent critical analysis (and debunking) of the arguments against naturalistic social sciences. Richards (1996), in a useful anthology edited by Rhoads and Thorn, discusses the role of the case study and its methodological import. An interesting example of the historiography of physical geography can be found in Sack's (1992) article where she discusses the debate about Davisian and Gilbertian geomorphology. A readable account of the nature of science and scientific activity, and the relationship of science with society, can be found in Woolgar (1988), *Science: The Very Idea*.

*Note*: Full details of the above can be found in the references list below.

## References

Bird, J. (1989) *The Changing Worlds of Geography: A Critical Guide to Concepts and Methods*. Oxford: Oxford University Press.

Davidson, D.A. (1978) *Science for Physical Geographers*. London: Edward Arnold.

Hacking, I. (1983) *Representing and Intervening*. Cambridge: Cambridge University Press.

Harré, R. (1981) *Great Scientific Experiments*. Oxford: Oxford University Press.

Latour, B. (1987) *Science in Action: How to Follow Scientists and Engineers through Society*. Cambridge, MA: Harvard University Press.

Pawson, R. (1989) *A Measure for Measures: A Manifesto for Empirical Sociology*. London: Routledge.

Phillips, D.C. (1992) *The Social Scientist's Bestiary: A Guide to Fabled Threats to, and Defences of, Naturalistic Social Sciences*. Oxford: Pergamon Press.

Richards, K.S. (1996) 'Samples and cases: generalisation and explanation in geomorphology', in B.L. Rhoads and C.E. Thorn (eds) *The Scientific Nature of Geomorphology*. Chichester: Wiley, pp. 171–90.

Richards, K.S., Brooks, S.M., Clifford, N.J., Harris, T.R.J. and Lane, S.N. (1997) 'Theory, measurement and testing in "real" geomorphology and physical geography', in D.R. Stoddart (ed.) *Process and Form in Geomorphology*. London: Routledge, pp. 265–92.

Richards, K.S., Hughes, F.M.R., El-hames, A.S., Harris, T., Pautou, G., Peiry, J.-L. and Girel, J. (1996) 'Integrated field, laboratory and numerical investigations of hydrological influences on the establishment of riparian tree species', in M.G. Anderson et al. (eds) *Floodplain Processess*. Chichester: Wiley, pp. 611–35.

Sack, D. (1992) 'New wine in old bottles: the historiography of a paradigm change', *Geomorphology*, 5: 251–63.

Stoddart, D.R. (1987) 'To claim the high ground: geography for the end of the century', *Transactions, Institute of British Geographers*, 12: 327–36.

Underwood, A.J. (1997) *Experiments in Ecology: Their Logical Design and Interpretation using Analysis of Variance*. Cambridge: Cambridge University Press.

Woolgar, S. (1988) *Science: The Very Idea*. Chichester: Ellis Horwood.

# 3 Geography and the Social Science Tradition

**Ron Johnston**

## Definition

Social science is the study of human society and activity. It includes such disciplines as sociology, economics and political science. The field expanded rapidly in the postwar period, using scientific analysis to solve real-world problems. Before the 1970s few human geographers identified the discipline as a social science, but many now do so. Initially this move was associated with the adoption of 'scientific' methods and positivist ontology, but today many geographers who see themselves as social scientists challenge this scientific orthodoxy and draw on a more diverse range of theories and approaches, including Marxism, feminism and postmodernism.

## INTRODUCTION

Geography in general, and human geography in particular, has moved among the major divisions of academic life within the universities over the last century. Before the 1970s, very few human geographers identified their discipline as a social science: two decades later, most did. That shift was neither 'natural' nor necessarily obvious: it resulted from conflict over the discipline's identity and over the willingness of 'the social sciences' to accept geographers within their orbit. This chapter traces some of those conflicts and the changes in geography that they involved, with particular reference to the situation in the UK and, to a lesser extent, North America.[1]

## GEOGRAPHY'S ORIGINS

Geography as both an intellectual and a practical activity has a long history (Livingstone, 1992); geographical material was being taught in the ancient British universities by the late sixteenth century (Cormack, 1997; Withers, 2002; Withers and Mayhew, 2002) and in several American colleges in the early nineteenth century (Koelsch, 2002). But it only became a recognized segment of the academic discipline of labour – with separate university departments of, and degrees in, geography – in the early twentieth century in both countries. In the UK, almost every university had a geography department (and a professor, indicative of the discipline's status) by the Second World War, but most of these were small, with few graduates each year and no more than five staff members (Johnston, 2003a); in the USA, there was no time when there was a geography department in the majority of universities.

Much of the early pressure for the discipline's establishment in British universities came from the Royal Geographical Society (founded in 1830), whose major concerns were with the promotion of British imperialism and associated notions of citizenship (Ploszajska, 1999; Driver, 2001; on early geography in North America, see Schulten, 2001): it focused attention on Cambridge and Oxford. Elsewhere, the demand for geographical teaching came from a variety of sources. In some universities its introduction was linked to a major donor's wishes. In others, the case for geography courses came from economists, who wanted teaching in commercial geography (Chisholm, 1886; Wise, 1975; Barnes, 2001a):[2] indeed, the first professorship in geography was held by L.W. Lyde (a classical scholar and author of numerous school texts with sales of over 4 million) who taught courses for the Department of Economics at University College London.[3] Elsewhere, a separate geography presence emerged from the Department of Geology to cover the study of contemporary landscapes. And in most, the main rationale for geography degrees was to train students who would then teach the discipline in the country's public and grammar schools. In these ways, geography was established by individuals from a variety of backgrounds; most – especially those who taught human geography – were arts graduates.[4] In the USA, there was no central body pressurizing universities to introduce geography courses and departments, and those established reflected local demands – with most departments originating with the appointments of geographers to teach courses for either geology or economics students.

There was a very strong symbiosis between the secondary schools and the universities in promoting geography in the UK – as in Germany (Schelhas and Hönsch, 2002): the schools provided the students, very many of whom became schoolteachers after graduation. This symbiosis was enhanced by the work of the Geographical Association, founded in 1893 to promote geographical education at all levels, but which focused on geography in schools. It remains an extremely important pressure group (Balchin, 1993): without it (spearheaded for much of its existence by a few senior academics) it is very unlikely that geography would be as large a discipline in the country's universities as it is now. In this, the UK situation contrasts with that in the USA where, although there was early pressure for some training of geography teachers in the 'normal schools' in some states, the discipline was not in the high-school curricula and thus very few proceeding to university had much knowledge of it.[5] Student interest there had to be captured by professors offering attractive and interesting courses within much broader curricula than was the case with the much narrower UK honours degrees; indeed, it is still the case that very few American undergraduates go to university with the specific intention of majoring in geography.

Geography's origins were reflected in how it was practised for the first half of the twentieth century. The roots in geology were the basis for the development of physical geography – especially geomorphology, as with the influence of the Harvard geologist, William Morris Davis (Chorley et al., 1973). Those in economics stimulated interest in patterns of economic activity – of agriculture, industry and trade[6] – and links with other disciplines (such as anthropology, very strong at Aberystwyth and Belfast in the UK) generated work on less developed societies. This was further enhanced by the creation of geography departments in the universities of many of the countries of the then British Empire, which were staffed by expatriates – many on fixed-term contracts – who did research on the local area; a number of British geographers also developed regional interests based on their experiences during the Second World War.

These divergent scholars shared concerns with the interrelations between the physical environment and human activity. For many, the environment was a determining influence on human activity; to others, increasingly the majority, it was a strong constraint, but the ultimate determinant was human free will. Whichever position was taken, however, the outcome was the same: a mosaic of areas with particular environmental characteristics and human activities. Such areas were regions, separate areas with distinct landscapes (both natural and human) that distinguished them from their neighbours. Geographers saw the main rationale of their discipline as identifying and describing these areas (at a variety of scales and on a range of criteria; see also Chapter 1). The region was the core geographical concept; defining regions – largely through map comparison techniques – occupied the heart of the discipline's methodology; and studying regions was the ultimate purpose of a training in geography. (Many honours degree courses in the UK, and especially their final years, were dominated by regional courses until the 1960s.)[7]

## SOCIAL SCIENCE ORIGINS

This orientation of the discipline meant that there was little contact with the social sciences. In the first half of the twentieth century only economics from that group of disciplines was established in most universities, but there were few links between its theoretical approaches and geographers' empirical concerns. Neoclassical economics sought to account for the market operations through deductive model-building: geographers mapped patterns of economic activity and related them to the physical environment. With few exceptions, geographers made no use of professional economists' tools in their research and teaching (as illustrated by Rawstron, 2002) – even though one of the strongest UK geography departments was at the London School of Economics and Political Science – the LSE (see Barnes, 2001a, on the history of economic geography in both the UK and the USA).

The other social sciences were but minor presences in universities before the Second World War; only anthropology was firmly established, reflecting the British imperial role and the desire to understand and 'improve' the lives of 'native peoples'.[8] Indeed, geography was institutionalized into UK academic life long before either sociology or political science in the twentieth century. The Royal Geographical Society was founded in 1830, for example; the Geographical Association in 1893; and the Institute of British Geographers in 1933:[9] the Political Studies Association was founded in 1950, and the British Sociological Association in 1951. There were just two sociology departments in British universities in the mid-1950s and, of the 54 academic sociologists across 16 universities a decade later, fully 16 of them were at the LSE.[10]

In the 1950s, therefore, there were some links between geographers and anthropologists, and a few with economists, but geography largely existed outside the social sciences, instead occupying a claimed bridging-point between the arts and the sciences, combining the study of human activity within its environmental context through a focus on regions. Its nearest academic neighbour, according to one much-referenced North American scholar at the time (Hartshorne, 1939), was history. Both employed 'exceptionalist' approaches: historians studied particular time periods whereas geographers studied particular places. Both provided explanatory accounts of their periods/places through a synthesis of available material: both eschewed generalization; and they came together – especially in the UK where a strong tradition developed around one scholar, H.C. Darby (Prince, 2000) – in the study of historical geography.

And then – from the mid-1950s on – came change: why? The first reason was the rapid growth of the social sciences, reflecting their perceived relevance and applied worth. Economics became increasingly important as states became larger actors in and regulators of economies, especially during wartime, and as individual businesses became more professional in their operations with ownership and management shifting from individuals and families to shareholders in corporations. Economists played major roles in wartime governments, for example, and remained important thereafter, as the expanded state took on wider peacetime roles in economic management.

Economists' roles also increased within the growing state apparatus because of the growth of the welfare state, which provided economic and social protection for the vulnerable, invested in the future through universal schooling and widening university education, and redistributed wealth to produce a more equal society – a dominant ideological force of the times. Sociologists played an important role in this too, providing intellectual foundations for the more applied disciplines of 'social administration' and 'social policy' as well as through the importance of their core concept of class to those promoting a redistributive state.

The applied relevance of political science, which emerged as a separate discipline from roots in history and philosophy, came through desires to understand the working of the state apparatus and ensure the efficient operation of state bureaucracies – public administration (both national and local). And as globalization increased, with all the associated political tensions and conflicts, and with the Cold War stimulated by the ideological gulf between east and west, so the study of international relations increased in importance.

These three disciplines at the core of the social sciences – economics, sociology and political science – became major components of the academic world from the 1950s on, therefore. Anthropology failed to expand at the same pace, however, as interest shifted from 'primitive' to 'modern' societies and the stimuli to studying the former were reduced with decolonization. Other disciplines which overlapped the social sciences similarly increased in academic importance – notably psychology, which assumed increased importance in understanding and managing human behaviour in a range of contexts.

This demand for the social sciences, from users and potential students, stimulated growth at the universities – though less so in England's ancient establishments than elsewhere. (Sociology has only recently achieved departmental status at Oxford, for example, and there is no separate political science department. A major centre for postgraduate research in those disciplines was established at Nuffield College in the 1940s, however.) The LSE became a major UK centre for social science teaching and research, having been a pioneer in those areas for more than half a century.[11] Furthermore, almost all the new UK universities established in the early 1960s invested heavily in the core social science disciplines.[12] And they were the core disciplines whose postgraduate training and research activities were funded when a Social Science Research Council (SSRC) was established in the UK in the mid-1960s.

## BELATED MEETING: HUMAN GEOGRAPHY AND SOCIAL SCIENCE

Where was geography when all this was going on? What was its contribution to the war effort and to the burgeoning demand for social science expertise thereafter? With regard to the former, geographers were involved in a range of intelligence-gathering and provision activities – much of it associated with mapping and air-photo interpretation and the production of handbooks on countries where military operations were likely (Balchin, 1987; Clout, 2003). An American geographer – Ackerman – claimed that geographers' contribution there was not of very high quality, which led him to campaign for changes in the nature of the discipline (Ackerman, 1945; 1958). But what about postwar reconstruction and the Welfare State?

One potential area for geographers was the growing activity of town and country planning (called city and regional planning in the USA). There were increasing concerns for the most efficient use of land, for example, to ensure an adequate food supply during wartime. Subsequently, attention shifted to the need to distribute economic activities efficiently rather than allowing an overconcentration in certain areas, which would make them vulnerable to air attack. The need to protect high-quality agricultural land, to prevent urban sprawl, to reduce concentration on certain regions and to distribute land uses within cities efficiently (notably though their transport systems) stimulated planning legislation. This was enacted in the late 1940s in the UK, with both national planning (extending the regional policies of the 1930s enacted to deal with the problems of industrial 'depressed areas') and a requirement for all local authorities to produce local plans within a national framework.

Geographers' knowledge and expertise about regions could provide information for the production of national and local land-use plans – and a major Land Utilisation Survey mounted by Dudley Stamp at the LSE in the 1930s (by far the largest geographical 'research project' in the first half of the century) provided both valuable information and a template for such data-gathering exercises (Stamp, 1946). But could geographers be more than just data-gatherers and displayers? From the early 1930s, one of the country's first urban geographers, Robert Dickinson, argued for geographers focusing attention not on 'formal regions' (the separate parts of the landscape mosaic defined largely by their physical characteristics) but on 'functional regions', the tributary areas of towns and cities which were the basic structure within which society was organized. He argued that local government should be restructured to fit this pattern of functional organization, and that intra-urban planning should recognize the 'natural areas' of cities. In 1947 he published a major book based on US and European as well as British sources – *City Region and Regionalism* – which promoted these goals. City regions, according to his Preface, are 'aspects of the inherent spatial or geographical structure of society upon which planning must be based' and he presented the book as '. . . not about planning. It is concerned with certain aspects of the inherent spatial or geographical structure of society on which planning must be based' (Dickinson, 1947: xiii). By the end of the 1950s, many geography graduates were entering the planning profession but, although some occupied leadership roles (see Willetts, 1987), most were at the level of data-gatherers and displayers: the leadership in spatial planning of the 'brave new world' was provided by architects, surveyors and engineers (see Hall, 2003).

Had geographers missed the boat? A new set of disciplines had come to the fore, from which they were largely excluded – although a few, seeing the potential, allied themselves closely to the new disciplines and transferred their allegiance accordingly.[13] In the 1950s and 1960s a new generation of geographers sought to reorientate their discipline towards the social sciences. Much of the early impetus occurred in the USA – where many British geographers went for postgraduate training and other experiences, bringing the new ideas back to their country in the early 1960s (Johnston, 1997a). The 'exceptionalism' of regional geography was rejected – as providing 'mere description' – and the newly emerging social sciences were lauded, in part at least because their approaches and methods were closer to those of the natural sciences than to the arts. Promotion of this new approach occurred in a number of centres virtually simultaneously. Martin and James (1993: 372) call it spontaneous, although networks linking the various groups were soon established.

Three aspects of the new work were especially attractive to postwar generations of scholars:

1  *Its concern for scientific rigour.* Much current geographical practice was portrayed as theoretically weak and lacking the objectively neutral approach associated with both the 'natural sciences'. Schaefer (1953) published a damning critique of the 'Hartshornian orthodoxy', arguing that geographers should focus on identifying the laws that underpin spatial arrangements. This involved adopting the hypothetico-deductive 'scientific method', fully explored by Harvey's (1969) pioneering examination of the philosophy of

science, *Explanation in Geography*. This concluded with the state-ment 'by our theories you will know us' (Harvey, 1969: 486): expla-nation and prediction were to be the goals of human geographic research.

2  *An argument that quantitative methods formed a necessary com-ponent of this more rigorous approach* to the portrayal and analysis of information, including geographical information, although not all the early proponents of this cause necessarily tied it directly to the philosophical claims regarding 'scientific method'. The adop-tion of standard statistical procedures was seen by some simply as the proper way to use data (as in Gregory's 1962, *Statistical Methods and the Geographer*). To be rigorous, geographers had to be quantitative.

3  *A realization that rigorously obtained research results could be applied to a wide range of 'real world' problems*. Many geo-graphers were concerned that their discipline did not have the status it deserved among decision-makers (see Coppock, 1974; Steel, 1974). The social science disciplines were much more influ-ential because they took a more rigorous approach to problem-solving associated with the 'scientific method' and 'quantification'. Geographers should promote their expertise in the creation of spatial order – increasingly needed with the growth in spatial plan-ning – but should do this as scientists (which increasingly physical geographers were becoming too).

Within this context, what should be geographers' substantive con-cerns? Those attracted to the cause explored the literature (past and present) for inspiration. They found it in the general concept of spatial organization, the spatially ordered arrangement of human activities. Exceptional among those stimuli was the work of a German geo-grapher, Walter Christaller (1966), who developed central place theory in the early 1930s to understand settlement patterns. Individuals who journey to shops and offices for goods and services want to minimize the time and cost involved: the needed facilities should be as close to their homes as possible and clustered together so that they can make as many of their purchases as possible in the same place. And the owners of businesses want to maximize their turnover – with people spending as much as possible in the shops and offices and as little on transport. Thus an efficient distribution of services was in the inter-ests of both suppliers and customers. Christaller showed that this would result in a distribution of service centres across a uniform plane (i.e. with no topographical barriers) in an hexagonal arrangement, with the smaller centres (providing fewer services) nested within the mar-ket areas of the larger – though the details of that arrangement would depend on whether the goal was to minimize the number of settlements or the total length of roads. (On central place theory and its early influence, see Berry, 1967; Barnes, 2001b.)

Other works – all by non-geographers – provided similar stimuli. Economists such as Hoover, Palander, Lösch and Weber, for example, suggested that manufacturing industries would be located so as to minimize their input costs (among which a major variable element was the costs of transporting them to the plant from a range of sources) as well as their distribution costs (getting the final goods to the market): least-cost location was the goal, and this could be modelled as a form of spatial economics. And a nineteenth-century German landowner-economist (von Thünen) derived a similar model for the location of agricultural production, suggesting a zonal patterning of different activities consistent with the costs of transporting the output to markets. Economists built on this to suggest similar zonal organization of land uses within cities, which would be correlated with the pattern of land values.

Work on spatial patterns was complemented by studies of flows, of movements of goods, people and information. Their modelling was also based on principles of least effort: people wish to minimize their travel costs. The Newtonian gravity model was adapted, using the analogy that the larger the places of origin and destination the greater the movement between them, but that this would decrease, the greater the distance separating them. The various models of patterns and flows were brought together – and a new discipline, regional science, was launched, though it failed to gain the status that its founder (Walter Isard) sought. These location–allocation models integrated locations and flows, suggesting both optimum locations for facilities and efficient flows between them.

Geographers – especially those trained after the Second World War – were attracted to these models. They used them as the foundations for hypotheses that could be tested, using rigorous, quantitatively based procedures to show that locational decision-making was economically rational, and that planning for new facilities and routes could be based on such models. In addition, they rediscovered models of the internal spatial organization of cities into 'natural areas' developed by sociologists and others at the University of Chicago (Dickinson was the first to notice them, in the 1930s), and models of voting behaviour built on the spatial flow of information. These were all brought together in innovative and influential textbooks which discussed both the patterns and the methods for analysing them – such as Haggett's (1965) *Locational Analysis in Human Geography*, Chorley and Haggett's (1967) *Models in Geography*, Morrill's *The Spatial Organization of Society* (1970) and Abler, Adams and Gould's *Spatial*

*Organization* (1971): in different ways these emphasized the theme earlier pronounced by Watson (1955) that 'geography is a discipline in distance'. Cox (1976) argued that this new orientation brought geographical interests in line with contemporary society: in the pre-industrialized world, 'vertical' relationships between society and nature predominated as influences on regional patterns; in the industrial world, the horizontal relationships between and within societies were salient – and their study involved geography joining the social sciences.

Over the next couple of decades, the volume of work in this mould expanded greatly, applying and modifying the 'classic models', developing statistical and mathematical procedures for analysing spatial organization, exploring the underlying philosophy of the 'scientific method' (positivism: Harvey, 1969), and arguing that their models could be used as planning tools for cities and regions (Wilson, 1974). And the substantive interests expanded: a Swedish geographer – Torsten Hägerstrand (1968) – developed models of spatial spread and diffusion which were widely applied to patterns of change (notably the spread of disease); and a subfield of 'behavioural geography' evolved to embrace the 'scientific' study of human spatial behaviour and decision-making through the quantitative analysis of data obtained from questionnaires and similar instruments (Johnston, 2003b).

## GAINING RECOGNITION

Human geography was very substantially remade during the 1950s to the 1970s, therefore, though not without considerable conflict with those who sought to defend the status quo in, especially, regional and historical geography (Johnston, 1997a). As such, the remodelled discipline presented itself as a social science, claiming a clear niche within that area of activity with its focus of location and space (identifying itself as spatial science or locational analysis). But it was too late to gain entry to most of the UK's new universities of the 1960s: of them, only Sussex had a (relatively small) geographical presence virtually from the outset, and one was added at Lancaster in the early 1970s (because geography departments could attract students). An attempt by the RGS to promote geography with the founding bodies for the new institutions was unsuccessful; its claims for the discipline failed to match the scientific mood of the times (Johnston, 2003a). A few of the others (East Anglia, Lancaster, Stirling and Ulster) included geographers within multidisciplinary environmental science schools, but human geographers were in the minority compared to their physical geography colleagues (who had also been seduced by the three characteristics of the 'scientific method' listed above, and were remodelling their part of the discipline too).

Geography was also excluded from the SSRC when it was established. A group challenged this, presenting a case based on the 'new' geography (which was contested by some heads of geography departments and others, who wanted to maintain the status quo and did not identify with the social sciences). This was accepted, but geography, unlike the original disciplines, was not accorded separate committee status within the SSRC; instead it was linked with planning (Chisholm, 2001). Having achieved that status, the chief author of the case published a number of books promoting the new view of the discipline (Chisholm and Manners, 1973; Chisholm and Rodgers, 1973; Chisholm, 1971; 1975). Similar attempts were made in the USA, and two ad hoc committees made the case for recognition of geography both within the country's main research academy (NAS-NRC, 1965) and its social science community (Taaffe, 1970): these were bolstered by a further attempt to sustain and enhance their position three decades later (NRC, 1997). Even so, several geography departments with graduate schools were closed in the last third of the twentieth century (including prestigious institutions such as the Universities of Chicago, Michigan and Pennsylvania, plus Columbia, Northwestern and Yale Universities), by the end of which only one of the country's Ivy League universities – Dartmouth – had a geography department. As Koelsch (2002: 270) expressed it: 'the closing of geography in the major private universities sent a powerful signal that geography is no longer valued by academic administrators in institutions that traditionally have turned out the country's economic decision-makers and its cultural and political élite'.

Although social scientific recognition has been achieved, nevertheless geography is still considered peripheral to some aspects of academic life. In almost every country there is one or more national academy, an elected body of the country's main scholars. In the UK the two main bodies are the Royal Society (for the sciences) and the British Academy for the humanities and social sciences. Only three geographers have ever been elected to the Royal Society (Fleure, Wooldridge and Rhind). And no geographer became a Fellow of the British Academy until 1977, when the historical geographer, Clifford Darby, was elected: today there are some 20 Fellows. In the USA, there are two major comparable institutions. The National Academy of Sciences currently has some 1854 active members, of whom just nine are geographers (with five previous members now deceased); only one geographer – Brian Berry – has served on its council. (There is also one foreign member who is a geographer, a Nigerian – Akin Mabogunje.) The other body is the American Academy of Arts and Sciences, which currently has 3700 fellows, of whom only six are geographers.

## OPENING OUT

But things did not stand still. The social sciences were changing fast during the last three decades of the century, and geographers were changing with them. They discovered stimuli in aspects of the core disciplines that they had previously largely ignored: in economics, for example, there were both welfare (Chisholm, 1966) and Marxian (Harvey, 1973; 1982; the first of these books – *Social Justice and the City* – was extremely influential in stimulating a new focus to much Anglo-American human geography) approaches to be explored; sociologists, including the Chicago School, had studied a much broader range of subjects, with a wider range of methods, than those initially identified and adopted by geographers (as Jackson and Smith, 1984, cogently argued); and a range of multidisciplinary approaches – such as world-systems analysis – offered new arenas within which a spatial perspective could be crafted (Taylor, 1982).

At the same time the quantitative/positivist 'revolution', which many welcomed for its 'conceptual rigour' (Davies, 1972), itself came under attack. By reducing all decision-making to economic criteria, subject to immutable 'laws' regarding least-costs, profit-maximization and distance-minimizing, geographers, it was claimed, were ignoring (even denigrating) the role of culture and individuality in human conditioning and behaviour. By suggesting the use of those 'laws' as the bases for spatial planning, they were simply seeking to reproduce the status quo – of capitalist domination. And by assuming universal patterns of behaviour they were patronizing those who chose to operate differently.

Out of these arguments grew three main strands of work. One was *Marxist-inspired* (often termed radical), which explored not only the workings of the economy from that perspective, and added a spatial dimension to it (notably in Harvey, 1982, and Smith, 1984), but also the class conflict which underpins Marxian analyses of the economy and is central to a major area of sociological and political science literature. For such work, the positivist 'scientific method' was irrelevant since it assumes constant conditions within which economic decisions are taken whereas, for Marxist scholars, continuous change is the norm. Among alternative methods, the most popular (either explicitly or implicitly) was critical realism (Sayer, 1984). This accepts that there are general (or immanent) tendencies operating within capitalism, but that they are implemented by individual human agents making decisions in context – as illustrated by Massey's (1984) classic study of the changing geography of economic activity in Great Britain. Since by those decisions they change the context – in Massey's analogy, a new round of decision-making imposes a new layer on the map of locational activity – then the contingent circumstances within which future decisions are made must change too. Furthermore, the decision-makers themselves change as they learn from the making and consequences of previous decisions. There is a continuous interplay between structure and agency, or context and decision-maker, which Giddens (1984) termed structuration in a major contribution to sociological theory that was also influential among geographers. Thus for realists it is possible to explain why an event occurred – why a factory was located in a particular site, for example – but not as an example of a general law of location: explanation refers to specific events in context when decision-makers react to circumstances in order to meet certain imperatives (such as making a profit) within the constraints of their particular situations (what they know; what they believe their competitors will do; how they manipulate that knowledge, etc.).

The second strand drew particularly on work in sociology, especially though not exclusively work on gender and the growth of *feminist scholarship*. The core of the argument was that individuals occupy multiple positions within society, not just the class position which is at the core of Marxian analyses. Feminist geographers argued that not only was geography a male-dominated discipline but also that its concerns reflected masculine positions (Rose, 1993). Women were subordinated and ignored, and the goal was both to remove that ignorance and demonstrate that gender divisions in society could not be reduced to class position. From this emerged a wider concern with 'positionality', which embraced not only gender divisions within society but also ethnic, racial and national, plus sexual orientation and other criteria on which individuals' identities were based – such as the position of those living in postcolonial situations. Thus even gender had to be subdivided recognizing, for example, the different positions (and politics) of white and black women, of women in developed and developing world contexts, in various religions and so forth (McDowell, 1993). Appreciating those divisions, and people's positionality within them – and the many hybrids that emerge through, for example, mixing in multi-ethnic cities – cannot be achieved by the abstract theorizing of either positivist spatial science or Marxian analysis. It calls for interpretative methodologies aimed at understanding through empathy, which can be gained through a variety of methods developed in other social sciences – such as participant observation, focus groups, in-depth interviewing, the examination of archived resources (novels, diaries, biographies, works of art, landscapes and homescapes, etc.) – which give access to how people interpret their place(s) in the world, and how they act accordingly. This was the case, for example, with the burgeoning subdiscipline of critical geopolitics in the 1990s which, through links with parallel developments in international relations, sought to appreciate how influential political thinkers and politicians develop and propagate mental maps of the world as structures for action (Dodds and Atkinson, 2000).

Much of this work came to be associated, more explicitly in some cases than others, with what become known as postmodernism, again a major development in the social sciences (outside economics). This argues that there are no absolute truths and therefore no grand theories that can provide both explanations and guides to action (political or otherwise). Truths are the beliefs on which people act, so there are multiple truths – none of which can claim primacy over others; although the 'value' of competing truths can be assessed ethically (Smith, 2000). People learn their truths from others – either directly or through indirect sources (such as books). Such learning is context dependent and, since most live relatively spatially constrained lives, the spaces within which they learn are their homes, their neighbourhoods, their workplaces, the formal organizations they participate in and so on. Appreciation of the role of context has brought places back to centre stage in much human geographical research, not in the former regional tradition with contexts defined by environmental features but, rather, in a much more plastic way: places are made, remade and dissolved; they may overlap – or they may be bounded and defended (see Chapter 9).

This revived interest in places, and a shift of focus away from space within the discipline, is a feature also of the third strand. Geographers are playing a central role within a burgeoning field of *cultural studies*, which brings together scholars from the humanities and social sciences in new ways of approaching the study of human behaviour in context (see Chapter 4 for a discussion of geography and the humanities). Their work ranges over many aspects of behaviour, including the micro-scale of the individual body, seeking to understand the meanings that underpin actions – many of which are never recorded during the processes of everyday life (see Chapter 5). The relationships between people and nature are also being reconsidered, breaking down the perceived artificial boundaries between these long-considered binary opposites (Whatmore, 2002). Here again, new approaches are being explored for the interrogation of actions, including places as their arena: indeed, such is the geographical contribution to cultural studies that some identify a 'spatial turn' within the humanities (Anderson et al., 2002); other geographers continue to explore the interactions between humans and their environment in more 'traditional' ways (Turner, 2002).

## CONCLUSIONS: HUMAN GEOGRAPHY – SOCIAL SCIENCE AT LAST

Geography came late to the social sciences, therefore, and by the time that human geographers sought to ally with them they found they had been excluded. In response, while remaking their own discipline they also had to make strong claims that their discipline was changing and was now clearly a social science. To do this, they emphasized a particular aspect of the social sciences, privileging economic over other forces as determinants of human behaviour, and emphasizing models of spatial behaviour – of organization and flows – in which those forces dominated. They achieved some success in this strategy. A stream of work was introduced which remains strong, although it has changed over the last four decades. Rigorous analysis of quantitative data remains at the core of what is known as the spatial analysis tradition (Johnston, 2003b), but formal models based on idealized spatial patterns derived from oversimplified principles have largely been jettisoned – though interestingly they were taken up by a school of economists in the 1990s, in a 'new economic geography' which geographers (with some exceptions) claim they disowned 20 years ago (Clark et al., 2000).[14]

Alongside the spatial analysts, with their increased technical sophistication and reliance on advanced technology, other geographers discovered a wide range of approaches to explanation and understanding within the social sciences. Some have adopted approaches to explanation which differ from the positivism on which the original spatial analysts relied: others have argued that explanation is not feasible and only understanding is possible. They interact with very different areas of social science from the spatial analysts, and they too have won recognition and regard among their interdisciplinary peers.

'Positionality' is as central to academic life as to all other areas of society. Individual academics are schooled in particular approaches to the overall goal of understanding and changing society, within their own context – their own 'place'. Geographers have their collective 'place' – a perspective based around the key concepts of place, space, environment and scale (Massey et al., 1999) – which they promote, and within the discipline different groups of geographers emphasize different concepts. From those bases, some located in 'real places' (particular graduate schools, for example), they interact with other social scientists, bringing separate perspectives to bear on shared subject-matter. Interactions among the practitioners create wholes that are greater than the sums of the parts, communities with new hybrid perspectives on worlds and how they should be studied. For the last three decades at least, human geographers have been party to these negotiations, having largely abandoned their origins as a discipline built on firm foundations in the physical sciences.

## Summary

- Over the last century human geography has evolved from its position on the boundary between the arts and the physical sciences to become firmly established as a social science.
- The social sciences (economics, sociology and politics) grew rapidly after the Second World War because of their relevance in understanding the emerging global economy and changing social and political relations.
- In the 1950s and 1960s a new generation of human geographers sought to reorientate the discipline towards the social sciences.
- These geographers were concerned with scientific rigour and adopted quantitative methods to analyse spatial patterns and to develop models of spatial organization.
- Subsequently, this scientific orthodoxy has been challenged and now human geographers who identify themselves as social scientists draw on a more diverse range of theories and approaches, including Marxism, feminism and postmodernism.

## Further reading

For overviews of the history of geography, see Martin and James's (1993) *All Possible Worlds*, Livingstone's (1992) *The Geographical Tradition*, Johnston's (1997a) *Geography and Geographers: Anglo-American Human Geography since 1945* and the essays in Dunbar (2002), *Geography: Discipline, Profession and Subject since 1870*. Much of the discipline's nature and development is charted, and its terminology outlined, in the many essays and entries in Johnston et al.'s (2000) *The Dictionary of Human Geography*. A useful anthology of relevant materials is Agnew et al.'s (1996) *Human Geography: An Essential Anthology*.

*Note*: Full details of the above can be found in the references list below.

### NOTES

1  This concentration on the UK in part reflects my own knowledge and in part the greater volume of material on the history of geography there relative to the USA. There is, of course, no single unified discipline of geography: how it has developed in different countries – even different universities – reflects local conditions and personalities.

2  Chisholm's book appeared in 20 separate editions, the last in 1980, having been rewritten by Sir Dudley Stamp.

3  There was briefly an earlier chair in geography at that college, in the 1830s, occupied by Alexander Maconochie (Ward, 1960).

4  There were exceptions of course: the first professor at Oxford (Mason, appointed in 1933) had no academic training.

5  One of the leading American geographers of the early twentieth century – Mark Jefferson – worked in a 'normal school': see Martin (1968).

6   The University of Melbourne had two geography departments until the late 1960s. One, the oldest, was a Department of Economic Geography in the Faculty of Commerce; the other, established in the early 1960s, was the Department of Geography in the Faculty of Arts.

7   As an undergraduate between 1959 and 1962, the courses I took in the first year were all compulsory; none were regional in orientation. The second year included compulsory courses on Great Britain and on Ireland, and there was one optional course – I did the regional geography of India. And then in the final year, in addition to a dissertation, there was one compulsory course (on the geography of France and Germany), one major option (I did applied geography) and one minor option (I did the regional geography of southwest Asia): in addition, there were two papers in the final exams (on map interpretation – using French and German maps – and a general essay) for which there were no courses. There were some systematic courses in physical geography, but none in human geography (e.g. nothing on urban geography or industrial geography, etc.).

8   On the history of anthropology, see Kuper (1996). On geography's link to the anthropologists' project – what he terms 'European subject; world object' – see Taylor (1993).

9   The RGS is a general society with the goal of advancing the study of geography in all walks of life; the GA focuses on geographical education, especially in schools; and the IBG (which merged with the RGS in 1995) was a learned society for researchers – mainly university staff and postgraduate students. The comparable organizations in the USA are the National Geographical Society, the National Council for Geographical Education and the Association of American Geographers (which merged with the Association of Professional Geographers in the 1940s).

10  Developments were somewhat earlier in the USA: Chicago had a very strong Sociology Department by the 1920s, for example.

11  Interestingly, Sir Halford Mackinder, who founded the School of Geography at Oxford University in 1887, was Director of the LSE at the beginning of the twentieth century prior to developing a career in politics and diplomacy.

12  The Universities were East Anglia, Essex, Kent, Lancaster, Stirling, Sussex, Ulster, Warwick and York.

13  This was especially the case with sociology, and several geographers – such as Bill Williams at Swansea, Ray Pahl at Kent and Duncan Timms at Stirling – were appointed to chairs in that discipline.

14  In the USA, a report seeking to enhance the case for geography within the country's research structure – called *Rediscovering Geography* (NRC, 1997) – concentrated almost entirely on the spatial analysis approach, in human as well as physical geography, clearly believing that this was the best way to promote the discipline (see Johnston, 1997b; 2000).

## References

Abler, R.F., Adams, J.S. and Gould, P.R. (1971) *Spatial Organization: The Geographer's View of the World*. Englewood Cliffs, NJ: Prentice Hall.

Ackerman, E.A. (1945) 'Geographic training, wartime research, and immediate professional objectives', *Annals of the Association of American Geographers*, 35: 121–43.

Ackerman, E.A. (1958) *Geography as a Fundamental Research Discipline. Research Paper 53.* Chicago, IL: Department of Geography, University of Chicago.

Agnew, J., Livingstone, D.N. and Rogers, A. (eds) (1996) *Human Geography: An Essential Anthology.* Oxford: Blackwell.

Anderson, K., Domosh, M., Pile, S. and Thrift, N.J. (eds) (2002) *Handbook of Cultural Geography.* London: Sage.

Balchin, W.G.V. (1987) 'United Kingdom geographers in the Second World War', *The Geographical Journal,* 153: 159–80.

Balchin, W.G.V. (1993) *The Geographical Association: The First Hundred Years, 1893–1993.* Sheffield: The Geographical Association.

Barnes, T.J. (2001a) 'In the beginning was economic geography: a science studies approach to disciplinary history', *Progress in Human Geography,* 25: 521–44.

Barnes, T.J. (2001b) 'Lives lived and lives told: biographies of geography's quantitative revolution', *Environment and Planning D: Society and Space,* 19: 409–29.

Berry, B.J.L. (1967) *The Geography of Market Centers and Retail Distribution.* Englewood Cliffs, NJ: Prentice Hall.

Chisholm, G.G. (1886) *Handbook of Commercial Geography.* London: Longman.

Chisholm, M. (1966) *Geography and Economics.* London: Bell.

Chisholm, M. (1971) *Research in Human Geography.* London: Heinemann.

Chisholm, M. (1975) *Human Geography: Evolution or Revolution.* London: Penguin Books.

Chisholm, M. (2001) 'Human geography joins the Social Science Research Council: personal recollections', *Area,* 33: 428–30.

Chisholm, M. and Manners, G. (eds) (1973) *Spatial Policy Problems of the British Economy.* Cambridge: Cambridge University Press.

Chisholm, M. and Rodgers, B. (eds) (1973) *Studies in Human Geography.* London: Heinemann.

Chorley, R.J., Beckinsale, R.P. and Dunn, A.J. (1973) *The History of the Study of Landforms. Volume II. The Life and Work of William Morris Davis.* London: Methuen.

Chorley, R.J. and Haggett, P. (eds) (1967) *Models in Geography.* London: Methuen.

Christaller, W. (1966) *Central Places in Southern Germany* (trans. C.W. Baskin from 1933 original in German). Englewood Cliffs, NJ: Prentice Hall.

Clark, G.L., Feldman, M.P. and Gertler, M.S. (eds) (2000) *The Oxford Handbook of Economic Geography.* Oxford: Oxford University Press.

Clout, H. (2003) 'Place description, regional geography and area studies: the chorological inheritance', in R.J. Johnston and M. Williams (eds) *A Century of British Geography.* Oxford: Oxford University Press.

Coppock, J.T. (1974) 'Geography and public policy: challenges, opportunities and implications', *Transactions, Institute of British Geographers,* 63: 1–16.

Cormack, L. (1997) *Charting an Empire: Geography at the English Universities, 1580–1620.* Chicago, IL: University of Chicago Press.

Cox, K.R. (1976) 'American geography: social science emergent', *Social Science Quarterly,* 57: 182–207.

Davies, W.K.D. (1972) 'The conceptual revolution in geography', in W.K.D. Davies (ed.) *The Conceptual Revolution in Geography.* London: University of London Press, pp. 9–17.

Dickinson, R.E. (1947) *City Region and Regionalism*. London: Routledge & Kegan Paul.

Dodds, K.J. and Atkinson, D. (eds) (2000) *Geopolitical Traditions: A Century of Geopolitical Thought*. London: Routledge.

Driver, F. (2001) *Geography Militant: Cultures of Exploration in an Age of Empire*. Oxford: Blackwell.

Dunbar, G.S. (ed.) (2002) *Geography: Discipline, Profession and Subject since 1870*. Amsterdam: Kluwer.

Giddens, A. (1984) *The Constitution of Society*. Cambridge: Polity Press.

Gregory, S. (1962) *Statistical Methods and the Geographer*. London: Longman.

Hägerstrand, T. (1968) *Innovation Diffusion as a Spatial Process*. Chicago, IL: University of Chicago Press.

Haggett, P. (1965) *Locational Analysis in Human Geography*. London: Edward Arnold.

Hall, P. (2003) 'Geographers and the urban century', in R.J. Johnston and M. Williams (eds) *A Century of British Geography*. Oxford: Oxford University Press.

Hartshorne, R. (1939) *The Nature of Geography*. Lancaster, PA: Association of American Geographers.

Harvey, D. (1969) *Explanation in Geography*. London: Edward Arnold.

Harvey, D. (1973) *Social Justice and the City*. London: Edward Arnold.

Harvey, D. (1982) *The Limits to Capital*. Oxford: Blackwell.

Jackson, P. and Smith, S.J. (1984) *Exploring Social Geography*. London: Allen & Unwin.

Johnston, R.J. (1997a) *Geography and Geographers: Anglo-American Human Geography since 1945*. London: Arnold.

Johnston, R.J. (1997b) 'Where's my bit gone? Reflections on *Rediscovering Geography*', *Urban Geography*, 18: 353–9.

Johnston, R.J. (2000) 'Intellectual respectability and disciplinary transformation? Radical geography and the institutionalisation of geography in the USA since 1945', *Environment and Planning A*, 32: 971–90.

Johnston, R.J. (2003a) 'The institutionalisation of geography as an academic discipline', in R.J. Johnston and M. Williams (eds) *A Century of British Geography*. Oxford: Oxford University Press.

Johnston, R.J. (2003b) 'Order in space: geography as a discipline in distance', in R.J. Johnston and M. Williams (eds) *A Century of British Geography*. Oxford: Oxford University Press.

Johnston, R.J., Gregory, D., Pratt, G. and Watts, M. (eds) (2000) *The Dictionary of Human Geography* (4th edn). Oxford: Blackwell.

Koelsch, W. (2002) 'Academic geography, American style: an institutional perspective', in G.S. Dunbar (ed.) *Geography: Discipline, Profession and Subject since 1870: An International Survey*. Dordrecht: Kluwer, pp. 281–316.

Kuper, A. (1996) *Anthropology and Anthropologists*. London: Routledge.

Livingstone, D.N. (1992) *The Geographical Tradition: Episodes in the History of a Contested Enterprise*. Oxford: Blackwell.

Martin, G.J. (1968) *Mark Jefferson: Geographer*. Ypsilanti, MI: Eastern Michigan University Press.

Martin, G.J. and James, P.E. (1993) *All Possible Worlds: A History of Geographical Ideas*. New York, NY: Wiley.

Massey, D. (1984) *Spatial Divisions of Labour: Social Structures and the Geography of Production*. London: Macmillan.

Massey, D., Allen J. and Sarre, P. (eds) (1999) *Human Geography Today*. Cambridge: Polity Press.

McDowell, L. (1993) 'Space, place and gender relations (two parts)', *Progress in Human Geography*, 17: 157–79 and 305–18.

Morrill, R.L. (1970) *The Spatial Organization of Society*. Belmont, CA: Wadsworth.

NAS-NRC (1965) *The Science of Geography*. Washington, DC: NAS-NRC.

NRC (1997) *Rediscovering Geography*. Washington, DC: NRC.

Ploszajska, T. (1999) *Geographical Education, Empire and Citizenship: Geographical Teaching and Learning in English Schools*. London: Historical Geography Research Group of the Royal Geographical Society.

Prince, H.C. (2000) *Geographers Engaged in Historical Geography in British Higher Education 1931–1991. Historical Geography Research Series 36*. London: Historical Geography Research Group of the Royal Geographical Society.

Rawstron, E.M. (2002) 'Textbooks that moved a generation', *Progress in Human Geography*, 26: 831–6.

Rose, G. (1993) *Feminism and Geography*. Cambridge: Polity Press.

Sayer, A. (1984) *Method in Social Science: A Realist Approach*. London: Hutchinson.

Schaefer, F.K. (1953) 'Exceptionalism in geography: a methodological examination', *Annals of the Association of American Geographers*, 43: 226–49.

Schelhas, B. and Hönsch, I. (2002) 'History of German geography: worldwide reputation, and strategies of nationalisation and institutionalisation', in G.S. Dunbar (ed.) *Geography: Discipline, Profession and Subject since 1870: An International Survey*. Dordrecht: Kluwer, pp. 9–44.

Schulten, S. (2001) *The Geographical Imagination in America, 1880–1950*. Chicago, IL: University of Chicago Press.

Smith, D.M. (2000) *Moral Geographies: Ethics in a World of Difference*. Edinburgh: Edinburgh University Press.

Smith, N. (1984) *Uneven Development: Nature, Capital and the Production of Space*. Oxford: Blackwell.

Stamp, L.D. (1946) *The Land of Britain: Its Use and Misuse*. London: Longman.

Steel, R.W. (1974) 'The Third World: geography in practice', *Geography*, 59: 189–97.

Taaffe, E.J. (1970) *Geography*. Englewood Cliffs, NJ: Prentice Hall.

Taylor, P.J. (1982) 'A materialist framework for human geography', *Transactions, Institute of British Geographers*, 7: 15–34.

Taylor, P.J. (1993) 'Full circle, or a new meaning for the global?' in R.J. Johnston (ed.) *The Challenge for Geography. A Changing Word; A Changing Discipline*. Oxford: Blackwell, pp. 181–97.

Turner, B.L. (2002) Contested identities: human–environment geography and disciplinary implications in a restructuring academy', *Annals of the Association of American Geographers*, 92: 52–74.

Ward, R.G. (1960) 'Captain Alexander Maconochie, R.N., 1787–1860', *The Geographical Journal*, 126: 459–68.

Watson, J.W. (1955) 'Geography: a discipline in distance', *Scottish Geographical Magazine*, 71: 1–13.

Whatmore, S. (2002) *Hybrid Geographies*. London: Sage.

Willetts, E.C. (1987) 'Geographers and their involvement in planning', in R.W. Steel (ed.) *British Geography 1918–1945*. Cambridge: Cambridge University Press, pp. 100–16.

Wilson, A.G. (1974) *Urban and Regional Models in Geography and Planning.* Chichester: Wiley.

Wise, M.J. (1975) 'A university teacher of geography', *Transactions, Institute of British Geographers*, 66: 1–16.

Withers, C.W.J. (2002) 'A partial biography: the formalization and institutionalization of geography in Britain since 1887', in G.S. Dunbar (ed.) *From Traveller to Scientist: The Professionalization and Institutionalization of Geography in Europe and North America since 1870.* Amsterdam: Kluwer, pp. 79–119.

Withers, C.W.J. and Mayhew, R.J. (2002) 'Rethinking "disciplinary" history: geography in British universities, c 1580–1887', *Transactions, Institute of British Geographers*, 27: 11–29.

# 4  Geography and the Humanities Tradition

**Alison Blunt**

## Definition

The humanities encompass the study of human creativity, knowledge, beliefs, ideas, imagination and experience. Such work has inspired a wide range of humanistic, cultural and historical geographical research. At the same time, ideas about space, place and imaginative geographies have inspired work across the humanities. This chapter explores the creative interfaces between geography and the humanities.

## INTRODUCTION

Geography and the humanities are intimately connected, although it is only relatively recently that such connections have been explicitly explored in theory and in practice, and both by geographers and by those working in other disciplines. The term 'humanities' refers to the study of human life and humanity more widely and is usually associated with particular subjects, approaches and methods (see Williams, 1976, for more on 'humanity' and related terms). Subjects usually (but not exclusively) located within the humanities include literary studies, languages, history, art, philosophy, archaeology and cultural studies. Research within the humanities is often concerned with human creativity, interpretation, meanings, values and experience, in both historical and contemporary contexts and over both imaginative and material terrains. Rather than view the humanities and social sciences as clearly distinct from each other, it is more helpful to think of a series of interfaces between them. Indeed, the diverse traditions and approaches within human geography have

shaped a wide range of research and teaching located at these interfaces, making productive connections that span sources, methods and
ideas both within and between the humanities and social sciences.
Much has changed since Denis Cosgrove wrote in 1989 that 'the idea
of human geography as a *humanity* is scarcely a mature or fully
developed one' (p. 121).

In the years since Cosgrove made this argument – and in large part
inspired by his own research on visual images, iconography and
landscape – the idea of human geography as a humanity has taken root
to an unprecedented extent. Early attempts to draw creative connections between geography and the humanities have long antecedents
(particularly in the work of Carl Sauer on premodern cultural landscapes), but are usually attributed to the work of humanistic geographers in the 1970s. As D.W. Meinig (1983: 315) explains:

> while we have long had a 'human geography', we have never before had
> an explicitly 'humanistic geography' with such a self-conscious drive to
> connect with that special body of knowledge, reflection, and substance
> and human experience and human expression, about what it means to
> be a human being on this earth.

(For an overview of humanistic geography, see Cloke et al., 1991).
Writing to counter positivist spatial science and structural Marxism
(see Chapter 3), humanistic geographers sought to replace 'rational
economic man' with a fully human subject, whose thoughts, experiences, values, emotions, agency and creativity made him a unique
individual within a wider humanity (and it *was* usually 'him' in
humanistic work; see below). David Ley and Marwyn Samuels, for
example, wanted 'man put back together with all the pieces in place,
including a heart and even a soul, with feelings as well as thoughts,
with some semblance of secular and perhaps transcendental meaning'
(1978: 2–3). At the same time, humanistic geographers claimed that
part of what made people human was an intense, sensual, and often
passionate, attachment to place. As Yi-Fu Tuan (1976: 269) wrote:

> How a mere space becomes an intensely human place is a task for the
> humanistic geographer; it appeals to such distinctively humanist inter-
> ests as the nature of experience, the quality of the emotional bond to
> physical objects, and the role of concepts and symbols in the creation of
> place identity.

The work of humanistic geographers was very influential in fore-
grounding human experience and a sense of place to counter the
abstractions of spatial science. Alongside Marxist geographical work
and the emerging discipline of cultural studies, humanistic geography
laid important foundations for a revitalized cultural geography, particularly from the 1980s onwards (Jackson, 1989; Crang, 1998; Cook
et al., 2000). But many critiques of humanistic geography centred on

its limited engagement not only with the humanities but also with the very idea of humanity itself. In relation to the humanities, Stephen Daniels (1985: 150) writes that 'The naïve approach of most humanistic geographers to the arts, especially their neglect of the artistic form, reveals their poor working knowledge of the humanities, especially literary and art criticism and history'. Not only did humanistic geography largely lack a theoretical and methodological engagement with other research in the humanities – particularly in humanistic studies of geography and literature and the surprising disregard of the visual arts – but humanistic geography also offered little to other scholars in the humanities seeking to understand literary texts and visual images in material as well as imaginative contexts. In relation to humanity more widely, Gillian Rose (1993) critiques the masculinism of humanistic geography that privileges but does not interrogate the primacy of the male subject. In particular, Rose shows how humanistic geographers have celebrated the home as the site of authentic meaning and value but failed to analyse the home as a gendered space shaped by different and unequal relations of power. Alongside such feminist critiques that show the importance of understanding the gendered dynamics of human subjectivity, postcolonial geographical work critiques ethnocentric – and often transparently white – visions of the 'human' in human and humanistic geography (for more on challenging the assumed transparency of whiteness, see Bonnett, 1999; for more on postcolonial geographies, see Blunt and McEwan, 2002).

Emerging, in part, from such critiques of humanistic geography, cultural and historical geographers today engage with the humanities in more critical and diverse ways. Cultural and historical geographers (and, as will become clear, the work of many geographers is *both* cultural and historical) have developed a more critical and theoretical engagement with other research in the humanities, often in the light of poststructuralism, feminism and postcolonialism. There has also been a greater recognition and interrogation of the complexities of humanity itself, both in terms of the politics of identity and also in terms of the moral and ethical considerations that are a crucial part of understanding the human world (Proctor and Smith, 1999; Smith, 2000). At the same time, rather than view space merely as a container for human life and experience, or merely to celebrate a sense of place and individual perceptions of place, there has been an increasingly critical understanding of the production of space and place mediated by different relations of power (see Chapters 5 and 9). More than ever before, scholars working in other disciplines in the humanities are thinking and writing in explicitly spatial terms, most notably in terms of imaginative geographies and the multiple and contested spaces of identity, which are often articulated through spatial images such as mobility, location, borderlands, exile and home. The rest of this chapter addresses these points more fully in relation to three main interfaces between geography and the humanities: geography, text and writing; geography, landscape and the visual arts; and geography, embodiment and performativity.

## GEOGRAPHY, TEXT AND WRITING

If 'geography' means 'writing the world', geographers write the world and think about the world *and* writing in different ways. Geographers have increasingly turned their attention not only to 'writing' and the 'world' being written about but also to the wider politics and poetics of representation (Barnes and Duncan, 1992; Duncan and Ley, 1993; Barnes and Gregory, 1997). For many humanistic geographers, the study of 'geography and literature' offered one way of reading human attachments to, and perceptions of, place (Tuan, 1978; Pocock, 1981; Mallory and Simpson-Housley, 1987). For Tuan (1978: 205):

> Literary art serves the geographer in three principal ways. As though experiment on possible modes of human experience and relationship, it provides hints as to what a geographer might look for when he studies, for instance, social space. As artifact it reveals the environmental perceptions and values of a culture . . . Finally, as an ambitious attempt to balance the subjective and the objective it is a model for geographical synthesis.

Seeking to foster a 'meeting of the disciplines' between geography and literature, Mallory and Simpson-Housely (1987: xii) hoped to 'bridge the gap between the geographer's factual descriptions and the writer's flights of imagination, hence giving the world – both in geographical and literary terms – a more unified shape'. But in both these examples, geography and literary studies – sometimes seeming to parallel 'reality' and 'imagination' – remained largely separate from each other. While literature (particularly novels) provided a new source for geographical analysis, the analysis itself often remained social scientific and paid little attention to literary theory and criticism. Rather than study literary forms, conventions and language, humanistic geographers often regarded 'literature simply [as] perception' (Pocock, 1981: 15). As Daniels (1985: 149) writes, 'For a humanist to abstract the perceptive insight of an artist from the language in which it is embodied is paradoxically to diminish an artist's humanity'.

Such early encounters between geography and literature also led to an increased awareness of geographical writing itself. Many geographers stressed the importance of vivid, compelling descriptions of the world, infusing an understanding of place with human meanings, values and experiences. So, for example, Pierce Lewis described his love of the Michigan sand dunes in sensual and embodied prose:

My love affair with [the] Michigan dunes . . . had everything to do with violent immediate sensations: the smell of October wind sweeping in from Lake Michigan, of sand blown hard against bare legs, the pale blur of sand pluming off the dune crest against a porcelain-blue sky, Lake Michigan a muffled roar beyond the distant beach, a hazy froth of jade and white. As I try to shape words to evoke my feelings, I know why the Impressionists painted landscapes as they did – not literally, but as fragments of color, splashes of pigment, bits of shattered prismatic light. One is meant to feel those landscapes, not to analyze them. I loved those great dunes in my bones and flesh. It was only much later that I learned to love them in my mind as well (1985: 468; also see Meinig, 1983).

As well as such 'thick' description, a closer attention to creative geographical writing also took a more experimental form, best shown in the work of Gunnar Olsson (1980; 1981) who focused attention on the slippages between language and meaning. While such experimental writings revealed the instabilities and ambiguities of meaning, the elision of creativity with experimentation has been critiqued for its opacity. As Daniels (1985: 148) writes: 'You don't have to be a linguistic puritan to conclude that the literary styles of humanistic geographers are no less gratuitous than the quantitative or algebraic flourishes of some positivist geographers'.

Since the late 1980s, a number of geographers have engaged more directly and fully with literary theory and have developed a more critical understanding of text, language, reading and writing (Brosseau, 1994; Sharp, 2000). Often inspired by poststructuralism – particularly in terms of discourse analysis and deconstruction – cultural geographers have turned their attention more fully to the politics of representation. Rather than analyse texts as unproblematic representations of the world or merely in terms of perception, cultural geographers ask more critical questions about texts and the contexts in which they are written and read. A number of geographers have also developed a more metaphorical understanding of text, as shown by the work of James and Nancy Duncan on reading landscapes as texts rather than reading texts merely as 'literary artifacts' (Duncan and Duncan, 1988; Duncan, 1990). Geographers have also begun to analyse a much wider range of writing, including travel writing (Blunt, 1994; Duncan and Gregory, 1999), tour guides (Gilbert, 1999; Howell, 2001), geography textbooks (Ploszajska, 2000), children's fiction (Phillips, 1997; 2001), letters, memoirs and diaries (Blunt, 2000a; Gowans, 2001), as well as novels (Sharp, 1994; Brosseau, 1995). One key theme has been an attempt to explore geographies of writing in both imaginative and material contexts and in the very form of writing itself. So, for example, diaries can be read as 'spatial stories', detailing events and feelings not only at the time but also in the place where they occur, while travel writing is inherently geographical in its depiction of mobility within and between the spaces of home and away. Much of this more recent geographical work on writing has been influenced by postcolonialism and feminism, interrogating identity in terms of gender and race and, to a lesser extent, class and sexuality. Most importantly, geographers increasingly recognize that writing – and representation more broadly – is located within a nexus of power relations.

While geographers are increasingly inspired by work across the humanities in the study of writing, a number of literary theorists have explored the imaginative geographies produced in part through writing. Perhaps most famously in this context, the literary critic and postcolonial theorist, Edward Said, has written about the imaginative geographies of 'self' and 'other' in his study of Orientalism (Said, 1995; for more on imaginative geographies and Said's work, see Gregory, 1995; Rose, 1995; Driver, 1999). Said shows the interplay of power, knowledge and representation through written and visual depictions of an exotic East. Focusing on texts such as travel writings, scholarly accounts and novels, Said argues that Orientalism produced knowledges about colonized people and places as inferior, irrational and 'other' to a powerful, rational, Western 'self'. As he shows, such writings, and the imaginative geographies they helped to shape, had material consequences in the exercise and legitimation of imperial rule. More recently, Said has written a memoir of his early life in Jerusalem, Cairo, Lebanon and the USA. Aptly titled *Out of Place* (Said, 1999), geography lies at the heart of Said's memoir. As he explains: 'Along with language, it is geography – especially in the displaced form of departures, arrivals, farewells, exile, nostalgia, homesickness, belonging and travel itself – that is at the core of my memories' (Said, 1999: xvi). Imaginative and material geographies are clearly central to interpreting text and writing not only within but also far beyond the discipline of geography.

### GEOGRAPHY, LANDSCAPE AND THE VISUAL ARTS

While humanistic geographers saw literature as an important source for describing human emotion, experience and a sense of place, they paid far less attention to the visual arts (Daniels, 1985). More recently, cultural geographers have increasingly recognized that vision, visual representation and 'the gaze' – in other words, practices of looking and observing – are inherently spatial. Interpreting the spatiality of the gaze means paying attention to *where* observation takes place as well as what is being observed; thinking critically about the distance between the observer and the subject of observation; seeking to locate

and embody the gaze and to challenge claims for objective detach-
ment; and analysing various framing strategies that put boundaries
around what is being observed (and far more on how to do this in
practice, see Rose, 2001). Geographers have always been interested in
visual representations of the world, particularly through maps and
mapping. Rather than view maps in positivist terms as objective
reflections of reality, many geographers now argue that all maps are
socially constructed forms of knowledge that are partial, infused with
different meanings and shaped by different relations of power (Harley,
1988; Woods, 1994). Geographers have, for example, studied the ways
in which maps have been produced and used not only as objects of
imperial power but also of postcolonial resistance (see Jacobs, 1996,
for an example of Aboriginal 'counter-cartography' in Australia). In his
work on avant-garde visions of the city, David Pinder (1996) explores
Utopian remappings of urban space as a space of human freedom and
creativity, in part by overturning conventions not only of cartography
but also of urban planning and design. Such critical, interpretive and
often deconstructive readings of maps are closely related to the visual
and textual analysis associated with work in the humanities.

Unlike the textual descriptions of landscape in humanistic geography, many cultural geographers have studied landscape as a 'way of
seeing' and have analysed visual depictions of landscape in painting
and photography and, to a lesser extent, in film (Cosgrove and Daniels,
1988; Ryan, 1997; Nash, 1999a; Schwartz and Ryan, 2003). As Cosgrove and Jackson (1987: 96) explain, 'the landscape concept is itself a
sophisticated cultural construction', and one that is closely tied to
power relations. Denis Cosgrove (1985) traces the 'idea of landscape'
to landscape painting and property ownership in the Renaissance,
arguing that visual depictions of landscape – revolutionized by the
geometric representation of space – were bound up with the interests
of the powerful. Geographers study the iconography of landscape,
which involves studying the symbolic significance of the landscape
itself and analysing symbolic markers within a landscape, such as
monuments and place-names, in their social and political context
(Cosgrove and Daniels, 1988; Johnson, 1995; Nash, 1999b). Geographers also study the ways in which ideas of landscape are closely
bound up with imaginative and material geographies of nation and
empire. David Matless (1998; see also Daniels, 1993), for example,
analyses a diverse range of visual and textual sources in his study of
landscape and Englishness from the early twentieth century to the
present day. Focusing on debates about nation, citizenship and heritage, he shows how ideas of Englishness were intimately bound up
with ideas of landscape. Other work shows how such ideas of Englishness and landscape have had far-reaching effects over imperial space.

In her discussion of the picturesque style of painting and viewing landscape, Catherine Nash explains that:

> [a] sense of the superiority of English landscape aesthetics was linked to a broader certainty that English ways were the best ways of doing things and of their natural superiority and authority over other people and places. Ideas of landscape were involved in the multiple ways in which European expansion within imperial trade and colonization was naturalized and legitimated (1999a: 220; also see Seymour, 2000).

A number of feminist geographers have argued that ideas of landscape, and the geographical analysis of visual images more widely, are profoundly gendered both in theory and practice. Gillian Rose critiques both 'social-scientific' and 'aesthetic' masculinism, with the latter claiming 'complete sensitivity to a mysterious yet crucial world' (1993: 61). Focusing on the work of 'new' cultural geographers, Rose argues that there are uneasy tensions between pleasure in, and knowledge about, landscape that reflect and repeat profoundly gendered distinctions between a feminine 'nature' and a masculine 'culture'. As she writes: 'Pleasure in the landscape is often seen as a threat to the scientific gaze, and it is argued that the geographer should not allow himself to be seduced by what he sees' (Rose, 1993: 72). Feminist geographers argue that not only are vision, visual representation, 'the gaze' and ideas about landscape inherently spatial, but that they are also gendered in important, but often unacknowledged, ways. As Rose (1993: 109) puts it, 'cultural geography's erotics of knowledge' are masculinist and heterosexist, gazing on a feminized landscape in voyeuristic, distanced and disembodied ways that render the specificity of its own gaze unmarked and invisible. Countering such a masculinist gaze, Catherine Nash has studied the gendered and sexualized interplay of landscape, body and nation in modern Ireland (Nash, 1994; 1996), while Rose has more recently written about photographs of, and by, women in the nineteenth and early twentieth centuries (1997; 2000).

I want to contrast two images to illustrate the gendered geographies of the gaze in more detail. Both images date from 1857 and portray gendered spaces of home, nation and empire through English eyes. The first is a painting called 'The sinews of old England' by George Elgar Hicks (Figure 4.1). It shows a family positioned on the threshold between home and world. The title and the painting represent English national identity in clearly embodied terms that centre on the strength and work of the man. But in this image such an embodied, masculine and working-class national identity also depends on feminized domesticity. The man is pictured with his wife and child, and the viewer can look into the tidy and well ordered home beyond the threshold. While the man gazes into the distance

*Source*: Unknown (thought to be in a private collection)

**FIGURE 4.1**    'The sinews of old England', by George Elgar Hicks, 1857

away from home and beyond the gaze of both his wife and the viewer, his wife looks up at him lovingly. While the man does not return this gaze, we, as viewers of the image, see the family as a whole within its pastoral and domestic context. This mid-Victorian painting is a classic representation of public and private space and the white, heterosexist gender order on which such spatial and social divisions relied.

Compare Hicks' painting with another image from 1857 (Figure 4.2). This second image is a cartoon entitled 'How the Mutiny came to English homes' and is located in India at the start of the uprising that threatened to overthrow British rule (for more on the 'mutiny', see Blunt 2000a and 2000b). While Hicks' painting offered a glimpse over the threshold into a tidy and ordered home of peace and calm, this image represents the home as a place of danger and violation. With a young child playing next to her and a baby at her breast, a British woman is located at the centre of domestic and familial calm that has just been shattered by the invasion of two Indian insurgents. The violent presence of these men is the only indication that the home is in India rather than Britain. The rebels appear set to destroy the defenceless woman, her children and the home itself. Moreover the box labelled 'England' on the chaise longue suggests the vulnerability of national and imperial power. The absent British husband and father – present only in a portrait on the wall – is unable to protect his wife

*Source*: Unknown

**FIGURE 4.2** 'How the Mutiny came to English homes', 1857

and family. This image stands in stark contrast to Hicks' painting of national identity embodied by a family on the threshold between home and world. It represents the threat to domestic, national and imperial life during the uprising, with the severity of the threat shown by the vulnerability of a British wife, mother and children. Both images show the political importance of the home in terms of national and imperial identity, and both images also show the gendered subjects and spaces shaping national and imperial identity. If, as Hicks' painting suggests, the 'sinews of old England' were embodied by masculine strength and feminine domesticity, the cartoon shows such sinews – and the security and identity of 'old England' itself – being torn apart.

While geographers now study visual images in more critical ways, a number of art historians and film theorists have begun to make the spatiality of visual images central to their work. Feminist work has been particularly influential in seeking to embody and contextualize the gendered spatiality of spectatorship and the gaze. As one example of such work, Griselda Pollock (1988) analyses the paintings of two female Impressionists in explicitly spatial ways. Focusing on the work of Berthe Morisot and Mary Cassatt in late nineteenth-century Paris, Pollock explores the spaces of their art in three connected ways. First, she writes about the spaces represented in their paintings, which were often interior, domestic spaces. Then she explores the spatial order within their work, showing how spaces are often separated by a balcony or a balustrade, marking spaces in gendered ways and positioning both the painter and the viewer in close proximity to the subject of the painting. Finally, she examines the social spaces of representation in terms of where and what a woman was able to paint. She quotes from the diary of Marie Baskirteff, another artist in Paris in the nineteenth century, who wrote that her lack of independence and mobility constrained her work:

> What I long for is the freedom of going about alone, of coming and going . . . of stopping and looking at the artistic shops, of entering churches and museums, of walking about old streets at night; that's what I long for; and that's the freedom without which one cannot become a real artist. Do you imagine that I get much good from what I see, chaperoned as I am, and when, in order to go to the Louvre, I must wait for my carriage, my lady companion, my family? (quoted in Pollock, 1988: 70).

Imaginative and material geographies are also clearly central to interpreting landscape and the visual arts not only within, but also far beyond, the discipline of geography.

## GEOGRAPHY, EMBODIMENT AND PERFORMATIVITY

Unlike humanistic geographical work that celebrated human creativity, agency and individuality, recent geographical work has engaged more critically with questions of identity by interrogating both the politics of identity and the political intersections of place and identity (Rose, 1995). A recent and important theme within such work has been to understand and interpret identities in embodied terms (Butler, 1999; Longhurst, 2000; McDowell, 1999; Valentine, 2001). Ideas about embodiment have important implications not only for studying identity but also for the production of knowledge more widely. Resisting disembodied claims to objectivity, many scholars argue that knowledge itself should be embodied. In the context of geographical knowledge more specifically, James Duncan and Derek Gregory (1999) argue that studies of travel writing should not remain solely textual but should also focus on the embodied practices of travel itself, while Felix Driver (2001) argues that fieldwork and exploration also need to be understood as embodied practices rather than solely read in textual terms. Retelling a famous story of exploration, John Wylie (2002a) studies the Antarctic expeditions of Scott and Amundsen in terms of their embodied understanding of the landscape and, more specifically, of ice. Contrasting English and Norwegian encounters with Antarctic ice, Wylie shows how the landscape itself came to shape different mobilities across it. Elsewhere, Wylie (2002b) discusses the different modes of dwelling and moving embodied by Scott and Amundsen in their polar expeditions. By paying attention to their 'sensualities and sensibilities' (2002b: 263), Wylie (p. 259) vivdly describes exploration as an embodied practice, as shown by the following passage:

> Each night in the tents, eyes blinded by the luminous intimacy of the landscape are treated with zinc sulphate and cocaine, then swaddled in rags and tea leaves. Frozen feet and hands are placed upon the warm chests and stomachs of consenting companions in a series of awkward embraces, unlikely arrangements of bodily parts. Antarctica demands, above all, that the frontiers of one's body be rigorously established and maintained.

Ideas about embodied practices are often closely tied to ideas about performance and performativity, as shown by geographical studies of city workplaces (McDowell, 1997) and gay and lesbian identities (Bell and Valentine, 1995). Geographical studies of embodiment and performativity offer important new ways of thinking about geography in relation to the humanities. At the same time, such studies also represent one of the most productive interfaces between social scientific and humanities-based geographical approaches, methods and analysis, whereby interviews, ethnographic research and focus group discussions can be combined with textual, visual and aural analysis of different sources and practices. Geographers have begun to engage with cultural forms and practices beyond literature and the visual arts that include creative fields such as dance, theatre and music (Leyshon et al., 1998; Malbon, 1999; Thrift, 1997; 1999). Geographers and many others working in the arts and humanities have also paid more critical attention to the spaces of display and performance, both within and beyond spaces that include art galleries, museums, music venues and nightclubs. Underlying this critical engagement with space, embodiment and performativity is a growing interest in cultural *practices* as well as cultural forms, artifacts, texts and images. The diversity of such interest is reflected in a section in each issue of the journal *Cultural Geographies* (formerly *Ecumene*) dedicated to 'Cultural geographies in practice', which 'offers a space for critical reflection on how practices within the artistic, civic and policy fields inform and relate to the journal's cultural geographic concerns'. Often closely bound up with this growing interest in cultural practices is a concern to articulate and explore sensual and emotional geographies. While humanistic geographers sought to replace 'rational economic man' with a thinking, feeling, human subject, but neglected the differences and inequalities between people, many human geographers today explore the politics of identity in embodied, sensual and emotional terms. I want to discuss two examples that illustrate this interest in cultural practices and sensual geographies before turning to some of the theoretical and methodological challenges posed by studying embodiment and performativity.

The first example concerns an artwork that takes place beyond the space of the gallery, and one that is concerned with sound, vision, movement and embodiment. In his study of an audio walk through east London by artist Janet Cardiff, David Pinder (2001: 2) shows that 'The artwork literally takes place in the streets, finding its meaning through its embodied enaction. In effect it is performed or co-created by participants'. While the artwork is 'a highly specific experience [that] is different according to mood, circumstances, events as well as the identity of the individual walker' (p. 15), it also raises wider questions about the city itself. As Pinder argues, the audio walk 'raises critical questions about reading and representing urban ambiences, about the interweaving of memories and urban space, and about the construction of senses of self through urban space-times' (p. 15). Addessing similar themes about embodied engagement and performance, but in a different context and both within and far beyond the space of the gallery, Ian Cook (2000) discusses the art installation by Shelley Sacks called 'Exchange values'. As Cook (p. 338) begins:

> They're in your face. Banana skins. Dried. Cured. Blackened. Flattened.
> Sewn together in a panel. Stretched. Taut. On a frame. Right in your
> face. And it smells. It's rich. Gorgeous. You can't move too far away. Get
> too distanced. If you want to keep the headphones on. The ones that are
> attached to the little metal box below the panel and the frame. The one
> with the number on. E490347.

As Cook explains, the number E409347 identifies Vitalis Emmanuel, a
banana farmer from St Lucia. He grows bananas for export, and this
number appears on the boxes that protect them as they are transported
around the world. Some of his bananas were handed out to people in
Nottingham on condition that they ate them on the spot and gave
back the skins. The banana skins form the basis of the installation,
and the metal box below contains a tape that plays the voices of 20
farmers including Vitalis Emmanuel. Cook (2000: 342) describes the
installation in terms of connective and collective creativity between,
for example:

> Shelley Sacks, the people who helped her to give out those bananas in
> Nottingham and to collect the skins, the people who helped to dry, cure
> and stitch those skins onto panels, the 20 St Lucian banana growers, and
> all those people – politicians, business people, extension officers and
> others – who helped her to find them. Add to that the representative
> groups campaigning for changing relationships between producers and
> consumers.

Such connections not only stretch over space but also create 'a
reflective space. A space of possibility. Where connections can be
seen. Felt. Thought through' (p. 342).

As these two examples show, ideas about embodiment and
performativity, cultural practices and sensual geographies are inspir-
ing a diverse and exciting body of geographical work. These ideas also
pose some important theoretical and practical challenges. A central
task for geographers working within a humanities tradition is to
examine their own research in practice by engaging more fully and
directly with methodological debates. Nigel Thrift's ideas about non-
representational theory – and the challenges of putting such ideas into
practice – are particularly important in this context. Thrift (1999: 318)
argues that cultural geography 'often seems to have taken representation as its central focus to its detriment' and proposes a 'non-
representational theory' of dance and other embodied practices. As
Catherine Nash (2000: 656) writes: 'Thrift is advocating a new and
demanding direction for cultural geography, away from the analysis of
texts, images and discourses, and towards understanding the micro-
geographies of habitual practices, departing from deconstructing representations to explore the non-representational'. Nash ends her review
of work on 'performativity in practice' with some wariness about 'a

retreat from feminism and the politics of the body in favour of the individualistic and universalizing sovereign subject' (2000: 662). Echoes of a humanistic geography clearly remain important and contested in humanities-based geographical work today.

## CONCLUSION

Much has changed since Cosgrove (1989) lamented that the idea of geography as a humanity was still largely undeveloped and immature. From the work of humanistic geographers in the 1970s, through the revitalization of cultural geography since the 1980s, the idea of geography as a humanity has taken much stronger root. The three parts of this chapter show the different ideas, sources and approaches involved in geographical studies of text and literature, landscape and the visual arts, and embodiment and performativity. In each case, several key themes emerge as geographers today engage with the humanities in more critical and diverse ways. First, cultural and historical geographers have developed a closer theoretical engagement with other research in the humanities, particularly in relation to studies of literature and the visual arts and, more recently and, to a lesser extent, in relation to studies of dance, theatre and music. Secondly, there has also been a greater recognition and interrogation of the complexities of humanity itself, both in terms of the politics of identity and in envisaging a humane geography. Finally, there has been an increasingly critical understanding of the power-laden production of space and place. Geographical ideas about the spatiality of identity and the politics of place have become increasingly important in other disciplines across the humanities. While the humanities continue to inspire a vibrant range of geographical research, geographical concerns are also increasingly important in stimulating and articulating other research across the humanities today.

## Summary

- Humanities-based geographical research has become increasingly important since the work of humanistic geographers in the 1970s and cultural geographers since the 1980s.
- More than ever before, scholars working in other disciplines in the humanities are engaging with geographical ideas.
- Examples of humanities-based geographical work include historical and cultural research and studies of literature, the visual arts, and other cultural forms and practices.

- Geographers working within a humanities tradition address critical questions about the production of space, the politics of representation and embodied knowledges and identities.

## Further reading

A good place to start is with some early humanistic geographical writings: Ley and Samuels' (1978) edited collection, *Humanistic Geography: Prospects and Problems*, Daniels' (1985) critique of humanistic geography and Cosgrove's (1989) argument about the importance of thinking about geography as a humanity. Two edited collections – Barnes and Duncan (1992) and Duncan and Ley (1993) – and an article in *Area* by Sharp (2000) are good starting points for more on geography, text and writing. Read Cosgrove (1985) in *Transactions*, Matless (1998) *Landscape and Englishness* and Nash (1996) in *Gender, Place and Culture* for more on geography, landscape and the visual arts. Then read Catherine Nash's (2000) review article about geography, embodiment and performativity, and see David Pinder's (2001) study of embodied geographies in the city and Ian Cook's (2000) article on the exhibition 'Exchange values'. Also see other papers in the regular section 'Cultural geographies in practice' in the journal *Cultural Geographies*.

*Note*: Full details of the above can be found in the references list below.

## References

Anderson, K., Domosh, M., Pile, S. and Thrift, N. (eds) (2002) *Handbook of Cultural Geography*. London: Sage.

Barnes, T. and Duncan, J. (eds) (1992) *Writing Worlds: Discourse, Text and Metaphor in the Representation of Landscape*. London: Routledge.

Barnes, T. and Gregory, D. (eds) (1997) *Reading Human Geography: The Poetics and Politics of Inquiry*. London: Arnold.

Bell, D. and Valentine, G. (eds) (1995) *Mapping Desire: Geographies of Sexualities*. London: Routledge.

Blunt, A. (1994) *Travel, Gender and Imperialism: Mary Kingsley and West Africa*. New York, NY: Guilford.

Blunt, A. (2000a) 'Spatial stories under siege: British women writing from Lucknow in 1857', *Gender, Place and Culture*, 7: 229–46.

Blunt, A. (2000b) 'Embodying war: British women and domestic defilement in the Indian 'mutiny', 1857–8', *Journal of Historical Geography*, 26: 403–28.

Blunt, A. and McEwan, C. (eds) (2002) *Postcolonial Geographies*. London: Continuum.

Bonnett, A. (1999) *White Identities*. Harlow: Prentice Hall.

Brosseau, M. (1994) 'Geography's literature', *Progress in Human Geography*, 18: 333–53.

Brosseau, M. (1995) 'The city in textual form: *Manhattan Transfer*'s New York', *Ecumene*, 2: 89–114.

Butler, R. (1999) 'The body', in P. Cloke et al. (eds) *Introducing Human Geographies*. London: Arnold: 238–45.

Cloke, P., Philo, C. and Sadler, D. (1991) *Approaching Human Geography.* London: Paul Chapman.

Cook, I. (2000) 'Social sculpture and connective aesthetics: Shelley Sacks's "Exchange values" ', *Ecumene*, 7: 337–43.

Cook, I., Crouch, D., Naylor, S. and Ryan, J. (eds) (2000) *Cultural Turns/ Geographical Turns.* London: Prentice Hall.

Cosgrove, D. (1985) 'Prospect, perspective and the evolution of the landscape idea', *Transactions of the Institute of British Geographers*, 10: 45–62.

Cosgrove, D. (1989) 'Geography is everywhere: culture and symbolism in human landscapes', in D. Gregory and R. Walford (eds) *Horizons in Human Geography.* Basingstoke: Macmillan: 118–35.

Cosgrove, D. and Daniels, S. (eds) (1988) *The Iconography of Landscape.* Cambridge: Cambridge University Press.

Cosgrove, D. and Jackson, P. (1987) 'New directions in cultural geography', *Area*, 19: 95–101.

Crang, M. (1998) *Cultural Geography.* London: Routledge.

Daniels, S. (1985) 'Arguments for a humanistic geography', in R.J. Johnston (ed.) *The Future of Geography.* London: Methuen: 143–58.

Daniels, S. (1993) *Fields of Vision: Landscape Imagery and National Identity in England and the United States.* Cambridge: Polity Press.

Driver, F. (1999) 'Imaginative geographies', in P. Cloke et al. (eds) *Introducing Human Geographies.* London: Arnold: 209–16.

Driver, F. (2001) *Geography Militant: Cultures of Exploration and Empire.* Oxford: Blackwell.

Duncan, J. (1990) *The City as Text: The Politics of Landscape Interpretation in the Kandyan Kingdom.* Cambridge: Cambridge University Press.

Duncan, J. and Duncan, N. (1988) '(Re)reading the landscape?', *Environment and Planning D: Society and Space*, 6: 117–26.

Duncan, J. and Gregory, D. (eds) (1999) *Writes of Passage: Reading Travel Writing.* London: Routledge.

Duncan, J. and Ley, D. (eds) (1993) *Place/Culture/Representation.* London: Routledge.

Gilbert, D. (1999) 'London in all its glory – or how to enjoy London: guidebook representations of imperial London', *Journal of Historical Geography*, 25: 279–97.

Gowans, G. (2001) 'Gender, imperialism and domesticity: British women repatriated from India, 1940–1947', *Gender, Place and Culture*, 8: 255–69.

Gregory, D. (1995) 'Imaginative geographies', *Progress in Human Geography*, 19: 447–85.

Harley, B. (1988) 'Maps, knowledge and power', in D. Cosgrove and S. Daniels (eds) *The Iconography of Landscape.* Cambridge: Cambridge University Press: 277–312.

Howell, P. (2001) 'Sex and the city of bachelors: popular masculinity and public space in nineteenth-century England and America', *Ecumene*, 8: 20–50.

Jackson, P. (1989) *Maps of Meaning: An Introduction to Cultural Geography.* London: Unwin Hyman.

Jacobs, J. (1996) *Edge of Empire: Postcolonialism and the City.* London: Routledge.

Johnson, N. (1995) 'Cast in stone: monuments, geography and nationalism', *Environment and Planning D: Society and Space*, 13: 51–65.

Lewis, P. (1985) 'Beyond description', *Annals of the Association of American Geographers*, 75: 465–77.

Ley, D. and Samuels, M. (eds) (1978) *Humanistic Geography: Prospects and Problems*. London: Croom Helm.

Leyshon, A., Matless, D. and Revill, G. (eds) (1998) *The Place of Music*. New York, NY: Guilford.

Longhurst, R. (2000) *Bodies: Exploring Fluid Boundaries*. London: Routledge.

Malbon, B. (1999) *Clubbing: Dancing, Ecstasy, Vitality*. London: Routledge.

Mallory, W.E. and Simpson-Housley, P. (eds) (1987) *Geography and Literature: A Meeting of the Disciplines*. Syracuse, NY: Syracuse University Press.

Matless, D. (1998) *Landscape and Englishness*. London: Reaktion.

McDowell, L. (1997) *Capital Culture: Gender at Work in the City*. Oxford: Blackwell.

McDowell, L. (1999) *Gender, Identity and Place: Understanding Feminist Geographies*. Cambridge: Polity Press.

Meinig, D.W. (1983) 'Geography as an art', *Annals of the Association of American Geographers*, 8: 314–28.

Nash, C. (1994) 'Remapping the body/land: new cartographies of identity, gender, and landscape in Ireland', in A. Blunt and G. Rose (eds) *Writing Women and Space: Colonial and Postcolonial Geographies*. New York, NY: Guilford: 227–50.

Nash, C. (1996) 'Reclaiming vision: looking at landscape and the body', *Gender, Place and Culture*, 3: 149–69.

Nash, C. (1999a) 'Landscapes', in P. Cloke et al. (eds) *Introducing Human Geographies*. London: Arnold: 217–25.

Nash, C. (1999b) 'Irish placenames: post-colonial locations', *Transactions of the Institute of British Geographers*, 24: 457–80.

Nash, C. (2000) 'Performativity in practice: some recent work in cultural geography', *Progress in Human Geography*, 24: 653–64.

Olsson, G. (1980) *Birds in Egg/Eggs in Bird*. London: Pion.

Olsson, G. (1981) 'On yearning for home: an epistemological view of ontological transformations', in D. Pocock (ed.) *Humanistic Geography and Literature*. London: Croom Helm: 121–9.

Phillips, R. (1997) *Mapping Men and Empire: A Geography of Adventure*. London: Routledge.

Phillips, R. (2001) 'Politics of reading: decolonizing children's geographies', *Ecumene*, 8: 125–50.

Pinder, D. (1996) 'Subverting cartography: the situationists and maps of the city', *Environment and Planning A*, 28: 405–27.

Pinder, D. (2001) 'Ghostly footsteps: voices, memories and walks in the city', *Ecumene*, 8: 1–19.

Ploszajska, T. (2000) 'Historiographies of geography and empire', in B. Graham and C. Nash (eds) *Modern Historical Geographies*. Harlow: Longman: 121–45.

Pocock, D. (ed.) (1981) *Humanistic Geography and Literature*. London: Croom Helm.

Pollock, G. (1988) *Vision and Difference: Femininity, Feminism and the Histories of Art*. London: Routledge.

Proctor, J. and Smith, D. (eds) (1999) *Geography and Ethics: Journeys in a Moral Terrain*. London: Routledge.

Rose, G. (1993) *Feminism and Geography*. Cambridge: Polity Press.

Rose, G. (1995) 'Place and identity: a sense of place', in D. Massey and P. Jess (eds) *A Place in the World?* Oxford: Oxford University Press: 87–132.

Rose, G. (1997) 'Engendering the slum: photography in East London in the 1930s', *Gender, Place and Culture*, 4: 277–300.

Rose, G. (2000) 'Practising photography: an archive, a study, some photographs and a researcher', *Journal of Historical Geography*, 26: 555–71.

Rose, G. (2001) *Visual Methodologies*. London: Sage.

Ryan, J. (1997) *Picturing Empire*. London: Reaktion.

Said, E. (1995, first published 1978) *Orientalism*. London: Penguin Books.

Said, E. (1999) *Out of Place: A Memoir*. London: Granta.

Schwartz, J. and Ryan, J. (eds) (2003) *Picturing Place: Photography and the Geographical Imagination*. London: IB Tauris.

Seymour, S. (2000) 'Historical geographies of landscape', in B. Graham and C. Nash (eds) *Modern Historical Geographies*. Harlow: Longman: 193–217.

Sharp, J. (1994) 'A topology of "post" nationality: (re)mapping identity in the *Satanic Verses*', *Ecumene*, 1: 65–76.

Sharp, J. (2000) 'Towards a critical analysis of fictive geographies', *Area*, 32: 327–34.

Smith, D. (2000) *Moral Geographies: Ethics in a World of Difference*. Edinburgh: Edinburgh University Press.

Thrift, N. (1997) 'The still point: resistance, expressive embodiment and dance', in S. Pile and M. Keith (eds) *Geographies of Resistance*. London: Routledge: 124–51.

Thrift, N. (1999) 'Steps to an ecology of place', in D. Massey et al. (eds) *Human Geography Today*. Cambridge: Polity Press: 295–322.

Tuan, Y.-F. (1976) 'Humanistic geography', *Annals of the Association of American Geographers*, 66: 266–76.

Tuan, Y.-F. (1978) 'Literature and geography: implications for geographical research', in D. Ley and M. Samuels (eds) *Humanistic Geography: Prospects and Problems*. London: Croom Helm: 194–206.

Valentine, G. (2001) *Social Geographies: Space and Society*. Harlow: Pearson.

Williams, R. (1976) *Keywords: A Vocabulary of Culture and Society*. London: Fontana.

Woods, D. (1994) *The Power of Maps*. New York, NY: Guilford.

Wylie, J. (2002a) 'Earthly poles: the Antarctic voyages of Scott and Amundsen', in A. Blunt and C. McEwan (eds) *Postcolonial Geographies*. London: Continuum: 169–83.

Wylie, J. (2002b) 'Becoming-icy: Scott and Amundsen's South Polar voyages, 1910–1913', *Cultural Geographies*, 9: 249–65.

# Key Concepts

# 5    Space: The Fundamental Stuff of Human Geography

**Nigel Thrift**

## Definition

As with terms like 'society' and 'nature', space is not a commonsense external background to human and social action. Rather, it is the outcome of a series of highly problematic temporary settlements that divide and connect things up into different kinds of collectives which are slowly provided with the means which render them durable and sustainable.

## INTRODUCTION

'Space' is often regarded as the fundamental stuff of geography. Indeed, so fundamental that the well-known anthropologist, Edward Hall, once compared it to sex: 'It is there but we don't talk about it. And if we do, we certainly are not expected to get technical or serious about it' (cited in Barcan and Buchanan, 1999: 7). Indeed, it would be fairly easy to argue that most of the time most geographers do tend to get rather embarrassed when challenged to come out with ideas about what the supposed core of their subject is, and yet they continue to assert its importance. Rather like sex, they argue, without space we would not be here. So is all this just mass disciplinary hypocrisy? Not really. It is more about the extreme difficulty of describing certain aspects of the medium which is the discipline's message.

This brief introduction to the topic of space aims to tell you what space is and why we as human geographers need to study it (see Chapter 6 on space and physical geography). It will do this as straightforwardly as possible but it is important to point out that one of the problems that geographers have with space is that something that appears as though it really ought to be quite straightforward very often isn't – after all, we all have trouble at times in getting from A to B.

Even nowadays, of course, some geographers still persist in believing that it ought to be possible to explain space and other like concepts in such simple terms that you should be able to understand what is going on straight off. But increasingly this kind of simpleminded approach has come to be understood as more likely to be a part of a desperate attempt to try to reduce the wonderful complexity and sheer richness of the world in ways which mimic the predictable worlds of those privileged few who have the ability to make things show up in the way they want them to (Latour, 1997). In this piece, in contrast, while I will certainly attempt to write about space clearly, you should not think that this will be the end of the matter. You will need to read more and think more really to start to get a grip on the grip that space exerts on all our lives – and, as we shall see, the ways that we can alter that grip in order to make new spaces.

Space has been written about in lots of ways. There are, for example, books upon books which document the different kinds of conceptions of space that can be found in disciplines like philosophy or physics (e.g. Crang and Thrift, 2000). But I want to keep well away from most of these accounts for now, though they will figure indirectly in quite a lot of what I have to say. Rather, I want to write about how modern geography thinks about space. That could cover pages and pages and so I will have to condense these thoughts into a manageable form. I will therefore make what some will regard as the outrageously simple claim that currently human geographers are chiefly writing about four different kinds of space. However different the writings about these different kind of spaces may appear to be, they all share a common ambition: that is to abandon the idea of any pre-existing space in which things are embedded for an idea of space as undergoing continual construction exactly through the agency of things encountering each other in more or less organized circulations. This is a *relational* view of space in which, rather than space being viewed as a container within which the world proceeds, space is seen as a co-product of those proceedings. To begin with, I will artificially separate these four spaces out but, as I will point out in the conclusion, the exciting thing about geography today is that we are learning how to put them together in combinations that are beginning to produce unexpected insights.

## FIRST SPACE: EMPIRICAL CONSTRUCTIONS

Talking of putting things together, let's start with the empirical construction of space. It only takes a few minutes of reflection to start listing down all the things that we rely on to keep our spaces going – houses, cars, mobiles, knives and forks, offices, bicycles, computers, clothes and dryers, cinemas, trains, televisions, garden paths – but because these things are usually so mundane we tend to overlook them. So we often forget just what an extraordinary achievement the fabric of our daily lives actually is. Indeed, it is only recently that geographers have started to think systematically about the humble texts, instruments and drills that make up so much of what we are.[1] Let's take just one example of the kind of space that we make every day: the space of measurement. We are so used to looking at road signs measured out in terms of metres and kilometres or consulting a map or looking up an address or working out how long a journey will take that we forget what an extraordinary historical achievement these very ordinary practices are. They didn't suddenly come into existence overnight but were the subject of progressive standardizations and co-ordinations that have taken centuries to put in to place. And they required extraordinary investments too. They required the invention of specialized devices that could measure the same things at the same places, culminating in today's satellite-based global positioning system (GPS). They required a whole knowledge of measurement that itself had to be able to be transported around the world in devices, books and journals. They required, latterly, endless boring committees that were able to agree that the same measures would be measured in the same way in different places and then integrated with each other. And they demanded a good deal of brute force. After all, many of the ways space is measured out around the world were imposed by imperial conquest, not prettily negotiated. Nevertheless, it is important to realize the sheer load of human effort that has gone in to making measured space and the often near to insane enterprises that have made this space possible. For example, let us remember, with a certain amount of awe, the attempts to give birth to a new unit of measure, the metre, under the first French Republic (Guedj, 2001; Alder, 2002). Between 1792 and 1799 the astronomers Pierre Méchain and Jean-Baptiste Delambre travelled from one end of France to the other measuring the length of the Paris meridian in order to determine the exact length of the standard metre, which the National Assembly had decreed would be one ten millionth of the quarter meridian. The enterprise was an extraordinary one, involving the dragging of large pieces of equipment up hill and down dale but it laid the basis for the whole decimal metric system which is now so familiar.[2]

What is remarkable about the present time is the way in which this empirical construction of space is currently taking another leap forward. In the late nineteenth century, there was a widespread standardization of time. Driven by the increasing speed of transport and communications and more exact timekeeping instruments, states agreed on a common standard of time (based on the Greenwich meridian) and on a set of time zones spanning the globe in each of which time would be agreed to be uniform. Now, as the twenty-first century begins, something very similar is taking place in space. Driven by the demands of modern logistics and new, more exact ways of registering space (most especially the combination of GPS, geographical information systems (GIS) and radio frequency identifier tags (RFID)), it will soon be possible to locate everything – yes, everything – using standards of measurement, some of which (as we have seen) were already being laid down in the eighteenth century. Through the standardization of space made possible by these technologies (and the large bureaucracies that employ them), each object and activity taking place on the globe will, at least in principle, be able to be exactly located. The result will be that we will live in a world of perpetual contact, in which it will be possible to track and trace most objects and activities on a continuous basis, constantly adjusting time and space in real time, so producing what is now called micro or hyper-coordination (Katz and Aakhus, 2002). Numerous examples of hyper-coordination already exist in the logistics industry where it is necessary continually to adjust delivery schedules, but they are also becoming common in our daily lives – for example, in the way in which we use mobile phone text messaging continually to adjust meetings with friends.

## SECOND SPACE: UNBLOCKING SPACE

The second way we need to think of space is as a series of carefully worked-up connections through which what we know as the world interacts. These connections consist of pathways which bind often quite unlike things together, usually on a routine, circulating basis. They can range all the way from the movements of office workers around offices to the movements these office workers themselves order – of trade, of travel, even of arms. They can range all the way from the movements of a few already slightly drunk teenagers around the bars of Benidorm to the global flows of tourists of which they are a part. They can range all the way from the very restricted movements of prisoners let out of their cells only one hour in every 24 to the vast disciplinary apparatus of states dispensing laws and correction on an increasingly international scale. And so on. Trying to think about a world based on these flows of goods and people and information and money has occupied the attention of geographers to an increasing extent because their presence has become increasingly evident as the world has become increasingly knitted together by them, a tendency that sometimes goes by the name of globalization.

The problem is that these pathways are difficult to represent conceptually. We can map them, we can list them, we can write about them, all key means by which the pathways themselves are able to achieve order. But how can we go a little further and create representational spaces which are still attached to these mundane means of achieving order but also pack an added analytical bite? For a long time in geography, the accepted way was to mimic a standard means by which the world is organized and draw boundaries around areas which were assumed to contain most of a particular kind of action and between which there was interaction. Once geographers had drawn lines round and labelled these large blocks, they held them responsible for producing characteristic forces or powers. So, for example, we might say that this block of interaction was a capitalist space or an imperialist space, a neoliberal space or a dependent space, a city space or a community space, and that it had particular inherent qualities. Such a strategy of regionalization is obviously useful. It captures and holds still a particular aspect of the world and it is doubtful that we could ever do without it. But it is always an approximation and it has some serious disadvantages, most notably the tendency to assume that boundary equals cause, and the tendency to freeze what is often a highly dynamic situation. So geographers began to become more and more impatient with these kind of representations, not so much because they were wrong but because they seemed to leave so much out of contention.

Nowadays, therefore, geographers tend to look for representations that can take more of the world in. One way of doing this has simply been to disaggregate these bounded spaces into smaller subordinate ones, usually with some of the same qualities but also with other qualities that only operate (or operate more strongly) at that scale (see also Chapter 12 on scale). But it is questionable whether such a mode of proceeding does anything more than continuing the same method of drawing lines round and labelling blocks of interaction, though in slightly different form by allowing the possibility of the creation of new blocks, or the migration of powers from one block to another. So many geographers are now trying a different tack. Instead of trying to draw boundaries around flows they are asking 'what if we regarded the world as made up of flows and tried to change our style of thought to accommodate that depiction' (Urry, 2000: 23)? That is no easy task but we can now see a whole series of approaches that are trying to start with movement as origin rather than endpoint and which stress travelling identities over fixed notions of belonging. For example, there is so-called actor-network theory which tries to trace out circulations in which the actor is the 'network' itself; things moving together through networks have powers (including the power to make stable spaces) that they could never have when separated. There is work on commodity chains which tries to map out the way commodities are assembled along pathways that cross the world. There is work by feminist theorists like Luce Irigaray or Elizabeth Grosz which is searching for spatial figures which can convey the ambition to build different, more fluid kinds of space. And a new more expansive vocabulary is coming into being that can match these several ambitions: events as well as structures, lines of flight as well as lines, transformation and becoming as well as system and being – all means of freeing thought from the straitjacket of container thinking.

In turn, all kinds of new spaces of differentiation are being constructed, sometimes fleeting and sometimes concerted experiments in living different kinds of life which are a set of questions about what kinds of space can be rather than definitive answers. And the questions are, as Elizabeth Grosz (2001: 130) puts it: 'How then can space function differently from the ways in which it has always functioned? What are the possibilities of inhabiting otherwise? Of being extended otherwise? Of living relations of nearness and farness differently?' All around the world geographers are now both studying and taking part in the spatial experiments which can begin to answer these questions. These experiments range far and wide; all the way from the kinds of experiments that are associated with reworking what we mean by 'wild' to the kinds of experiments that are meant to perform everyday life differently to the kinds of experiments which are trying to do new kinds of global (Blunt and Wills, 2000). No one quite knows what they are doing. So?

## THIRD SPACE: IMAGE SPACE

The third kind of space consists of what we might call pictures or, perhaps better because of all the associations that word conjures up, images. In the past, mention of the image might well have conjured up the notion of a formal painting. But nowadays, images come in all shapes and sizes – from paintings to photographs, from portraits to postcards, from religious icons to pastoral landscapes, from collages to pastiches, from the simplest graphs to the most complex animations (see also Chapters 4 and 17 for a discussion of how and why geographers study images). What is certain is that images are a key element of space because it is so often through them that we register the spaces around us and imagine how they might turn up in the future. The point is even more important because increasingly we live in a world in which pictures of things like news events can be as or more important than the things themselves, or can be a large part of how a thing is constituted (as in the case of a brand or a media celebrity). Part of the reason for the pervasiveness of images is that we now live in a world populated by all kinds of screen which produce a continuous feed of images. These screens are now so pervasive that we hardly notice their existence (McCarthy, 2001). So we find television screens populating not just homes but also bars, airports, shops, malls and waiting rooms, while computer screens can be found in dealing rooms, offices, studies and bedrooms and, increasingly, as public access screens in airports and stations as well as in Internet cafés. This extraordinary proliferation of screens over the last 50 years has produced an image realm of exceptional richness which is changing how we do space.

This change can be linked to another way in which our thinking about image space is changing. In the past, particular kinds of image very often created spaces in their likeness. So, for example, a particular notion of spatial symmetry helped to produce the Palladian landscape while a particular kind of modernist sensibility helped to produce the kinds of strictly laid-out and ascetically ordered landscapes still to be found lingering on in many urban housing estates. But the proliferation of images has made it increasingly difficult to read off images straightforwardly on to space like this. And it has also pointed to an aspect of images which has often been heretofore neglected; that they are the result of complex processes of mediation which themselves bear meaning. For example, Bruno Latour (1998) shows how a finished piece of work like a religious painting can involve all kinds of intermediaries, each of which can be bearers of spatial meaning – varnishes, dealers, patrons, assistants, maps, measuring devices, graphs, charts, angels, saints, monks, worshippers – and each of which has its own complicated intervening geographies. Such an example also shows that there is no direct reference to the world contained in an image but rather a never-ending set of transformations – or what Latour calls 'cooking steps' – each of which can involve quite different ways of seeing and working on the image.

If there are now so many image spaces swirling through us in so many different ways, it is clear they must compete for our attention. And it is this aspect of image space that I want to point to in concluding this section. For what is clear is that the issue of attention is probably the most pressing one now facing the geography of images (Thrift, 2003). Caught in a snowstorm of images, why do we attend to some images rather than others? In the nineteenth century, the matter of attention was a key element of debates on space. It was subsequently taken up by writers like Walter Benjamin and Georg Simmel who argued that the constant barrage of images was causing people to grow a kind of mental carapace which would protect them from this continuous bombardment, a carapace which was showing up in cities as new and studied social styles (like cynicism and a blasé attitude), which constructed certain routinized kinds of attention. However, the growth of the mass media like the cinema had also provided the opportunity for new kinds of moving images to come into existence which, to some extent, undercut these social styles and produced new apprehensions of space.

In the twenty-first century we can see this debate being replayed as geographers consider the ways in which new image forms are again providing new social and cultural pathologies but also new opportunities, as we have seen in the case of the sheer pervasiveness of the screen and the images supplied by it. We can wrap all these new image forms up in one big package called 'postmodernism' (Harvey, 1989), making all the images add up to one vast capitalist spectacle, but better by far to do what geographers are doing now and try to look at all the cooking steps of different kinds of image and their geographies, testing each step for its various potentials to tell us something new about how we see the world. This is a much harder slog, of course, one which requires a lot of methodological expertise (Rose, 2001). It also makes it much more difficult to write in terms of one stable big picture like postmodernism. But then perhaps that is not such a bad thing. After all, one of the continuing dangers of work on images is to read too much significance into them, rather than considering them as just another set of mundane tools and practices of seeing which allow us to see some things and not others and so construct some spaces and not others.

### FOURTH SPACE: PLACE SPACE

The final kind of space is space understood as place: I say 'understood' loosely since the nature of place is anything but fully understood (see Chapter 10 for a full discussion of place). One reason for this is precisely that place so often seems to be caught up with the idea of a natural register. Whether it be the quiet glories of Thoreau's Walden Pond or the noisy cultural authenticity of an urban enclave, somehow place is more 'real' than space, a stance born out of the intellectual certainties of humanism and the idea that certain spaces are somehow more 'human' than others; these are the places where bodies can more easily live out (or at least approximate) a particular Western idea of what human being should be being. But other geographers are moving away from this kind of certainty both about what 'human' and 'being' might be. They are more interested in testing the limits of 'human' and 'being' through experiment and, in the process, are starting to point to new kinds of space.

Whatever the case, all of those working on place seem to agree that place consists of particular rhythms of being that confirm and naturalize the existence of certain spaces. Often they will use phrases like 'everyday life' to indicate the way that people, through following daily rhythms of being, just continue to expect the world to keep on turning up and, in doing so, help precisely to achieve that effect (Lefebvre, 1996). The problem is that rhythms of being can vary so enormously that such phrases often provide only the most tenuous hold on what happens. This problem of variation does not just exist because there are so many different rhythms of being but also because when the minutiae of everyday interaction are closely looked at what we see is not just routines but also all kinds of creative improvisations which are not routine at all (though they may have the effect of allowing that routine to continue). So, in everyday life, what is striking is how people are able to use events over which they often have very little control to open up little spaces in which they can assert themselves, however faintly. Using talk, gesture and more general bodily movement, they can open up pockets of interaction over which they can have control (see Chapter 13 on resilience and resistance). Clearly, an important part of this process is that spatial awareness we call place. For places not only offer resources of many different kinds (for example, spatial layouts which may allow certain kinds of interaction rather than others) but they also provide cues to memory and behaviour. In a very real sense, places are a part of the interaction.

One thing that does seem to be widely agreed is that place is involved with embodiment. It is difficult to think of places outside the body. Think, for example, of a country walk and place consists not just of eye surveying prospect but also the push and pull of walking up hill and down dale, the sound of birds and the wind in the trees, the touch of wall and branch, the smell of trampled grass and manure. Or think of a walk in the city and place consists not just of eye making contact with other people or advertising signs or buildings but also the sound of traffic noise and conversation, the touch of ticket machine and handrail, the smell of exhaust fumes and cooking food. Once we start to think of place in this kind of way we also start to take notice of all kinds of things which previously were hidden from us. So, there is now a thriving study of how sound (and especially music) conjures up place associations (Leyshon et al., 2000). And other senses like touch and smell are also beginning to receive their due.[3]

But there is a big problem here. What do we mean by 'body'? And this is where we get to the most intriguing prospect of all. For though it is possible to think of *the* body as flesh surrounded by an envelope of skin, all the current thinking suggests that this container thinking is too simple. It probably makes much more sense to think of the individual body as a part of something much more complex, as a link in a larger spatial dance with other 'dividual' parts of bodies and things and places which is constantly reacting to encounters and evolving out of them not individual awareness but dividual 'a-whereness'. And this larger dance is held together in particular by the play of 'affects' like love and hate, sympathy and antipathy, jealousy and despair, hope and disappointment and so on. Affect is often thought of as just a posh word for emotion but it is meant to point to something which is non-individual, an impersonal force resulting from the encounter, an ordering of the relations between bodies which results in an increase or decrease in the potential to act. Place (understood as a part of this complex process of embodiment) is a crucial actor in producing affects because, in particular, it can change the composition of an encounter by changing the affective connections that are made. Thus, as we all know, certain places can and do bring us to life in certain ways, whereas others do the opposite. It is this expressive quality of place which has recently lead to the emphasis on perform-ance in geography (see Chapter 4). For, through experiments with particular kinds of performance (from art to dance to drama), it may be possible to show some of this affective play and use the under-standings (or should it be stancings) of place thereby gained to make places which can help to produce the same sense of empowerment and general creative potential that we currently most often identify with, situations like standing on the top of a hill on a windy day drinking in the atmosphere or being moved by a great new piece of music. In other words, geographers working on place have started to join in a kind of politics which is intent on freeing up more of the potentials of place – and installing some new ones.

## CONCLUSIONS: JOINED-UP SPACING

What is fascinating about the present time is that geographers are now attempting to fit these four kinds of space together, partly because models for doing this are now erupting in the social sciences and humanities in a way that they never did before. In the past, many theoretical models of space that had an ambition to connect spaces of various kinds simply simulated the command and control models of the dominant systems around them. So, for example, many early Marxist models of capitalist space produced spatial connection by nesting some kinds of 'little' space in other 'big' spaces, declaring the 'little' spaces to be 'unique' and the big spaces to be 'general' (see also Chapter 12). But nowadays this simple 'size' distinction does not hold. We are no longer sure what is big or little or what is general or unique. Instead, as we have seen in the case of each of the four spaces we have examined, the hunt is on to think about space in quite different ways – ways which can prompt new 'a-where-nesses' (Massey and Thrift, 2002).

And this relates to the most important point that I want to make. This is that all these ways of thinking space are attempts to rethink what constitutes *power* once we can no longer think of power as simply command and control (Allen, 2002). So new thinking about the empirical construction of space involves considering the prolonged hard grind of actually putting viable pathways together, especially when, as nowadays, they can stretch around the world and back. New thinking about unblocking space involves the difficult task of re-describing the world as flow and continuous transformation. New thinking about image space involves reconsidering how images are circulated and kept stable when that circulation involves large numbers of intermediaries. And new thinking about place space involves trying to understand the gaps in the rhythms of everyday life through which new performances are able to pass. What we are seeing are new spaces being imagined into being by reworking the spatial technologies that we hold dear, and what is clear is that these acts of imagination are all profoundly political acts; what we often think of as 'abstract' conceptions of space are a part of the fabric of our being, and transforming how we think those conceptions means transforming 'ourselves'.

## Summary

- Space arises out of the hard and continuous work of building up and maintaining collectives by bringing different things (bodies, animals and plants, manufactured objects, landscapes) into alignment. All kinds of different spaces can and therefore do exist which may or may not relate to each other.
- For the purpose of simplification, it is possible to identify four different kinds of these constructed spaces: empirical, block, image and place.
- Empirical space refers to the process whereby the mundane fabric of daily life is constructed.
- Block space refers to the process whereby routine pathways of interaction are set up around which boundaries are often drawn.
- Image space refers to the process whereby the proliferation of images has produced new apprehensions of space.
- Place refers to the process whereby spaces are ordered in ways that open up affective and other embodied potentials.

## Further reading

Space has been written about in lots of ways. One book which documents some of the different conceptions of space that are drawn upon by different disciplines is Crang and Thrift's (2000) *Thinking Space*. Different takes on the nature of space within the discipline of geography are evident in Anderson et al.'s (2003) *The Cultural Geography Handbook*,

Gregory's (1994) *Geographical Imaginations*, Harvey's (1989) *The Condition of Post-modernity*, Thrift's (1996) *Spatial Formations* and Massey's (1992) *Space, Place and Gender*. Most recently, some geographers have begun to draw on actor-network theory in thinking about the complexity and richness of space. See, for example, Latour's (1997) *Common Knowledge* paper in the special issue of *Environment and Planning D: Society and Space* edited by Hetherington and Law (2000) as an introduction to these ideas.

*Note*: Full details of the above can be found in the references list below.

## NOTES

1   This is a little bit unfair. The exceptions to this rule include work by those who have been interested in the history of cartography and navigation, such as the late Eva Taylor (see Taylor, 1930).
2   I could have chosen many other examples – for example, the history of the surveying of Britain or the mapping of Switzerland (Gugerli, 1998).
3   This move also underlines how often space is bodied in various ways. And, if space is bodied then it will, for example, be actively gendered. Therefore, to return to the beginning of this essay, space has numerous sexual dimensions (see Pile and Nast, 2000).

## References

Alder, K. (2002) *The Measure of All Things. The Seven Year Odyssey that Transformed the World*. London: Little Brown.
Allen, J. (2002) *Hidden Geographies of Power*. Cambridge: Polity Press.
Anderson, K., Domosh, M., Pile, S. and Thrift, N.J. (eds) (2003) *The Cultural Geography Handbook*. London: Sage.
Barcan, R. and Buchanan, I. (eds) (1999) *Imagining Australian Space. Cultural Studies and Spatial Inquiry*. Nedlands: University of Western Australia Press.
Blunt, A. and Wills, J. (2000) *Dissident Geographies*. London: Longman.
Crang, M. and Thrift, N.J. (eds) (2000) *Thinking Space*. London: Routledge.
Gregory, D. (1994) *Geographical Imaginations*. Oxford: Blackwell.
Grosz, E. (2001) *Architecture from the Outside. Essays on Virtual and Real Space*. Cambridge, MA: MIT Press.
Guedj, D. (2001) *The Measure of the World*. Chicago, IL: Chicago University Press.
Gugerli, D. (1998) 'Politics on the topographer's table: the Helvetic triangulation of cartography, politics and representation', in T. Lenoir (ed.) *Inscribing Science. Scientific Texts and the Materiality of Communication*. Stanford, CA: Stanford University Press, pp. 91–118.
Harvey, D. (1989) *The Condition of Postmodernity*. Oxford: Blackwell.
Hetherington, K. and Law, J. (2000) 'After networks', special issue of *Environment and Planning D: Society and Space*, 28.
Katz, J. and Aakhus, M. (eds) (2002) *Perpetual Contact. Mobile Communication, Private Talk, Public Communication*. Cambridge: Cambridge University Press.
Latour, B. (1997) 'Trains of thought: Piaget, formalism and the fifth dimension', *Common Knowledge*, 6: 170–91.

Latour, B. (1998) 'How to be iconophilic in art, science and religion?' in C.A. Jones and P. Galison (eds) *Picturing Science, Producing Art.* New York, NY: Routledge, pp. 418–40.

Lefebvre, H. (1996) *Writings on Cities.* Oxford: Blackwell.

Leyshon, A., Matless, D. and Revill, G. (eds) (2000) *The Place of Music.* New York, NY: Guilford.

Massey, D. (1992) *Space, Place and Gender.* Cambridge: Polity Press.

Massey, D. and Thrift, N.J. (2002) 'The passion of place', in R.J. Johnston and M. Williams (eds) *A Century of British Geography.* Oxford: Oxford University Press

McCarthy, A. (2001) *Ambient Television.* Durham, NC: Duke University Press.

Pile, S. and Nast, H. (eds) (2000) *Places outside the Body.* London: Routledge.

Rose, G. (2001) *Visual Methodologies.* London: Sage.

Taylor, E.G.R. (1930) *Tudor Geography 1485–1583.* London: Methuen.

Thrift, N.J. (1996) *Spatial Formations.* London: Sage.

Thrift, N.J. (2003) 'Bare life', in H. Thomas and J. Ahmed (eds) *Cultural Bodies.* Oxford: Blackwell.

Urry, J. (2000) *Sociology beyond Societies. Mobilities for the Twenty-first Century.* London: Routledge.

# 6 Space: Making Room for Space in Physical Geography

**Martin Kent**

## Definition

Geographers are poor at defining space. The *Oxford English Dictionary* defines space in two ways: (1) 'A continuous extension viewed with or without reference to the existence of objects within it'; and (2) 'The interval between points or objects viewed as having one, two or three dimensions.' The geographer's prime interest is in the objects within the space and their relative position, which involves the description, explanation and prediction of the distribution of phenomena. The relationships between objects in space is at the core of geography.

## INTRODUCTION

The importance of the concept of space in geography has always been controversial (Gatrell, 1983; Unwin, 1992; Holt-Jensen, 1999) and whether geography and geographers should primarily focus on, or at the very least give some recognition to, the importance of space remains a fundamental question for the discipline. This chapter examines the concept of space in the context of physical geography (see Chapter 5 for a human geography perspective on space). The chapter begins with the suggestion that physical geographers have neglected the vital spatial dimension of their subject over the past few decades and explores the possible reasons for this. The spatial units and approaches to mapping that are recognized by the major sub-disciplines are then examined and, in turn, the relatively poor spatial synthesis across physical geography is considered. Finally it is argued that new technologies and developments in related disciplines, like ecology, present physical geographers with exciting possibilities for stimulating a new awareness of space. Recent developments in bio-geography provide early indications of this 'spatial reawakening' in physical geography.

## SPACE NEGLECTED

There is no doubt that physical geographers have generally been less willing or able than their human counterparts to adopt a predominantly spatial emphasis to their studies and research. A search through the most recent texts on the nature and philosophy of (physical) geography (e.g. Haines-Young and Petch, 1986; Stoddart, 1986; Haggett, 1990; Rogers et al., 1992; Unwin, 1992; Rhoads and Thorn, 1996; Slaymaker and Spencer, 1998; Holt-Jensen, 1999; Gregory, 2000) reveals that remarkably little is written on the importance of space as a concept in physical geography. Gregory (2000) lists 11 'Tenets for Physical Geography' of which the first is 'Emphasise the spatial perspective' (p. 287) yet, paradoxically, the word 'space' is not mentioned in the index and there is just one reference to 'spatial analysis'.

Spatial analysis provided a unifying theme for geography during the 1960s (Unwin, 1992) and emphasized that geographers should 'pay attention to the spatial arrangement of the phenomena in an area and not so much to the phenomena themselves' (Schaefer, 1953: 228). Most developments took place in human geography rather than in physical, although some attempts were made to demonstrate the potential significance of spatial analysis to the physical side of the subject, notably by Doornkamp and King (1971) and Chorley (1972). Within spatial analysis, concepts of distance are fundamental, which traditionally in the three-dimensional Euclidean view of space is expressed as a straight line between any two points or a three-dimensional volume that may be occupied by any phenomena. Gatrell (1983) stresses that this is the concept of *absolute space*. Human geographers quickly developed new ideas about space and distance – for example, 'time-distance', 'cost or economic distance', 'cognitive or perceptual distance' and 'social distance' (Unwin, 1992) – all of which treat space as a relative phenomenon (*relative space*). Ultimately, these developments have led to widespread rejection of spatial analysis by human geographers, many of whom now regard space as a social construct.

In comparison, within physical geography, concepts of distance have remained strongly absolute and defined in terms of Euclidean and metric space. Perhaps physical geographers, with their ongoing emphasis on absolute space, could have been expected to take up the challenge and to focus more on spatial analysis. However, they have generally also chosen to reject this pathway and downgrade or neglect spatial analysis. This downgrading of space by physical geographers can be linked to three alternative views that, rather than space, human–environment relationships, process and the time dimension are at the core of the discipline.

Thus, the first widely held view is that physical geography is primarily about human relationships with and impact on the *environment* and the resulting need for environmental management (Briggs, 1981; Pacione, 1999). While there is absolutely no doubt concerning the impact of human activity on all physical and environmental systems, to place this at the core of geography raises some interesting questions about the difference between physical geography and the much more recently evolved discipline of environmental science. Is environmental science simply physical (and some human) geography without a spatial emphasis? To be provocative, are physical geographers really geographers at all if they downgrade or neglect a spatial emphasis to the subject?

It is interesting that one of the most widely sold textbooks in physical geography since its publication in 1985 (David Briggs et al.'s *Fundamentals of Physical Geography*) reappeared in 1997 as a second edition but under the revised title, *Fundamentals of the Physical Environment*. Other core texts aimed at and written by geographers are now predominantly presented as 'environmental' rather than 'geographical'. While this is partly attributable to the wiliness of both publishers and authors in exploiting the market within both geography and environmental science degrees, it also can perhaps be seen as emphasizing the ambivalence of many physical geographers towards placing space and spatial analysis at the centre of physical geography and their tendency simply to equate geography with the environment.

Secondly, a commonly quoted idea is that physical geographers study 'pattern and *process*'. The word 'pattern' clearly has spatial connotations and is closely linked to ideas of geographers being concerned with the spatial description of phenomena. The word 'process', on the other hand, is linked with explanation of (spatial) patterns and distributions and relates to the understanding of how patterns are derived, models of their underlying functionality and how patterns may change over time (Stoddart, 1997). Recently, physical geographers have tended to overemphasize the importance of process and hence function and explanation at the expense of pattern and the spatial approach.

This point is accentuated further when one considers the third alternative view and the recent emphasis that has also been placed on environmental change and the importance of *time* by many physical geographers. Time and process are inextricably linked. An excellent introduction to the general approach is presented in Slaymaker and Spencer (1998), who demonstrate the relevance of an understanding of past climatic and environmental variability and the increasing impact of human activity to key present and future issues of environmental management and climate change. Quaternary science and the study of palaeoenvironments have come to the core of physical geography over the past 30 years and studies of environmental change over the Holocene period have achieved particular significance (Roberts, 1998). Although it is implicit that environmental change is expressed in terms of change in spatial patterns of the various components of physical geography through time, the main emphasis has been primarily on temporal changes and understanding the processes that underlie them. Such research is vital in informing the future predictions of climate and environmental change in response to human forcing. This emphasis on time, important though it is, has perhaps been at the expense of space, and as long ago as 1987, Clark et al. (1987: 384), in the conclusion to their edited volume on physical geography, concluded that 'Spatial concepts are certainly of importance, but seem at the moment to be subsidiary to the elucidation of temporal behaviour'.

Space is, of course, intimately linked to time and physical geographers are acutely aware of this (Schumm and Lichty, 1965; Harrison and Warren, 1970; Lane, 2001). Massey (1999: 273), writing on the relationships between physical and human geography, also concurs with the general view that space has been comparatively ignored 'In contrast to the prominence of time and historicity in the debates that I have explored so far, space has had a very low profile. It is denigrated as a simple absence of history and/or not accorded the same depth of intellectual treatment as time.' Completely to divorce space and time is impossible, and any attempts to do so are artificial. The problem, however, remains that the time dimension has tended to receive greater attention than space in much of physical geography.

To summarize, general disinterest in spatial analysis, with the particular exceptions discussed below, and the rise of environment, process and time as core foci in physical geography mean that physical geographers have not worried greatly about the relevance and importance of the spatial basis of their discipline.

## CONCEPTS OF SPACE AND SPATIAL UNITS IN PHYSICAL GEOGRAPHY

It is instructive to review the conventional spatial units and approaches to mapping that are recognized by the major subdisciplines of physical geography (Figure 6.1). The problem of scale is, however, immediately important since, for all subdisciplines, spatial units and approaches to mapping for research and management can be defined and completed at a range of scales (see Chapter 11 on scale). Spatial units are also often nested within each other across this range of scales. Of the various components of physical geography, perhaps biogeography can be shown to have identified most strongly with the spatial aspects of physical geography over the past 30 years.

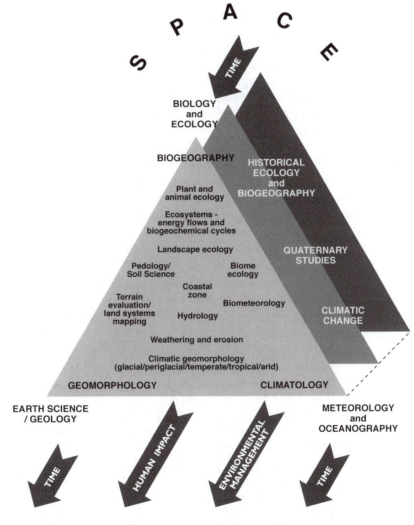

**FIGURE 6.1**  The various components of physical geography and their relationships in space and time

## *Biogeography*

### THE ECOSYSTEM AND PLANT COMMUNITY CONCEPTS

Two of the major paradigms of biogeography are, first, the ecosystem concept, originally described by Tansley (1935). The spatial expression of the ecosystem concept is, however, linked to the second key paradigm, which is that of the plant community or, more generally, the habitat. In unmodified environments, vegetation or habitat is 'the outward and visible sign' of a particular ecosystem type because it is within or underneath vegetation or habitat that all higher trophic levels live, feed and reproduce. Thus vegetation or habitat usually represent the key defining spatial concepts for ecosystems, and mapping or spatial representation is frequently completed on the basis of both physiognomic and/or floristic variability (Kent and Coker, 1992; Huggett, 1998). The question of whether plant community and ecosystem types can be mapped and thus given spatial expression at all and whether they are distributed as a continuum rather than as a set of mappable types is still heavily debated by both ecologists and biogeographers, and the idea that types can even be considered as 'concrete' mappable types, rather than 'abstract' categories, remains controversial (Kent and Coker, 1992; Kent et al., 1997). The recent completion of the National Vegetation Classification (NVC) for Britain (Rodwell, 1991–2000), with all major vegetation types now being recognized, demonstrates this problem clearly, with the main classification being essentially 'abstract' but with many examples now emerging of 'concrete' mapping of NVC types in different parts of Britain (e.g. Dargie, 1998). The nature of boundaries, transitions or ecotones between different plant community/vegetation types within a given region has tended to be conveniently ignored by both biogeographers and ecologists. Similarly, the application of multivariate analysis using techniques of numerical classification and ordination and the 'fitting' of vegetation samples that are 'transitional' between types, often using the MATCH and TABLEFIT computer programs of Malloch (1991) and Hill (1991), are also still highly contentious issues (Kent et al., 1997).

The process of defining and then mapping plant communities or habitats and hence ecosystems is thus fraught with difficulty. Often the level of detail that can be incorporated is low, and vegetation categories are reduced to generalized habitats and definition of boundaries between habitats is often arbitrary. The phase 1 habitat mapping programme for the mapping of plant communities in Britain during the 1980s and 1990s provides an excellent example. The level of generalization was often so high as to render the final maps as of comparatively little value, and for various practical reasons, the accuracy of many of the maps has subsequently been questioned

(Cherrill and McClean, 1999). This problem of boundary recognition is highlighted as a major new potential research area for physical geography in the conclusion to this chapter.

### THE CONCEPT OF 'ECOLOGICAL DISTANCE'

Plant and animal ecology and biogeography also represent one area of physical geography where alternative concepts of distance have been applied. The search for pattern in plant and animal community data routinely involves the application of multivariate analysis (Kent and Coker, 1992; Waite, 2000). As Gatrell (1983) points out, numerous (dis)similarity coefficients have been devised by ecologists to assess the degree of matching between samples of community composition. Many of these are described as having 'metric' properties (i.e. they are based on concepts of Euclidean space) – not least the coefficient of Euclidean distance itself (Kent and Coker, 1992; Waite, 2000). Numerous methods of 'mapping' samples in terms of their species composition expressed as 'ecological distances' have been devised under the general heading of 'ordination techniques' – for example, non-metric multidimensional scaling, principal components analysis and various forms of correspondence analysis. Since all samples of species composition are collected in space, it is possible to relate true spatial distance (absolute space) to change in species composition expressed as various forms of 'ecological distance' (relative space). However, despite their convenience, problems of spatial distortion abound in such analyses (Kent and Coker, 1992).

### MAPPING OF INDIVIDUAL SPECIES

The other key form of spatial expression in biogeography is the mapping of individual species distributions, often also with a view to understanding their tolerance ranges and environmental controls. In the case of rare and endangered species, changing species distributions on these maps are a vital tool for biological conservation. In Britain, the distribution mapping of individual species for most groups of organisms has now been completed on a 10 km$^2$ grid (Perring and Walters, 1982; Centre for Ecology and Hydrology, 2001; Preston et al., 2002), and similar mapping schemes are increasingly available elsewhere in Europe and across the globe. However, animal communities and species assemblages of higher trophic levels, as opposed to individual species, although recognized, are rarely mapped as such since they exist within the habitat provided by vegetation.

### CHANGING SPATIAL DISTRIBUTIONS THROUGH TIME – QUATERNARY AND HISTORICAL BIOGEOGRAPHY

Mapping of changes in species and community distributions through time introduces the fields of Quaternary science and historical ecology/biogeography (Figure 6.1). Most studies in this area again provide 'abstract' concepts of changes in both individual and community distributions through time rather than 'concrete' spatial mapping of those changes. Also, as suggested above, most interpretation and inference concentrates on temporal rather than spatial variation. The current trend towards fine-resolution description and analysis of microfossils and environmental indicators in cores from peat, lakes, sediments and the oceans, which assists in linking rapid short-term ecological changes to longer-term geological time, emphasizes this point further (Bennett, 1997).

## THE 'ERGODIC HYPOTHESIS' – SUBSTITUTION OF SPACE FOR TIME

Biogeographers and ecologists also often link space and time through the 'ergodic hypothesis' whereby, as an expedient research strategy, different areas in space are taken to represent different stages in time (Chorley and Kennedy, 1971; Bennett and Chorley, 1978). Classic examples are in the study of successional processes, where differing locations or spatial units within a local area at one point in time are taken to represent the sequences of species changes that would occur in that area over long periods of time. Similar approaches have been applied in geomorphology in the context of landform development (Chorley and Kennedy, 1971).

## SOILS AND PEDOLOGY

Soils and pedology are traditionally seen as a part of biogeography but are, of course, also closely linked to geomorphology. In soil mapping, similar problems to those of mapping vegetation exist in that a set of soil 'types' is typically described in the 'abstract' sense for any region but precise mapping of those types in 'concrete' terms is often difficult (White, 1997; Gerrard, 2000). The problems of soil classification to define and recognize types or series are very similar to those for vegetation classification, and use of multivariate analysis and numerical classification and ordination techniques in this context is still debated (Webster, 1977; 1985). Problems of boundary recognition and definition are thus once again emphasized.

## LANDSCAPE ECOLOGY – A NEW SPATIAL SCIENCE?

A major trend within the ecological side of biogeography over the past decade has been to adopt the approach and methodology of landscape ecology (Forman and Godron, 1986; Forman, 1995; Kupfer, 1995; Kent et al., 1997). Landscape ecology is concerned with the description, analysis and explanation of the spatial patterns of plant community and land-use types within a given landscape or region (see Chapter 10). Landscapes are composed of 'patches' that are distributed as a mosaic within any local area. The mosaic of patches at the patch scale within a landscape can be aggregated to give the landscape scale, and both higher and lower scales may be identified above the landscape scale and below the patch scale (Figure 6.2). These individual patches have

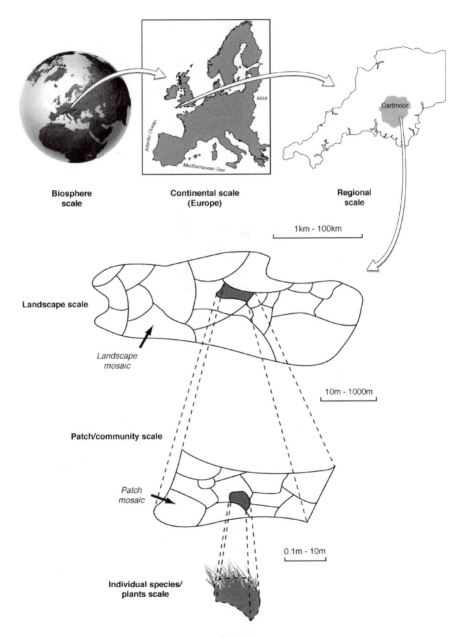

**FIGURE 6.2**    The key scales of the 'patch' and the 'landscape' within the overall spatial hierarchy of landscape ecology

varying degrees of 'naturalness', and a spectrum of patch types with varying intensity of human impact and modification, known as land-cover types, may be identified. Linear patches and features, such as river courses and hedgerows, are known as corridors and again boundaries between patches or corridors of different type are of considerable interest.

The methodology of landscape ecology is closely linked to recent developments in both remote sensing and geographic information systems (GIS) (Haines-Young et al., 1993; Bissonette, 1997; Klopatek and Gardner, 1999; Farina, 1998; 2000; Turner et al., 2001). Recognition of patches first requires classification of patches based on their ecological and land-use attributes into land-cover types. Once patches of a particular type have been defined, they can be examined in terms

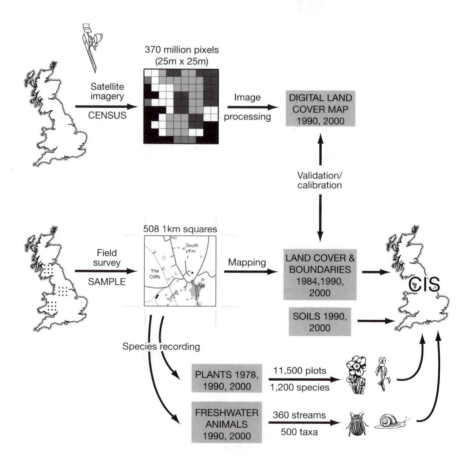

*Source*: Adapted from Department of Environment (1993); Department of Environment, Transport and the Regions/Centre for Ecology and Hydrology (2000)

**FIGURE 6.3** The methodological approach of the Countryside Surveys of Great Britain

of numerous parameters relating to size, shape and fragmentation. Various computer packages for spatial description and analysis of landscape patches are available, of which the best known is FRAG-STATS, based on the GIS programs, Arc-Info, Arc-View and ArcGIS (McGarigal and Marks, 1994/2000; ESRI, 2001).

This advent of landscape ecology, GIS and remote sensing has provided a valuable tool for habitat mapping in biogeography, with one of the best examples being the UK Countryside Surveys carried out between 1978 and 2000 (Department of Environment, 1993; Department of Environment, Transport and the Regions, 2000). Using remote sensing, field survey and geographic information systems, the Countryside Survey is now repeated every 10 years, thus representing a spatial census of habitat and land-cover change across Britain.

In the Countryside Information System (CIS) (Figure 6.3), the main categories derived from remote sensing are of generalized habitat types, principally because only major differences in vegetation and hence ecosystem physiognomy, rather than detailed variations in floristic composition, are picked out by remote-sensing imagery. However, this is counterbalanced by extensive repeat field surveys on the ground within all main landscape/habitat types, yielding data not only on changes in habitat extent but also in species richness and diversity of plants and freshwater organisms at 10-year intervals.

### Climatology

#### AIR MASSES AND FRONTS

As arguably dealing with the most dynamic part of the environment, the spatial units of climatology are represented by air masses and their numerous different characteristics and the fronts that characterize the boundaries between them. Both air masses and fronts are constantly changing in space and time. Nevertheless, the major air mass and frontal zones of the planet are widely recognized (Briggs et al., 1997; Barry and Chorley, 1998; Goudie, 2001). Similarly, climate types are recognized and classified (Oliver, 1991). Within major air masses, further dynamic variability and interaction occur at more detailed scales in response to relief and to relative proximity in space of land and sea. Any weather map or sequence of weather maps through time constitutes a representation of the changing spatial patterns of air masses and the nature of the fronts between them. Spatial and temporal patterns of extreme events are now a vital part of climatology and have significant social, economic and even political consequences (Perry, 1995). Magnitude and frequency studies of extreme climatic events and their spatial expression are now a key input to hazard studies in the whole of physical geography.

## METEOROLOGY

The processes behind the patterns described in climatology are studied in the subject of meteorology, and the prediction of changes in air mass and frontal patterns at a wide range of spatial and temporal scales provides the essence of weather and climate forecasting, which is central to the discipline. However, the brevity of this section serves to emphasize the point that the part of physical geography where spatial concepts are probably most difficult to work with is climatology, primarily because of the exceptionally dynamic nature of atmospheric processes.

### Geomorphology and hydrology

#### THE DRAINAGE BASIN AS A SPATIAL UNIT IN GEOMORPHOLOGY AND HYDROLOGY

Over the past 30 years, perhaps the most important spatial unit for geomorphology and hydrology has been the drainage basin, catchment or watershed, linked to the functioning systems model of the land phase of the hydrological cycle (Chorley, 1969; Gregory and Walling, 1973; Likens et al., 1977; Newson, 1995). In terms of hydrology, the analysis of both quantity and quality of water and changes through time, particularly in response to human impact, has been facilitated by the drainage basin approach, and much practical management is now based on the catchment concept (Gower, 1980; Newson, 1995). Studies of ecosystem biogeochemistry and nutrient cycling have also been primarily based on the catchment/watershed–ecosystem approach (Likens et al., 1977).

The catchment has also represented a spatial unit for comparison, with morphological, hydrological and geomorphological properties of different sets of drainage basins being compared (Doornkamp and King, 1971; Chorley, 1972). Stream ordering and the nesting of catchments are a particularly important concept that links across different spatial scales (Likens et al., 1977; Newson, 1995). The catchment is also a key unit for those physical geographers studying environmental change, with lakes and their sediments representing a valuable record of past hydrological, geomorphological and anthropogenic land-use change in upstream catchments (O'Sullivan, 1979; Lau and Lane, 2001).

#### (GEO)MORPHOLOGICAL MAPPING

Evans (1990: 97) expresses the comparative lack of interest in spatial aspects of geomorphology when he states:

> The excitement of work on process mechanisms, and the decline of the spatial tradition in geography, have pushed mapping to the periphery of academic concern, though it is important everywhere in applied work.

The production of geomorphological maps has a very low priority in Britain and North America, but many field-based publications contain maps of selected features relevant to their themes; these features may be the dominant landforms within the limited areas involved.

Morphological mapping in geomorphology developed from the work of Waters (1958) and Savigear (1965). Spatial units are defined using symbols for breaks of slope, and particular geomorphological features are shown using specific symbols (Bakker, 1963). Mapping may be completed in the field as well as from aerial photography (Verstappen, 1983). Further sophistication may be included, for example, by information on the genesis and nature of materials (Demek et al., 1972). However, problems occur in areas where breaks of slope are not obvious and slopes are smooth. The techniques have been extensively applied, for example, in the work of Sissons (1967; 1974), when mapping features of glacial geomorphology in Scotland. Some countries have produced national geomorphological maps (e.g. France, Belgium and Hungary, with maps at 1:25 000 and 1:50 000) but detail is often poor at this scale.

The advent of geographical positioning systems (GPS) and geographical information systems (GIS) has greatly improved accuracy and presentation of results (DeMers, 2000; Raper, 2000; Longley et al., 2001). Many geomorphological maps are produced for consultancy reports and in applied projects. These are often site and problem specific and may often not be in the public domain. Digital terrain modelling using GIS is now widely applied and is an important tool for geomorphologists (DeMers, 2000). Again GPS has proved invaluable in this respect (Longley et al., 2001).

As discussed below, geomorphological mapping, digital terrain modelling and GIS also provide the basis for the discipline of terrain evaluation and land-systems mapping, which provides a strong integrative spatial theme for the whole of physical geography.

## SPATIAL SYNTHESIS IN PHYSICAL GEOGRAPHY

The problem of the increasing fragmentation of physical geography has been highlighted by many authors (Unwin, 1992; Holt-Jensen, 1999; Gregory, 2000), but geography is also meant to be a subject of synthesis and some notable attempts at spatial synthesis in physical geography exist. Examples of spatial synthesis occur at two scales: first, at the landscape unit–ecosystem scale, in the disciplines of terrain evaluation and mapping (Townshend, 1981; Vink, 1983; Mitchell, 1991), which are linked to the more recent developments in landscape ecology discussed above (Forman and Godron, 1986; Forman, 1995;

Farina, 1998; 2000; Turner et al., 2001); and, secondly, at the biome scale, in what can be described as regional physical geography, of which the best contemporary examples are probably Briggs et al. (1997) and Goudie (2001).

### Terrain evaluation and land-systems mapping

Terrain evaluation and land-systems mapping evolved in those parts of the globe where assessments of land resources over very large areas were required, often for resource exploitation – for example, the Commonwealth Scientific and Industrial Research Organization (CSIRO) in Australia and the Canadian Government Land Surveys (Environment Canada) (Townshend, 1981; Vink, 1983; Mitchell, 1991). The importance of such surveys lay in their ideas on synthesis and that strong relationships existed among geology, geomorphology and slope form, overlying soils and vegetation (ecosystems) and, finally, land-use, where human modification was important. Particular types of slope form (slope facets) were linked to certain soil and vegetation (ecosystem) types and these in turn determined land-use activity, resulting in a form of habitat/ecosystem-type mapping. Recently, Warren (2001) has called for a re-evaluation of these ideas in relation to valley-side slope processes to provide a basis for conservation management. Aerial photography, remote sensing, GIS and digital terrain modelling are now an essential part of such evaluation and yet again problems of boundary delimitation between different terrain types or slope facets represent a major research focus for the subject (Townshend, 1981; Burrough and Frank, 1996; Burrough and McDonnell, 1998).

### Regional physical geography

The other potential area of spatial synthesis in physical geography is that of 'regional physical geography'. Although such ideas have taken a back seat over the past 30 years due to the 'process revolution' and fragmentation of subdisciplines in physical geography, there is no doubt that the possibility still exists. In theory, such spatial regional synthesis is possible at a range of scales, starting at the biome scale and progressively becoming more detailed at the regional and then local scales (Figure 6.2). Examples of physical synthesis at the biome scale still exist, most notably in the work of Goudie (2001), who takes a global perspective to physical geography but then in the second part of his text examines the physical geography of the 'major world zones', as he describes them. Regional and local studies are, however, virtually non-existent.

## THE FUTURE: BIOGEOGRAPHY AS AN INDICATOR OF A 'SPATIAL REAWAKENING' IN PHYSICAL GEOGRAPHY

Where does the future lie in terms of the awareness and degree of emphasis of the concept of space within physical geography? At the present time, the one area where a re-emergence of space as a key concept is clearly occurring is in biogeography and also within its related disciplines of biology and ecology. Fascinatingly, biologists and ecologists have become increasingly aware of the importance of space over the past decade, and the new subdiscipline of 'metapopulation ecology', which is concerned with research into species populations in 'spatially patchy' environments, has been at the forefront of this. The following quotation from the concluding chapter of Hanski (1999: 261) demonstrates this new focus:

> Ecologists are used to thinking about assemblages of interacting enti-
> ties, such as individuals in populations, but usually without an explicit
> reference to space. The novelty of spatial ecology is in the claim that the
> spatial locations of individuals, populations and communities can have
> equally significant consequences on dynamics as birth and death rates,
> competition and predation. Spatial ecology is one of the most visible
> developments in ecology and population biology in recent years; some
> regard it as a new paradigm arising towards the end of the twentieth
> century. New paradigm or not, it is true that never before has space been
> considered to be so pivotal to so many biological phenomena as today.

Metapopulation ecology has evolved from and is now replacing one of the few examples of spatial theory in physical geography, namely, the theory of island biogeography, originally proposed by MacArthur and Wilson (1963; 1967). The theory has been widely interpreted as having important implications for biological conservation (Diamond, 1975; Kent, 1987; Shafer, 1990) but more recently has been seriously questioned, particularly the underlying assumptions of equilibrium (Stott, 1998; Whittaker, 1998; 2000). Related to this, a whole new area called 'macroecology' is emerging (Gaston and Blackburn, 2000), which looks set to take the now well established subject of landscape ecology and blend it with metapopulation ecology and ideas of habitat fragmentation to give a whole new emphasis to the field of spatial ecology.

This new focus on space is expressed elsewhere in ecology. Dale (1999) has reinvigorated and revitalized the methodology of spatial pattern analysis for ecologists, and Dieckmann et al. (2000) show a new set of theoretical and conceptual ideas. Problems of boundary definition and mapping are assuming a new importance related to recent developments in GIS (Burrough and Frank, 1996). Kent et al. (1997) have reviewed the concepts of the plant community and the problems of defining and mapping plant community boundaries and gradients. Once again in this area, ecologists and mathematicians, rather than biogeographers, have made major contributions to boundary delimitation and mapping through new advances in geostatistics, including applications of fuzzy classification and crisp and fuzzy boundary detection (e.g. Fortin et al., 2000; Jacquez et al., 2000). A new user-friendly software package for boundary detection in spatially distributed ecological data – BoundarySeer (TerraSeer, 2001) – has recently been released (Figure 6.4), based on the work of the above researchers, none of whom, incidentally, are geographers. Many of these methods are also available within standard GIS applications such as ArcInfo, ArcView and ArcGIS (McGarigal and Marks, 1994/2000; ESRI, 2001) and there is no doubt that use of GIS as a tool

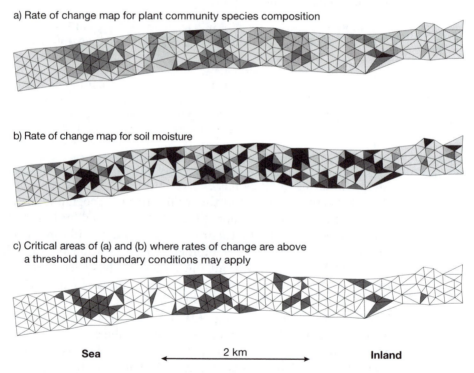

**FIGURE 6.4** Examples of mapping and boundary detection of plant communities and an environmental variable (soil moisture) using the 'BoundarySeer' package of Jacquez et al. (2000) and TerraSeer (2001). Data are from a 2 km × 200 m 'tranome' (a 2D transect) across the machair sand dunes of South Uist in the Outer Hebrides of Scotland. The first map shows rate of change in plant community composition derived from 'triangulation wombling' of scores from the first axis of an ordination of the species data from quadrats located at each of the 217 points in the triangular grid. The darker the tone, the greater the rate of change. The second map presents rates of change of soil moisture values on the same grid, while the third shows critical areas where rates of change are above a given threshold and thus boundary conditions may apply

Reproduced with the kind permission of Catherine Reid, University of Plymouth

of spatial description and analysis within all parts of physical geography will increase even further over the next decade (Burrough and Frank, 1996; Longley et al., 2001).

## CONCLUSION

As yet, these rapidly developing trends in biology and ecology are only just beginning to percolate into biogeography and, ironically, it seems almost as if everyone *but* geographers and biogeographers are at the forefront of this new revolution. Whether any similar revitalization of spatial concepts and spatial analysis is likely in the rest of physical geography remains unclear. While many physical geographers will always argue that space and a spatial emphasis are implicit in their work, perhaps now is an appropriate moment for them to re-examine the importance of space and spatial concepts and analysis in their various researches and in their teaching. The increasing importance of GIS and remote sensing within physical geography and related new developments in spatial analysis almost certainly represent a significant way forward in addition to the major trends in the subject identified at the outset of this chapter. The understanding of process is only as useful as the ability to apply it to the highly spatially variable nature of the earth's surface. Thus defining the geographic variation of the key input parameters and variables of all process models is essential, as is the geographic variability of their outputs. The above technologies can assist greatly with doing this. While the emphasis of physical geographers on processes and time will undoubtedly and rightly continue, in the future physical geographers may not be able to ignore the concept of space quite as easily as perhaps they have done over the past 30 years.

## Summary

- Spatial concepts have been neglected in physical geography over the past 30 years.
- Physical geographers are poor at explicitly defining and explaining spatial concepts in their subject.
- Physical geographers have perhaps emphasized process at the expense of pattern (space) in recent years.
- The time dimension in physical geography has received greater attention than the spatial dimension.
- Spatial units, such as the plant community, the air mass and the drainage basin, are fundamental to physical geography.
- Spatial synthesis across physical geography is poorly developed.

- The development of geographical information systems (GIS) and remote sensing offers exciting new possibilities for spatial analysis in physical geography.
- Recent developments in biology and ecology may stimulate a new awareness of the importance of space in physical geography through new approaches to spatial analysis in biogeography.

## Further reading

Gatrell's (1983) *Distance and Space: A Geographical Perspective* still provides an excellent introduction to concepts of space in geography. A valuable overview of the nature of physical geography is provided in *The Changing Nature of Physical Geography* (Gregory, 2000). The time element and time–space relationships are well covered in Slaymaker and Spencer's (1998) *Physical Geography and Global Environmental Change*, and the article by Schumm and Lichty (1965) is still essential reading for all physical geographers. Forman and Godron's (1986) *Landscape Ecology* remains the best introduction to landscape ecology as a truly spatial-based approach to biogeography. For a really challenging read on the awakening of spatial analysis in biology and ecology rather than physical geography, look at (but don't try to read in detail!) Dale's (1999) *Spatial Pattern Analysis in Plant Ecology* and Hanski's (1999) *Metapopulation Ecology*. Geographical information systems (GIS) are still insufficiently integrated into physical geography yet will be an even more vital tool in spatial description and analysis for the subject in the future. Burrough and McDonnell's (1998) *Principles of Geographical Information Systems* and Longley et al.'s (2001) *Geographic Information Systems and Science* provide the best introductions to the subject.

*Note*: Full details of the above can be found in the references list below.

## References

Bakker, J.P. (1963) 'Different types of geomorphological maps, problems of geomorphological mapping', *Geographical Studies*, 46: 13–31.

Barry, R. and Chorley, R.J. (1998) *Atmosphere Weather and Climate* (7th edn). London: Routledge.

Bennett, K.D. (1997) *Evolution and Ecology: The Pace of Life*. Cambridge: Cambridge University Press.

Bennett, R.J. and Chorley, R.J. (1978) *Environmental Systems: Philosophy, Analysis and Control*. London: Methuen.

Bissonette, J.A. (ed.) (1997) *Wildlife and Landscape Ecology: Effects of Pattern and Scale*. New York, NY: Springer.

Briggs, D.J. (1981) 'Editorial: the principles and practice of applied geography', *Applied Geography*, 1: 1–8.

Briggs, D.J. and Smithson, P. (1985) *Fundamentals of Physical Geography*. London: Hutchinson/Routledge.

Briggs, D., Smithson, P., Addison, K. and Atkinson, K. (1997) *Fundamentals of the Physical Environment*. London: Routledge.

Burrough, P.A. and Frank, A.U. (eds) (1996) *Geographic Objects with Indeterminate Boundaries. GISDATA* 2. London: Taylor & Francis.

Burrough, P.A. and McDonnell, R.A. (1998) *Principles of Geographical Information Systems* (2nd edn). Oxford: Clarendon Press.

Centre for Ecology and Hydrology (2001) Available at www.ceh.ac.uk/data/ EIC.htm.

Cherrill, A. and McClean, C. (1999) 'The reliability of "Phase 1" habitat mapping in the UK: the extent and types of observer bias', *Landscape and Urban Planning*, 45: 131–44.

Chorley, R.J. (1969) 'The drainage basin as a fundamental geomorphic unit', in R.J. Chorley (ed.) *Water, Earth and Man*. London: Methuen, pp. 77–100.

Chorley, R.J. (ed.) (1972) *Spatial Analysis in Geomorphology*. London: Methuen.

Chorley, R.J. and Kennedy, B.A. (1971) *Physical Geography: A Systems Approach*. London: Prentice Hall.

Clark, M.J., Gregory, K.J. and Gurnell, A.M. (eds) (1987) *Horizons in Physical Geography*. Basingstoke: Macmillan.

Dale, M.R.T. (1999) *Spatial Pattern Analysis in Plant Ecology*. Cambridge: Cambridge University Press.

Dargie, T.C.D. (1998) *Sand Dune Vegetation Survey of Scotland: Western Isles. Scottish Natural Heritage Research, Survey and Monitoring Report* 96 (3 vols). Battleby, Perth: Scottish Natural Heritage.

Demek, J., Embleton, C., Gellert, J.F. and Verstappen, H. (1972) *Manual of Detailed Geomorphological Mapping*. Prague: Academia.

DeMers, M. (2000) *Fundamentals of Geographic Information Systems* (2nd edn). London: Wiley.

Department of Environment (1993) *Countryside Survey 1990 – Summary Report*. London: HMSO/Department of the Environment.

Department of Environment, Transport and the Regions/Centre for Ecology and Hydrology (2000) *Accounting for Nature: Assessing Habitats in the UK Countryside*. London: Department of Environment, Transport and the Regions.

Diamond, J.M. (1975) 'The island dilemma: lessons of modern biogeographical studies for the design of nature reserves', *Biological Conservation*, 7: 129–46.

Dieckmann, U., Law, R. and Metz, J.A.J. (eds) (2000) *The Geometry of Ecological Interactions*. Cambridge: Cambridge University Press.

Doornkamp, J.C. and King, C.A.M. (1971) *Numerical Analysis in Geomorphology*. London: Edward Arnold.

ESRI (Environmental Systems Research Institute) (2001) *ArcGIS 8.1 Software*. Redlands, CA: ESRI Inc.

Evans, I.S. (1990) 'Cartographic techniques in geomorphology', in A. Goudie (ed.) *Geomorphological Techniques* (2nd edn). London: Unwin Hyman, pp. 97–108.

Farina, A. (1998) *Principles and Methods in Landscape Ecology*. Dordrecht: Kluwer Academic.

Farina, A. (2000) *Landscape Ecology in Action*. Dordrecht: Elsevier.

Forman, R.T.T. (1995) *Landscape Mosaics*. Cambridge: Cambridge University Press.

Forman, R.T.T. and Godron, M. (1986) *Landscape Ecology*. Chichester: Wiley.

Fortin, M.-J., Olson, R.J., Ferson, S., Iverson, L., Hunsaker, C., Edwards, G., Levine, D., Butera, K. and Klemas, V. (2000) 'Issues related to the detection of boundaries', *Landscape Ecology*, 15: 453–66.

Gaston, K.J. and Blackburn, T.M. (2000) *Pattern and Process in Macroecology*. Oxford: Blackwell Science.

Gatrell, A.C. (1983) *Distance and Space: A Geographical Perspective*. Oxford: Clarendon Press.

Gerrard, J. (2000) *Fundamentals of Soils*. London: Routledge.

Goudie, A. (2001) *The Nature of the Environment* (4th edn). Oxford: Blackwell.

Gower, A.M. (ed.) (1980) *Water Quality in Catchment Ecosystems*. Chichester: Wiley.

Gregory, K.J. (2000) *The Changing Nature of Physical Geography*. London: Arnold.

Gregory, K.J. and Walling, D.E. (1973) *Drainage Basin Form and Process*. London: Arnold.

Haggett, P. (1990) *The Geographer's Art*. Oxford: Blackwell.

Haines-Young, R., Green, D.R. and Cousins, S. (eds) (1993) *Landscape Ecology and Geographic Information Systems*. London: Taylor & Francis.

Haines-Young, R. and Petch, J. (1986) *Physical Geography: Its Nature and Methods*. London: Harper & Row.

Hanski, I. (1999) *Metapopulation Ecology*. Oxford: Oxford University Press.

Harrison, C.M. and Warren, A. (1970) 'Conservation, stability and management', *Area*, 2: 26–32.

Hill, M.O. (1991) *TABLEFIT – for Identification of Vegetation Types. Version 0.0*. Monks Wood: Institute of Terrestrial Ecology.

Holt-Jensen, A. (1999) *Geography: History and Concepts* (3rd edn). London: Sage.

Huggett, R.J. (1998) *Fundamentals of Biogeography*. London: Routledge.

Jacquez, G.M., Maruca, S.L. and Fortin, M.-J. (2000) 'From fields to objects: a review of geographic boundary analysis', *Journal of Geographical Systems*, 2: 221–41.

Kent, M. (1987) 'Island biogeography and habitat conservation', *Progress in Physical Geography*, 11: 91–102.

Kent, M. and Coker, P. (1992) *Vegetation Description and Analysis: A Practical Approach*. Chichester: Wiley.

Kent, M., Gill, W.J., Weaver, R.E. and Armitage, R.E. (1997) 'Landscape and plant community boundaries in biogeography', *Progress in Physical Geography*, 21: 315–53.

Klopatek, J.M. and Gardner, R.H. (eds) (1999) *Landscape Ecological Analysis: Issues and Applications*. New York, NY: Springer.

Kupfer, J.A. (1995) 'Landscape ecology and biogeography', *Progress in Physical Geography*, 19: 18–34.

Lane, S. (2001) 'Constructive comments on D. Massey "Space-time, 'science' and the relationship between physical geography and human geography"', *Transactions, Institute of British Geographers*, 26: 243–56.

Lau, S.S.S. and Lane, S.N. (2001) 'Continuity and change in environmental systems – the case of shallow lake ecosystems', *Progress in Physical Geography*, 25: 178–202.

Likens, G.E., Bormann, F.H., Pierce, R.S., Eaton, J.S. and Johnson, N.M. (1977) *Biogeochemistry of a Forested Ecosystem*. Berlin: Springer-Verlag.

Longley, P.A., Goodchild, M.F., Maguire, D.J. and Rhind, D.W. (2001) *Geographic Information Systems and Science*. Chichester: Wiley.

MacArthur, R.H. and Wilson, E.O. (1963) 'An equilibrium theory for insular zoogeography', *Evolution*, 17: 372–87.

MacArthur, R.H. and Wilson, E.O. (1967) *The Theory of Island Biogeography*. Princeton, NJ: Princeton University Press.

Malloch, J.C. (1991) *MATCH (VERSION 1.3) – a Computer Program to Aid the Assignment of Vegetation Data to the Communities and the Sub-communities of the National Vegetation Classification*. Lancaster: University of Lancaster.

Massey, D. (1999) 'Space-time, "science" and the relationship between physical geography and human geography', *Transactions, Institute of British Geographers*, 24: 261–76.

McGarigal, K. and Marks, B.J. (1994/2000) *Fragstats: Spatial Pattern Analysis Program for Quantifying Landscape Structure*. Corvallis, OR: Oregon State University (available at http://www.innovativegis.con/fragstatsarc/manual/manpref.htm).

Mitchell, C.W. (1991) *Terrain Evaluation. An Introductory Handbook to the History, Principles and Methods of Practical Terrain Assessment*. Harlow: Longman Scientific and Technical.

Newson, M. (1995) *Hydrology and the River Environment*. Oxford: Clarendon Press.

Oliver, J. (1991) 'The history, status and future of climatic classification', *Physical Geography*, 12: 231–51.

O'Sullivan, P. (1979) 'The ecosystem-watershed concept in the environmental sciences: a review', *Journal of Environmental Studies*, 13: 273–81.

Pacione, M. (1999) *Applied Geography: Principles and Practice*. London: Routledge.

Perring, F.H. and Walters, S.M. (1982) *Atlas of the British Flora* (3rd edn). Wakefield: E.P. Publishing.

Perry, A.W. (1995) 'New climatologists for a new climatology', *Progress in Physical Geography*, 19: 280–5.

Preston, C.D., Pearman, D.A. and Dines, T.D. (2002) *New Atlas of the British and Irish Flora*. Oxford: Oxford University Press.

Raper, J. (2000) *Multidimensional Geographic Information Science*. London: Taylor & Francis.

Rhoads, B.L. and Thorn, C.E. (1996) *The Scientific Nature of Geomorphology*. Chichester: Wiley.

Roberts, N. (1998) *The Holocene: An Environmental History* (2nd edn). Oxford: Blackwell.

Rodwell, J.S. (ed.) (1991–2000) *British Plant Communities. Vols 1–5*. Cambridge: Cambridge University Press.

Rogers, A., Viles, H. and Goudie, A. (1992) *The Student's Companion to Geography*. Oxford: Blackwell.

Savigear, R.A.G. (1965) 'A technique of morphological mapping', *Annals of the Association of American Geographers*, 55: 514–38.

Schaefer, F.K. (1953) 'Exceptionalism in geography: a methodological examination', *Annals of the Association of American Geographers*, 43: 226–49.

Schumm, S.A. and Lichty, R.W. (1965) 'Time, space and causality in geomorphology', *American Journal of Science*, 263: 110–19.

Shafer, C.L. (1990) *Nature Reserves: Island Theory and Conservation Practice*. Washington, DC: Smithsonian Institute Press.

Sissons, J.B. (1967) *The Evolution of Scotland's Scenery.* Edinburgh: Oliver & Boyd.

Sissons, J.B. (1974) 'A late glacial ice cap in the central Grampians, Scotland', *Transactions, Institute of British Geographers* OS, 62: 95–114.

Slaymaker, O. and Spencer, T. (1998) *Physical Geography and Global Environmental Change.* Harlow: Addison Wesley Longman.

Stoddart, D.R. (1986) *On Geography.* Oxford: Blackwell.

Stoddart, D.R. (ed.) (1997) *Process and Form in Geomorphology.* London: Routledge.

Stott, P. (1998) 'Biogeography and ecology in crisis: the urgent need for a new metalanguage', *Journal of Biogeography*, 25: 1–2.

Tansley, A.G. (1935) 'The use and abuse of vegetational concepts and terms', *Ecology*, 16: 284–307.

TerraSeer (2001) *BoundarySeer – Software for Geographic Boundary Analysis.* Ann Arbor, MI: Biomedware.

Townshend, J.R.G. (ed.) (1981) *Terrain Analysis and Remote Sensing.* London: George Allen & Unwin.

Turner, M.G., Gardner, R.H. and O'Neill, R.V. (2001) *Landscape Ecology in Theory and Practice.* New York, NY: Springer.

Unwin, T. (1992) *The Place of Geography.* Harlow: Longman.

Verstappen, H.T. (1983) *Applied Geomorphology: Geomorphological Surveys for Environmental Development.* Amsterdam: Elsevier.

Vink, A.P.A. (1983) *Landscape Ecology and Land Use* (trans. and edited D.A. Davidson). London: Longman.

Waite, S. (2000) *Statistical Ecology in Practice.* London: Prentice Hall.

Warren, A. (2001) 'Valley-side slopes', in A. Warren and J.R. French (eds) *Habitat Conservation: Managing the Physical Environment.* Chichester: Wiley, pp. 39–66.

Waters, R.S. (1958) 'Morphological mapping', *Geography*, 43: 10–17.

Webster, R. (1977) *Quantitative and Numerical Methods in Soil Classification and Survey.* Oxford: Oxford University Press.

Webster, R. (1985) 'Quantitative spatial analysis of soil in the field', *Advances in Soil Science*, 3: 1–10.

White, R.E. (1997) *Principles and Practice of Soil Science.* Oxford: Blackwell Science.

Whittaker, R.J. (1998) *Island Biogeography.* Oxford: Oxford University Press.

Whittaker, R.J. (2000) 'Scale, succession and complexity in island biogeography: are we asking the right questions?' *Global Ecology and Biogeography*, 9: 75–85.

### *Acknowledgements*

Martin Kent would like to thank Cheryl Hayward, Marzuki Haji, Brian Rogers and Tim Absalom of the Cartography Unit of the Department of Geographical Sciences, University of Plymouth, for drawing the figures.

# 7 Time: Change and Stability in Environmental Systems

## John B. Thornes

## Definition

Time is a framework in which geomorphological events are often placed to infer cause-and-effect relationships. Historic geomorphology uses this framework to reconstruct the past. Evolutionary geomorphology attempts to argue from process deductions about how landforms develop in time – what are the trajectories through time. Dynamical geomorphology attempts to explain this evolution through time in terms of non-linear behaviour. Time is not a variable, nor does it explain outcomes.

## INTRODUCTION

Since the appearance of *Geomorphology and Time* (Thornes and Brunsden, 1977), a number of key developments across the broad field of physical sciences have had a profound impact on the ways we view time. These changes, like those of the quantitative revolution that profoundly affected physical geography in the 1960s and 1970s, parallel the shift from the 'old science and old thinking' to the 'new science and new thinking' (Marshall and Zohar, 1997). They involve the paradigms of dynamical systems, non-linearities, chaotic behaviour and panarchy. The purpose of this chapter is to provide a readable and digestible introduction to those ideas and the impacts they are likely to have on physical geography (see Chapter 8 on time in human geography). It is not the objective to provide a history of the subject itself that has been admirably accomplished by Gregory (1985).

Although many of these ideas took root long before the 1960s, they have taken time to mature and become accepted. The rapid progress of quantitative ecology under the influence of May in the UK and the Princeton School in the USA, of non-linear geomorphology by

Schumm at Fort Collins and Favis-Mortlock at Oxford, and of dynamical climatology by Lorenz and Trenberth, have so much altered the view of how natural physical systems change in time that the moment has come for an overall reconstruction of what and how we research and teach in physical geography. This chapter attempts in some ways to provide a basis for this much needed review. It is consciously non-mathematical in approach. It reflects heavily the influences and originality of the author's own teachers: W.E.H. Culling, F.K. Hare and D.R. Harris.

## A VERY BRIEF HISTORY OF THE TIME PERSPECTIVE

Physical geography comprises geomorphology, biogeography and climatology – all the subjects that study the character of the face of the earth, its form, shape, variety and origins, its history and evolution. Because geomorphology took its roots in American geology and before that from the stratigraphers who roamed the British countryside, it had, from the outset, a heavily historical and evolutionary conceptual basis, as illustrated in the works of W.M. Davis. This was typified by Davis's model of the cycle of erosion (1899). In this model Davis characterized the landform sequence that would occur with the passage of time, following an initial disturbance in the form of a relatively sudden uplift (see also Chapter 15). Later geomorphologists used these landform characteristics to give a relative age to areas of country, notably in southeast England (Wooldridge and Linton, 1955) and in the Appalachians, just as geologists use fossils to date rocks and petrography to characterize the conditions in which rock formations were developed and therefore to infer a time sequence of the history of events. In the 1960s, the interest and emphasis lay in using these stratigraphic events to infer the historical sequence of events in the Quaternary period. Dating methods ranged from the simple rule of superimposition for sedimentary rocks (deepest are oldest) to radioisotope dating. As the methods of dating improved, so the precision of the historical sequence could also be improved. Emphasis on climatic dating (such as glaciations and warm and cold periods) coupled with pollen stratigraphy, confused the issue by shifting the emphasis to dating landforms away from the study of processes and sometimes leading to circularity of argument through lack of clarity of objectives.

The coupling of environmental change and historical dating methods flowered in the middle of last century when techniques were invented, developed, improved and applied to environmental reconstruction. Among the most important of these were radiocarbon dating (Libby, 1955) and oxygen-isotope temperature estimations and dating (Shackleton, 1977). These illustrate the point made by Harré (1969) that scientific progress is often made by technical innovation. Of course, the greatest example of this lies with the mid-century development and proliferation of computers. The second came with the great accessibility of remotely sensed images and processing capacity. The first ensured that early analytical models in geomorphology, ecology and climatology could be solved by digital numerical methods. The second, mainly through the openness and generosity of the US government, enabled serious and purposeful monitoring of the earth's surface at progressively higher frequencies. Data to test theories of change through time suddenly became abundant and the numerical capacity to model these changes mathematically ushered in the expansion of studies of global and regional climate changes and their impacts that were the precursors of the present preoccupation with global warming and its actual and potential circumstances for the earth's population.

Throughout most of the last century, the concept of succession prevailed in biogeographical thought. This involved the idea that, following disturbances, there was a progressive change leading to a new equilibrium state called the 'climax', in which the vegetation achieved its full potential, given the prevailing soil, climate and socio-economic conditions. This was able to rest empirically on the description and characterization of the vegetation across the globe that had been successively accomplished by the earliest physical geographers (notably von Humboldt) in their exploration of the globe. Classification of climates had also been practised for some decades before it became central to the problems of global change.

Already by 1920, ecology was experiencing the innovative development of the exploration of single-species population models and models of competition between competing species that would herald the eventual dismissal of the simplified conceptual basis of the theory of succession. By the 1940s, the climatologist, Edward Lorenz, had stumbled, through computational problems, on to the essential difficulties of non-linearity and chaos in climatic modelling, the so-called sensitivity-to-initial-conditions (see below). Schumm (1979) had upset the geomorphological apple-cart by pointing out that, contrary to the idea of smooth progressive change in geomorphological behaviour, many processes procede by a series of episodes of intensive change as 'intrinsic' thresholds were crossed. This brought about a major rethink in geomorphology when it was realized that geomorphological history depended not only on gradually changing extrinsic forces (mainly climatic and tectonic) but also on the robustness (resilience) of geomorphological systems to absorb the impacts of these changes. It was only by the end of the century that Phillips

(1999) collected together the material and consolidated the view of non-linear physical geography.

Another important landmark appeared about the same time. This is the *panarchy paradigm* that builds on the original work of Holling (1973) in developing the metaphor of stability and resilience, admitting the existence of multiple stable states and applying it to understanding transformations in human and natural systems. In particular, Holling and Gunderson (2002) developed the concept of adaptive cycles following disturbance. Together with newly developed ideas about spatial ecology (Tilman and Kareiva, 1997), these intellectual advances seem set to bring about a restructuring of the biogeography component of physical geography in the near future.

The main thrust of the quantitative revolution of the 1960s and 1970s was with empirical analysis of large datasets. As well as the spatial data, attention also focused on time structured data and the empirical analysis of time-series for both physical environmental series (such as water quality data) or social series (such as employment, demographic and epidemiological data). The work largely comprised the decomposition of time series into trend, periodic, persistence and noise components (for further details, see Thornes and Brunsden, 1977: 81–4). The methodology was essentially inductive, though epidemiological work also had to come to grips with modelling the wave-like behaviour of infectious outbreaks, and ecologists were already deducing non-linear competitive behaviour using difference and differential equations. Scheidegger (1960) and Kirkby (1972) opened up geomorphology to mathematical modelling through imaginative applications of standard mathematical procedures in hillslope morphology closely linked to the newly developed hillslope hydrology. By the early 1980s, the stage was set for the development of a dynamical geomorphology called 'evolutionary geomorphology', in which the emphasis is on the trajectories of behaviour through time rather than the description of the sequence of (dated) events in the historical approach we outlined earlier.

**THE NON-LINEAR PARADIGM**

Until the mid-twentieth century, scientific thinking was dominated by change as a linear phenomenon in which small forces produce proportionately small responses. Doubling the force doubled the response. Moreover, the phenomena were kept in check by negative feedbacks that tended to subdue strong departures from the norm in time. Natural systems were perceived to be well behaved, analysable, predictable and controllable. As Marshall and Zohar (1997: 248) describe it: 'Linear science complemented the dreams of a nineteenth century society that was rule-bound, reliable, predictable and unlikely to shock, a society that believed in eternal progress through the manipulation and exploitation of natural and human resources'.

In physical geography the linear world is represented by the idea of convergence on an equilibrium state. Equilibrium is a mathematical concept in which the rate of change is zero. It is generally taken to indicate a non-changing state of the system under observation. Thinking has been dominated by Le Chatalier's principle that systems react to changes in ways that tend to minimize the effect of the initial disturbance. This has been implicit for almost 100 years in the concepts of succession and climatic climax, usually attributed to F.E. Clements (1928). We shall see later in this chapter that this concept has now to be rejected. Similarly, geomorphology was preoccupied by progression towards stable equilibrium states of alluvial channel hydraulic geometry, stable network patterns and 'characteristic' slope forms. In climatology the global circulation, notably the circumpolar westerlies and their wave characteristics, was thought of as a stable equilibrium form of the earth's atmospheric heat engine, brought into being by advection of heat and momentum from the Equator to the poles.

In non-linear systems change is rapid and unexpected and can be triggered by small events. The response is usually out of all proportion to the causative event. This is usually named the Butterfly Effect, following the work of climatologist, Edward Lorenz (1963). It expresses the idea that the global circulation is so non-linear that a butterfly fluttering its wings on one side of the world would perturb the circulation and that, in a non-linear system, these perturbations would be amplified to interfere with the general circulation in another part of the world.

Big effects do not necessarily have big causes. Brunsden and Thornes (1979) developed this theme in geomorphology on the basis of Schumm's earlier work. They suggested that a key feature of geomorphological systems was their sensitivity to small perturbations. This is illustrated by Schumm's example of the shift from the meandering state to braided state in rivers. It had been shown that meandering and braided channels can be separated by a line on a graph of slope and discharge (Figure 7.1). When the line is crossed as a result of a change in either slope or discharge or both, the system changes from one river morphology to another. In other words it is not simply a linear response but the whole character of the system changes. If a point (e.g. A) is near to the line, only tiny changes in the controlling variables (slope or discharge) will push it across the threshold – for example, from a meandering to a braided habit, as shown by the arrows in Figure 7.1. At point B, the system is very stable because

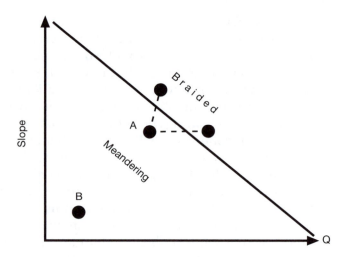

*Source*: Based on Schumm (1968)

**FIGURE 7.1**   The discriminant function between braiding and meandering, showing an unstable channel at position A and a stable channel at position B. Dotted lines show shifts in slope (vertical) and discharge (horizontal) from A

large shifts would be needed in either slope or discharge before any change in state could result.

Although this example illustrates the case of two controlling variable (slope and discharge), Figure 7.2 illustrates the problem of three controlling variables. In this case, the vertical axis is the

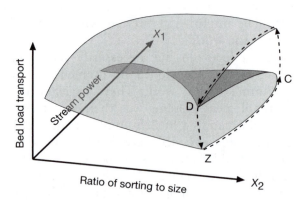

*Source*: From Thornes (1980)

**FIGURE 7.2**   Catastrophe theory representation of bed-load sediment transport. At C there is sudden high entrainment of bed material as stream power reaches a threshold value, leading to high bed-load transport. At D there is massive deposition as the threshold of transport for material of this sorting and grain size is passed as stream power falls. The dotted line represents the passage of a flood wave in an ephemeral channel

response (here the sediment transport in a river). The horizontal axes are the stream power ($x_1$) and the sorting of bed material ($x_2$). In all cases (represented by the shaded curved surface) sediment transport increases as stream power increases (going from front to back of the figure, parallel to $x_1$). At low values of $x_2$ (poorly sorted) there is always some sediment that matches the available stream power and therefore can be entrained (picked up), so the curve is relatively smooth. With better sorting (further along the $x_2$ axis) the grain size becomes more concentrated at one size group. Only when the stream power is sufficient to entrain that grain size can the sediment load increase. The threshold of entrainment (the value at which transport of a given particle size begins) increases as the ratio of mean grain size/sorting ($x_2$) increases. This produces a jump in the amount of sediment transported after the tiny increase in stream power that crosses the threshold. This jump is called a catastrophe (French for step), not to be confused with catastrophic, meaning disaster. The discovery of non-linear catastrophes has led to a whole area of non-linear science called catastrophe theory with its own specialized body of mathematics. This example is considered at length for ephemeral stream behaviour in Thornes (1980), and catastrophe theory has been the subject of application in human geography and ecology in Wilson's excellent text (1981).

Sometimes the system is completely transformed and a new system re-emerges from the change. This may occur where an ecosystem is completely ravaged by fire, climatic change or cultivation. After a fire, for example, Mediterranean scrub re-emerges as woodland after 12–15 years (Obando, 2002). After oil spills, the ecosystem reorganizes itself. Emergent phenomena form another special field of non-linear behaviour that attempts to explain how patterns emerge after a major disturbance. Other examples are the emergence of a drainage pattern in a former lake shore or an ancient sea plain after uplift. Rebeiro-Hargrave (1999) developed a model to account for the gully-dominated drainage system that has emerged since the middle Tertiary in the Guadix and Baza sedimentary basins in southeast Spain, east of Granada. He used a technique called cellular automata to model the evolution of the drainage of the basin under rules of gully extension on a sedimentary plane undergoing tectonic tilting. The emergent pattern produced in his model is very similar to that which exists today and emerged during the Quaternary era.

Non-linearity has become a way of life. It can be summarized as highly complicated responses to simple changes in inputs that may appear insignificantly small but lead the systems across thresholds that unexpectedly and rapidly shift the behaviour. This has been a major threat to historical sciences such as geology and geomorphology as they existed in the middle of last century. Large and sudden changes, such as the onset of glaciation, had been assumed to be produced by large and sudden changes in the controlling systems (Oerlermans and van der Veen, 1984). It is now recognized that the Dansgaard–Oeschlager events (large flows of cold meltwater into the North Atlantic) and even the onset of the Quaternary glaciations themselves are the manifestations of the non-linear behaviour of the ice–land–ocean–atmosphere system of interaction. Many of the phenomena that were attributed in the 1960s to climate change are now attributed to hidden non-linearities in natural systems.

## THE PANARCHY PARADIGM

The apparent importance of understanding the behaviour of environmental systems, particularly their non-linear behaviour with respect to stability, is that it appears to offer comfort to environmental managers. If the threshold of stream plan-form behaviour (meandering vs. braided) can be exactly identified then, for a given (engineered) discharge, it is theoretically possible to train a river to avoid the slopes across which instability is likely to be generated. By lengthening or shortening a channel between two heights, the slope can be changed. Even better, if the stable condition for pool and riffle and meander wavelength can be identified and if this attractor can be reached, the river system will find it hard to shift away from this equilibrium. This is the essential basis of river restoration:

• Identify the stable equilibria.
• Engineer the river into that equilibrium.
• Keep it there by minor adjustments (artificial bars and pools).

A natural system that can be tamed in this way is said to be resilient – that is, it responds to management perturbations by adopting a new equilibrium. There are abundant examples of this kind of philosophy, but there are two major types of uncertainty that undermine it. The first is that extreme events can obliterate the good intentions (a great flood can simply remove all the gravel bars that represent equilibrium conditions). Secondly, there are unknown and unforeseen complications (including hidden thresholds) that make an oversimplified view of the system exceptionally dangerous.

Such unforeseen complications are revealed in the concept of 'carrying capacity'. Carrying capacity is the number of animals that can graze grassland without damage to the rangeland ecology, without the consequent soil erosion and land degradation. Of course, a pig is not the same as a cow, so the environmental managers designed 'animal units' as the basis for agronomists to estimate the carrying capacity, and this could act as a management target that would enable them to keep the system close to the equilibrium state with respect to ecology and land degradation. Unfortunately this approach failed to accept that the target grass cover (the desired equilibrium to prevent erosion) and the perturbations that could shift the system away from the target equilibrium are unforeseeable – the magnitude and frequency of wetter or drier periods and their impact on grassland are discussed in a later section of this chapter. Even in the very short term, the internal dynamics of the competition between erosion and grass or water leads to unstable oscillations of the vegetation cover (see Brandt and Thornes, 1993). Therefore the push to develop 'carrying capacity' as an indicator of desertification in Mediterranean environments, long after it has been abandoned elsewhere, is based on false premises. A further problem is that this approach engenders a false sense of security because of its apparent basis in science and engineering and the attachment of hard targets. In an easily read and refreshingly honest appraisal of the problems of understanding transformations in human and natural systems, Gunderson and Holling (2002: xxii) attack the simple prescriptions of the concept of environmental systems control because 'They seem to replace inherent uncertainty with the spurious certainty of ideology, precise numbers or action. The theories implicit in these examples ignore multiple stable states'. They note the problem imposed by non-linearities of the types described above. They go on to say that 'The theories ignore the possibility that the slow erosion of key controlling processes can abruptly flip an ecosystem or economy into a different state that might be effectively irreversible'.

This is an absolutely crucial point in the evolutionary view of systems. It throws into serious doubt the belief that complex system management could or should be based on a few key indicators. There is an obsession on the part of the environmental agencies in Europe and the USA to develop single-figure indicators to monitor environmental management targets.

Later in their treatise, Holling et al. (2002) translate the discussion of equilibria into 'caricatures of nature' as four myths, as illustrated by Figure 7.3. The caricatures are as follows:

1 *Nature flat* – a system in which there are few or no forces affecting stability.
2 *Nature balanced* – a system at or near an equilibrium condition, the system at the bottom of a cup.
3 *Nature anarchic* – a view of the nature as globally unstable – dominated by hyperbolic processes of growth and collapse.
4 *Nature resilient* – a nature of multi-stable states, some of which become irreversible traps while others become natural alternating states that are experienced as part of the internal dynamics.

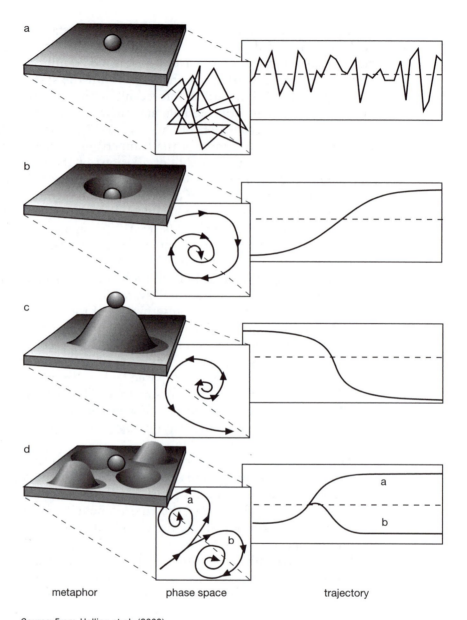

metaphor              phase space              trajectory

Source: From Holling et al. (2002)

**FIGURE 7.3**   Depictions of the four myths of nature by Holling et al. (2002): (a) nature flat;
(b) nature balanced; (c) nature anarchic; (d) nature resilient (multiple stable states)

This caricature recognizes that instabilities organize the behaviour as
much as stabilities do. These are the dynamical systems that prevail
in the subject-matter of physical geography in which non-linearities
dominate. Holling has argued this case since 1973. Holling et al.

(2002: 5) are seeking a theory of adaptive change 'to help us to understand the changes occurring globally'. They use the term panarchy to 'describe the cross-scale, interdisciplinary and dynamic nature of the theory whose essential focus is to rationalize the interplay between change and persistence, between the predictable and the unpredictable'.

Their general theory emerges as a model of the cycle of resilience and adaptation following change. This is envisaged as a Mobius strip trajectory through time, as shown in Figure 7.4. If we start following a major perturbation (say a shock) at the bottom left of the strip, the new situation is characterized by reorganization and the transient appearance or expansion of organisms – the pioneer stage (or $r$-phase in ecology). This leads to steady consolidation leading towards the relatively stable $k$-phase, which is characterized by conservative behaviour at equilibrium (again, see Obando, 2002). The system's connectedness increases until the system collapses, perhaps as a result of some small change leading to the crossing of an internal threshold. This in turn leads to what the authors call the release and eventually to a reorganization that starts again the adaptive cycle. The cycle, so described, is reminiscent of Davis's cycle of erosion (1899). Uplift leads to a new initiation of a drainage system on the raised surface (close to the end-surface of Penck) and this is consolidated through its control on runoff and sediment yield and feedback that lead to self-

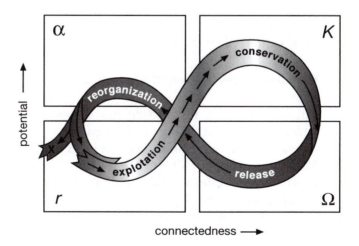

*Source*: From Holling et al. (2002)

**FIGURE 7.4**  A stylized representation of the four ecosystem states as an ecosystem adapts following change (Holling et al., 2002). The vertical axis is the capacity or potential that is inherent in the accumulated resources of biomass and nutrients. The horizontal axis is the connectedness. The cycle starts at the bottom left and the main (light) curve from $r$ to $k$ is the logistic curve discussed in the text

adaptation of the drainage network, as modelled by Rodriguez-Iturbe (1994). There is, however, no direct analogy with the 'release' phase of the equilibrium in the Davis model. Thornes (1990) showed that, through the competition between erosion and vegetation, there appeared a boom-and-bust cycle in Mediterranean *matorral* that was simulated by Brandt and Thornes (1993) and which shared a regular cycle with the adaptive behaviour of Holling et al.'s model.

The adaptive cycle model is still in its infancy and only time will tell if it is sufficiently robust. The weakness at the moment seems to be in the explanation of the release (collapse) phase which appeals more to external variables (wind, fire, drought) than to intrinsic internal dynamics. Periodic flips from one stable state to another are 'mediated by changes in the slow variables (the horizontal axis variables in the catastrophe theory) that suddenly trigger fast variable response, or escape' (Holling et al., 2002: 35). There seems little doubt that the panarchy paradigm will provide the focus for thinking and exploration of the systems control in environmental management philosophy in the future. If it overturns the idea that systems are generally in a stable equilibrium state and hence resilient in the engineering sense, this will represent a significant forward step.

## TWO TIME-BASED PROBLEMS

In this final section two examples have been chosen to illustrate the ideas expressed so far in this chapter.

### Global changes and vegetation shifts

The concept of a progressive move towards a potential has been widely explored in ecology and biogeography. Here the upper limit of biomass or species richness became fixed in the concept of succession. This was viewed by Odum (1969) as: (1) an orderly process that is reasonably directional and therefore predictable; (2) resulting from modification of the physical environment by the evolving plant community (for example, a soil develops); and (3) culminating in a stable (climax, mature) ecosystem with homeostatic properties. The upper limit is reached when all the resources available to the community (water, radiation and nutrients) are consumed in producing biomass. Likewise a lower stable limit would be fixed by any limitation on resources – Liebeg's Law. Any one of them might be 'limiting' and, in the 1960s and 1970s, they were reinterpreted in the 'new ecology' as the capacity or potential. Thus, in semi-arid environments, water is the limit to growth. The rate of growth increased or decreased according to the difference between the current state of the system and the capacity. This is modelled by the logistic growth equation. Because any of water, radiation and nutrients can be limiting, the succession concept of Clements (1928) has been interpreted as leading to the climate climax vegetation. The assumption that the vegetation reaches the capacity determined by the limits of growth is implicit in the climatic climax vegetation and the concept of succession. It forms the basis of the interpretation of historical evidence of vegetation cover in climatic terms (for example, in palynology) and the reconstruction of past vegetation shifts resulting from global climate change. Sometimes it is used to predict the potential of future climate changes.

Thornes (in press) has interpreted shifts of the bush–grass boundary of the Eastern Cape of South Africa in terms of the impact of soil moisture (reflecting precipitation changes). As precipitation changes, the biomass responds, usually with a lag in time. The vegetation cover is said to 'track' the precipitation. This tracking is of the boom-and-bust variety referred to above. In periods of heavy rain, vegetation 'overshoots' the potential and then dies back to the equilibrium or capacity value. In dry periods the vegetation dies back to a very sparse cover. Because rainfall is modelled as a random walk, the vegetation cover tracks it also as a random walk. Since South African rainfall is strongly influenced by oscillations due to ENSO oscillations (see below), wet and dry periods succeed each other. It was shown for both the Mediterranean and the Eastern Cape that, if a community is disturbed from its equilibrium trajectory by a sudden change of climate, it could be a very long time before equilibrium is restored for typical plant covers with a distinctive growth curve.

It is well known that these reactions to climate change are complicated by grazing activities (Noy-Meir, 1982) and that the impacts of grazing are non-linear, perhaps even chaotic in nature, with small effects pushing the system into new trajectories. There are three consequences of this work. It shows that there are two stable states, one with no vegetation, the other with the maximum possible amount. Small shifts of rainfall can move the vegetation system to one or another of them. Once a change is set in train, it moves towards one of the stable states. This also supports the arguments of Holmgren and Scheffer (2001) that, in managing the vegetation cover, it might be best to wait until the turns in rainfall create the conditions that will lead towards a new desired stable state. From the first point it also follows that the interpretation of shifts in geomorphological process (notably soil erosion) as a direct, linear consequence of climate change needs to be treated with caution, given the non-linear nature of vegetation responses.

### Circulation dynamics and complexity

Much of the chaotic dynamics paradigm, arising from Edward Lorenz's discovery of the Butterfly Effect, came from his attempt to model circulation. We should therefore not be surprised to find that the modelling of circulation change, at the very core of the debate on the impacts of global warning, is fraught with difficulty created by non-linearities that lead to chaotic outcomes and, indeed, this is the case. There are several aspects to this:

- Non-linearities mean that complicated and unexpected responses may be expected from quite small changes in inputs. Similarly, very complicated behaviour can be the outcome of simple causes. We do not have to seek complicated inputs to get complicated outputs from complex systems.
- Because of this chaotic behaviour, time-series of outputs, as discussed above, should be explored for indications of deterministic chaos (meaning oscillations caused by unstable numerical behaviour).
- Prediction becomes even more difficult as a result of chaos so that global climate change modelling is subject to even greater uncertainty. Not only do impacts of climate change reflect non-linearities but the inputs themselves (climate scenarios) are also the outcomes of chaotic dynamics.

Lorenz (1963) produced a drastically oversimplified model of thermal convection in a layer of fluid. The motion was driven by the temperature difference between the bottom of the fluid layer and the top. Subsequently the equations used by Lorenz to describe the speed of the moving air have attracted attention purely on their own mathematical merits, as demonstrated by Acheson (1997). Given three differential equations, Acheson demonstrates that there are three equilibrium states. He also shows that if the van der Pol equations are solved for the evolution through time, they can be seen to exhibit chaotic behaviour (Figure 7.5). This is shown by their irregular behaviour and their extreme sensitivity to initial conditions (Acheson, 1997). With an initial difference of just one part in 1000, the oscillations diverge as the difference becomes greater than about 13. The bold curve is for the starting $x$ $(x_0)$ = 5.000 and the lighter curve for $x_0$ = 5.005. Even when the difference is reduced to a factor of 100, to just one part in 10 000, the outcomes stay together for only a little longer, until $t$ = about 16. Lorenz realized that this extreme behaviour had profound implications. He concluded in 1963 that 'In view of the inevitable inaccuracy and incompleteness of weather observations, precise very long range forecasting would seem to be non-existent'

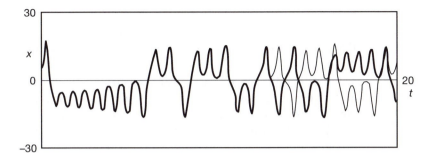

*Source*: Acheson (1997) Reprinted with permission of Oxford University Press

**FIGURE 7.5**   Numerical solution of the Lorenz equations for $r = 28$ and with $(x, y, z) = (x_0, 5, 5)$ at $t = 0$. The bold curve is for $x_0 = 5.000$, and the lighter curve for $x_0 = 5.005$. The two diverge after about 13 oscillations, showing sensitivity in the outcome to small differences in initial conditions $(x_0)$

(cited in Acheson, 1997: 159). Here 'long' means greater than three days!

Although some of these problems persist at the scale of global circulation, they are less acute. Prediction of climate (rather than weather) is more feasible because average conditions, not day-day variations, are being considered (Schneider, 1992). Imagine a billiard table with many balls. It would be difficult to predict the behaviour of a single ball if many are set in motion. Nevertheless, the average number of balls reaching a pocket in an average amount of time may be rather easier to predict. This statistical mechanics approach makes very helpful statements about the probability distribution of a large number of atoms in a gas (the ensemble) compared with the behaviour of individual atoms. Moreover, at the space and time scales of climate, the atmospheric changes are forced by other changes that respond only very slowly and by their own feedback systems, such as the effect of oceanic temperature. Generally the degree of predictability is related to the period of amplitude and forcing. Schneider goes on to point out that, after a long time period, the global-scale climate system will move to a unique equilibrium after a transient adjustment for these longer times, assuming the existence of such unique and stable equilibria.

The periodically changing climate response that illustrates internal working of the climate system is in the ENSO (El Niñō Southern Oscillation) circulation of the eastern Pacific and Australo-Indonesian regions (Cane, 1997). El Niño is a massive warming of the coastal waters off Peru and Ecuador that is accompanied by torrential rain and catastrophic flooding. El Niño events have been known since 1726 and there is evidence indicating occurrences for at least a millennium

before that (Quinn et al., 1987). The Southern Oscillation is a difference in the sea-surface pressures between the southeast tropical
Pacific and the Australo-Indonesian regions (between the stations of
Tahiti and Darwin). When the waters in the eastern tropical Pacific
are abnormally warm, sea-level pressure drops in the eastern Pacific
and rises in the west. The ENSO reveals characteristic features of
simple non-linear dynamics – regular oscillations and periodic switching between alternate quasi-stable states. Nevertheless, both the
physics and the mathematics are far too complicated for this chapter
and the reader is referred to Cane's review (1997). The first signs
appear in the Northern Hemisphere spring, build to a peak at the end
of the year and, by the following summer, the warm event is usually
over. The phenomenon occurs every four years, and modellers have
been able to predict it very successfully on the basis of Bjerknes'
(1969) hypothesis of the interactions among the Kelvin waves, the sea
temperature and the Rossby waves of the tropical Pacific. According
to Cane (1997: 586, emphasis added):

> ENSO is the premium example of variability, *stemming entirely from
> the internal* working of the climate system. Attempts to model and
> predict ENSO test our ability to model and predict larger time-scale
> variations and do it in a context with more observational data to verify
> against . . . It is the most advanced example of modelling two-way
> interactions between the atmosphere and ocean.

Another important characteristic is switching between quasi-stable
states. The switching is between the relative dominance of the Kelvin
and Rossby waves and this tension gives rise to the unstable oscillatory behaviour.

## CONCLUSIONS

In the last 25 years it has become obvious that the adoption of
classical scientific approaches in physical geographical investigations
will not necessarily provide a basis for explanation. The understanding of physical processes requires that we accept that, instead of
smooth linear behaviour over time, the real world is characterized by
suddenly changing oscillatory behaviour in which the systems tend to
switch between stable states. In the past there was a tendency to seek
externally varying forces to produce changes in a somewhat simple
cause-and-effect mentality. It was as if every geomorphological change
in the stratigraphical or morphological record should have a climate
change explanation. So, for example, the debate over the origin of
Dartmoor tors became a heated argument about alternative climatic
explanations rather than a discussion of relative rates of weathering

and erosion, where it rightfully belongs. The more the stratigraphy appeared to be periodic, the greater was the tendency to seek external forcing in the form of external controls that themselves vary periodically.

Decomposition of time series can be misleading and oscillatory behaviour that persists for a long period in one form or another should lead us to seek evidence of sensitivity to initial conditions and evidence of chaotic dynamics that is intrinsic to the system. The critical question when there is evidence of periodic behaviour should be 'Does it indicate dynamical instability?'

This chapter is not a call for the abandonment of historical studies in physical geography but, rather, a caution about the interpretation of unexpected behaviour.

## Summary

- In the last 25 years the importance of non-linearities, chaos, complexity and issues of the stability of equilibria have dominated the study of phenomena through time.
- Conventional time-series analysis is empirically based. Special techniques are needed to identify the non-linearities.
- Non-linearities are characterized by periodic behaviour, sensitivity to initial conditions and switching between main behaviours. The switching may be caused by external forcing, crossing internal thresholds or by small random perturbations near to critical points.
- There are families of new models to be used in the case of non-linear behaviour.
- All future physical geography will have to accept, identify and incorporate non-linear behaviour and identify its effects on the time-series of outputs from environmental systems.

## Further reading

Acheson's (1997) *From Calculus to Chaos: An Introduction to Dynamics* is a mathematical treatment of most of the concept at levels just within the reach of those of you who have studied A-level mathematics. Brunsden and Thornes' (1979) paper, 'Landscape sensitivity and change', illustrates the basic equilibrium approach to 1960s geomorphology and asks how the sensitivity of geomorphological systems to external perturbations can be identified and quantified through the transient-form ratio. The chapter by Chorley on Davis's cyclical model, in Chorley and Haggett (1965), broke the mould of mid-twentieth century historical geomorphology (denudation chronology) and opened the way to process studies and model building. Bennett and Chorley's (1978) *Environmental Systems. Philosophy, Analysis and Control* is the classical approach to systems analysis by two leading geographers and the first to recognize the stability problems that are discussed in this chapter. It is demanding reading. Gunderson and Holling's edited volume (2002), *Panarchy. Understanding Transformations in Human and Natural Systems*, provides a refreshing, up-to-date appraisal of oversimplified approaches to environmental management. It stresses the need for understanding the complexity of environmental behaviour and proposes new concepts to deal with it. *Who's Afraid of Schroedinger's Cat?*, by Marshall and Zohar (1997), is a remarkably readable collection of items on the new science from the arrow of time, through fuzzy logic, to the wave equation with complexity and chaos on the way. Introductory chapters are an absolute must. The most recent comprehensive review of complex behaviour in geomorphological systems is provided by Phillips (1999) in *Earth Surface Systems: Complexity, Order and Scale*. Roetzheim's (1994) *Enter the Complexity Lab: Where Chaos Meets Complexity* is a fun book that clearly explains the main concepts of complexity, stability and chaos. It includes a disk of computer programs (games?) that illustrate the principles, using 'birds', ants and even mice in mazes. Chaos, cellular automata, life models are all here. Although 30 years old, Wilson's (1981) *Catastrophe Theory and Bifurcation* is still the clearest and most accessible source for catastrophe, bifurcations and other aspects of non-linear behaviour in geographical systems.

*Note*: Full details of the above can be found in the references list below.

## References

Acheson, D. (1997) *From Calculus to Chaos: An Introduction to Dynamics*. Oxford: Oxford University Press.

Bennett, R.J. and Chorley, R.J. (1978) *Environmental Systems: Philosophy, Analysis and Control*. London: Methuen.

Bjerknes, J. (1969) 'Atmospheric teleconnections from the equatorial Pacific', *Monthly Weather Review*, 97: 163–72

Brandt, C.J. and Thornes, J.B. (1993) 'Erosion–vegetation competition in an environment undergoing climatic change with stochastic rainfall variations', in A.C. Millington and K.T. Pye (eds) *Environmental Change in the Drylands. Biogeographical and Geomorphological Responses*. Chichester: Wiley, pp. 306–20.

Brunsden, D. and Thornes, J.B. (1979) 'Landscape sensitivity and change', *Transactions, Institute of British Geography*, 4: 463–84.

Cane, M.A. (1997) 'Tropical Pacific ENSO models: ENSO – a model of the coupled system', in K.E. Trenberth (ed.) *Climate System Modelling*. Cambridge: Cambridge University Press, pp. 583–616.

Chorley, R.J. (1965) 'A re-evaluation of the geomorphic system of W.M. Davis', in R.J. Chorley and P. Haggett (eds) *Frontiers in Geographical Teaching*. London: Methuen: 21–38.

Clements, F.E. (1928) *Plant Succession and Indicators: A Definitive Edition of Plant Succession and Plant Indicators*. New York, NY: H.W. Wilson.

Davis, W.M. (1899) 'The geographical cycle', *Geographical Journal*, XIV: 481–504.

Gregory, K.J. (1985) *The Nature of Physical Geography*. London: Edward Arnold.

Gunderson, L.H. and Holling, C.S. (2002) 'Preface', in L.H. Gunderson and C.S. Holling (eds) *Panarchy. Understanding Transformations in Human and Natural Systems*. Washington, DC: Island Press, p. xxii.

Harré, R. (1969) *Scientific Thought 1900–1960*. Oxford: Clarendon Press.

Holling, C.S. (1973) 'Resilience and stability of ecological systems', *Annual Review of Ecology and Systematics*, 4: 1–24.

Holling, C.S. and Gunderson, L.H. (2002) 'Resilience and adaptive cycles', in L.H. Gunderson and C.S. Holling (eds) *Panarch: Understanding Transformations in Human and Natural Systems*. Washington, DC: Island Press, pp. 25–63.

Holling, C.S., Gunderson, L.H. and Ludwig, D. (2002) 'In quest of a theory of adaptive change', in L.H. Gunderson and C.S. Holling (eds) *Panarch: Understanding Transformations in Human and Natural Systems*. Washington, DC: Island Press, pp. 3–24.

Holmgren, M. and Scheffer, M. (2001) 'El Niño as a window of opportunity for the restoration of degraded arid ecosystems', *Ecosystems*, 4: 141–49.

Kirkby, M.J. (1972) 'Characteristic slope forms', in D. Brunsden (compiler) *Hillslope Form and Process*. London: IBG Special Publication: 15–31.

Libby, W.F. (1955) *Radiocarbon Dating* (2nd edn). Chicago, IL: Chicago University Press.

Lorenz, E.N. (1963) *The Essence of Chaos*. London: University College Press.

Marshall, I. and Zohar, D. (1997) *Who's Afraid of Schroedinger's Cat?* London: Bloomsbury Press.

McIntosh, R.P. (1981) 'Succession and ecological theory', in D.C. West et al. (eds) *Forest Succession: Concepts and Application*. Berlin: Springer-Verlag, pp. 10–23.

Noy-Meir, I. (1982) 'Stability of plant herbivore models and possible application to savanna', in B.J. Huntley and B.J. Walker (eds) *Ecology of Tropical Savannas*. Berlin: Springer-Verlag, pp. 591–609.

Obando, J.A. (2002) 'The impact of land abandonment on regeneration of semi-arid vegetation: a case study from the Guadalentin', in N.A. Geeson et al. (eds) *Mediterranean Desertification: A Mosaic of Processes*. Chichester: Wiley, pp. 247–68.

Odum, E.P. (1969) 'Generalization of successional-climax paradigm with over-shoot', *Science*, 164: 262.

Oerlermans, J. and van der Veen, C.J. (1984) *Glacial Fluctuations and Climate Change*. Dordrecht: Kluwer Academic.

Perry, J.N., Smith, R.H., Wolwod, I.P. and Morse, D.R. (2000) *Chaos in Real Data*. Dordrecht: Kluwer Academic.

Phillips, J.D. (1999) *Earth Surface Systems: Complexity, Order and Scale*. Oxford: Blackwell.

Quinn, W., Dopf, D., Short, K. and Kuo-Yang, R.W. (1987) 'Historical trends and statistics of southern oscillations, El Niño and Indonesian droughts', *Fisheries Bulletin*, 76: 663–78.

Rebeiro-Hargrave, A. (1999) 'Large scale modelling of drainage evolution in tectonically active asymetric intermontane basins, using cellular automata', *Zeitschrift für Geomorphologie*, Suppl. Bd, 118: 121–34.

Rodriguez-Iturbe, I. (1994) 'The geomorphological unit hydrograph', in K. Bevan and M.J. Kirkby (eds) *Channel Network Hydrology*. Chichester: Wiley, pp. 43–69.

Roetzheim, W.H. (1994) *Enter the Complexity Lab: Where Chaos Meets Complexity*. Indianapolis, IN: Sams Publishing/Prentice Hall.

Scheidegger, A.E. (1960) *Theoretical Geomorphology*. Berlin: Springer-Verlag.

Schneider, S.H. (1992) 'Introduction to climate modelling', in K.E. Trenberth (ed.) *Climate System Modelling*. Cambridge: Cambridge University Press, pp. 3–26.

Schumm, S.A. (1968) 'River adjustment to altered hydrologic regime: Murrumbridge River and palaeochannels Australia', *United States Geological Survey Professional Paper 598*: 68 pp.

Schumm, S.A. (1979) 'Thresholds in geomorphology', *Transactions, Institute of British Geographers*, 4: 485–515.

Shackleton, N.J. (1977) 'The oxygen isotope record of the late Pleistocene', *Philosophical Transactions of the Royal Society*, B280: 169–82.

Thornes, J.B. (1980) 'Structural instability and ephemeral channel behaviour', *Zeitschrift für Geomorphologie*, supplement band 39: 136–52.

Thornes, J.B. (1990) 'The interaction of erosional and vegetational dynamics in land degradation: spatial outcomes', in J.B. Thornes (ed.) *Vegetation and Erosion*. Chichester: Wiley, pp. 41–53.

Thornes, J.B. (in press) 'Exploring the grass–bush transitions in South Africa through modelling the response of biomass to environmental change', *Geographical Journal*.

Thornes, J.B. and Brunsden, D. (eds) (1977) *Geomorphology and Time*. London: Methuen.

Tilman, D. and Kareiva, P. (eds) (1997) *Spatial Ecology: The Role of Space in Population Dynamics and Interspecific Interactions. Monographs in Population Biology* 30. Princeton, NJ: Princeton University Press.

Wilson, A.G. (1981) *Catastrophe Theory and Bifurcation*. London: Croom Helm.

Wooldridge, S.W. and Linton, D.L. (1955) *Structure, Surface and Drainage in South East England* (2nd edn). London: George Philip & Son.

# 8 Time: From Hegemonic Change to Everyday Life

**Peter J. Taylor**

## Definition

Time and space form the basic physical dimensions of the universe. As such, time is used to measure change, including societal change. Social time indicates time with content: human phenomena in the process of change. Social time is invariably linked to social space as 'time-space'.

### INTRODUCTION: THE MEANINGS OF TIME IN HUMAN GEOGRAPHY

In human geography, time has been conceptualized in two distinct ways (see Chapter 7 on time in physical geography). One view of time is as a physical dimension, something that can be measured precisely. Thus a geographer might add a 'time dimension' to a model of, say, settlements to show how patterns have developed over a given period. Such models are called *dynamic models* (in contrast to 'static models' that describe a situation at one point in time) and rely on *time-series data and analysis*. The second view of time is as social change, where the emphasis is upon the 'content of time'. Thus a geographer might study the evolution of a particular settlement pattern as an outcome of industrialization. Such an approach focuses upon *social processes*: industrialization is a bundle of such process relating to shifting work practices with many concomitant economic, political and cultural changes. The form that such study takes depends upon the *social theory* that is used to define the nature of social change.

This chapter will focus on this second view of time. The argument consists of three parts: first, the ways in which geographers have studied time are described; secondly, different patterns of time are discussed; and, finally, we interpret how we use our ideas about time to interpret the present. We cannot begin the argument, however, without introducing two basic relations: time and space, and time and modernity.

For geographers, time cannot be studied independently of space. Like time, space can be viewed as either 'physical' space, which we think of as *geometry*, or 'social' space, which we think of as *place* – that is, a space with content (see Chapters 5 and 9 for more on the multiple definitions of space and place in human geography). In the former case space is viewed as three-dimensional so that time becomes the 'fourth dimension'. In the latter case, social processes are studied *in situ* so that we study the composition of places as in regional geography. Because space and time are so indelibly linked, in much of the discussion below there will be reference to *time-space* phenomena. In short, throughout their concern to understand time – whether measuring temporal trends or interpreting compositional changes – geographers will be interested in locations, from local to global, wherein the trends or changes occur.

We live in a world we call 'modern'. The idea of modern is a very time-laden one. Being modern is to 'move with the times', to be 'up to date', to be a user of the newest gadgets or ideas, to be a follower of the latest fashions in clothes, furnishings or games. Collectively a society of modern people defines a state of modernity. There is a sense that when either an individual or a society do not posses the newest artifacts then they are deemed to be somehow 'behind' in time. Note that this is pre-eminently a 'social time' idea; it makes no physical sense for contemporary individuals or societies to be in front or behind in 'real' time. However, a consequence of concerns for this social time in modernity is that the latter has come to define a society that experiences rapid and ceaseless social change. Thus the concept of time is central to the meaning of modernity and of geographers' attempts to understand this condition.

## HOW TIME HAS BEEN STUDIED BY HUMAN GEOGRAPHERS

The study of time in human geography has long been the domain of historical geographers. Their traditional concern for the development of landscapes and regions was challenged through the upheavals of the discipline in the 1960s. It is from this period that concepts of time began to be incorporated more generally throughout human geography in new systematic ways (Carlstein et al., 1978a; 1978b; 1978c). Five temporal models and concepts can be identified.

The first of these is *time-space convergence*. Through noting that the time it takes to transverse distances has fallen dramatically in recent centuries, it is sometimes suggested that the world is becoming somehow 'smaller'. This idea of a 'shrinking world' is not unique to geography but its precise measurement as time-space convergence is. This concept was devised by Janelle (1969). Drawing on data for the time it took to travel between particular pairs of towns from using stage-coaches to flying aeroplanes, Janelle constructed graphs showing the decline in length of time from the seventeenth to the twentieth century (Figure 8.1). As a particular example, he measured travel times between Edinburgh and London showing how they declined from taking weeks to complete the journal to just a few hours. More precisely, for the period 1776–1966, Janelle calculated an average 'convergence rate' between the two cities of 29 minutes per year. Subsequent researchers have used travel times between several cities to create new 'time maps' wherein physical distances between locations are replaced by 'time-distances' (Forer, 1978).

Hägerstrand's (1973) *time-geography*, the second of the concepts identified here, is arguably the most original contribution by human geography to the study of time. Using a two-dimensional space as a base map to which time was added as a vertical dimension, Hägerstrand attempted to trace the time-space paths of individuals 'upwards' and sideways through this three-dimensional diagram as

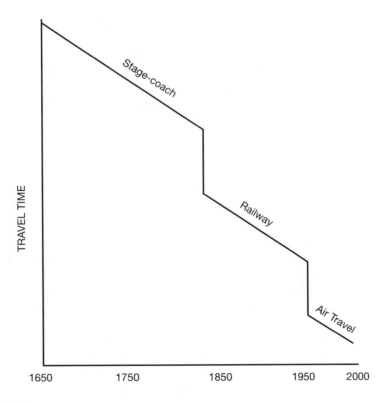

**FIGURE 8.1**  Time–space convergence between two cities

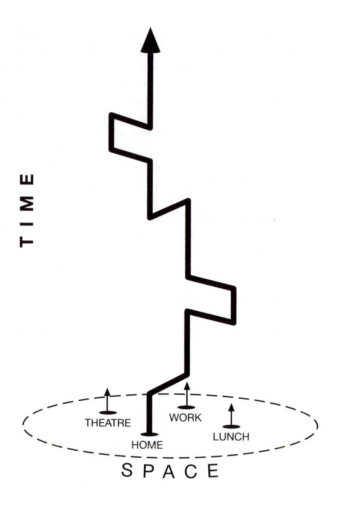

**FIGURE 8.2**  An individual time–space path

they carried out everyday tasks (Figure 8.2). For instance, during a single day a person would start on the base map at his or her residence, would then travel upwards (through time) and sideways (through space) to a workplace followed by further movements on upwards through time and across the map to a lunchtime meeting, and so on until returning at the 'top' of the diagram to the initial spatial starting point, the home. For each individual, depending on his or her access to travel facilities, there is a time-space prism that defines the boundaries of what activities are possible from his or her home base. This defines possible resources available to an individual: for instance, local schools but not multiligual international schools, or local general shops but not highly specialized retail outlets. In addition, movement through the prism is further constrained by necessary interactions with others, as at a work meeting or collecting a child from daycare, and where there is competition for time-space with others carrying out their everyday tasks. The classic case of the latter is a typical city 'rush hour' where commuters' time-space paths converge to create a time-space population concentration. This framework illustrates how space and time are resources routinely deployed in everyday life. The historical research of Pred (1986) is the exemplar for using the time-geography model for understanding how everyday life is altered through social changes.

Harvey's (1990) concept of *time-space compression* is undoubtedly the most influential contribution of human geography to the recent study of time. Fascinated by the nineteenth-century notion of the 'annihilation of space by time', Harvey develops an argument that links space and time to both economic necessities and cultural expressions. Starting with the rationalization of space in the cartographies of the sixteenth-century Renaissance and the eighteenth-century Enlightenment that expressed both the (spatial) power of state and property, Harvey argues that in order to 'conquer space' new space has to be produced, notably in transport and communications. This is the same reduction of 'spatial barriers' that produces the shrinking world of time-space convergence but here it is treated as much more than the changing dimensions of society. Within modern capitalist relations this shrinking world is part of a cyclical process of 'creative destruction' as new investments are required to resolve crises of overproduction. These new spaces create a general feeling of the world 'speeding up', leading to an 'overwhelming sense of compression'. It is the latter experience that is reflected in how the world is represented as cultural élites attempt to harness the maelstrom of incessant change that is modernity (Berman, 1982). The great artistic movements of modernism in the late nineteenth and early twentieth centuries, and postmodernism in the late twentieth century, are both interpreted as cultural reactions to this 'speeding up' of time. Time-space compression is, therefore, a powerful organizing concept that connects economy and culture.

World-systems analysis as developed by Wallerstein (1988; 1991) offers an alternative materialist interpretation of the relations between time and space in human geography. To emphasize the indissolubility of the two concepts he coins the word *TimeSpace* to describe his particular integration. Starting with Braudel's (1980) identification of three categories of social time – the short term, the medium term and the long term – Wallerstein adds a geographical component by attaching a spatial scale to each temporal span. Braudel's first time is episodic time, the time of traditional history that traces change through events, affairs, occasions and happenings. Wallerstein identifies geopolitical space, the immediate locales of events, as the spatial equivalent, thus creating episodic-geopolitical TimeSpace. Braudel's second span is a more general patterned time (trends and cycles) to which Wallerstein adds an ideological space – locales of different divisions of the world such as East/West during the cold war. This produces a cyclical-ideological TimeSpace. Braudel's third span is structural time which deals with the long-term slow movement of everyday life that underpins society. Wallerstein interprets this as the structures of historical systems such as the historical rise and demise of the great ancient empires. The structural space of this time is the boundary expansion of the system and its reproduction through core-periphery structures. This all defines a structural TimeSpace. Wallerstein's research focuses on the modern world system – our 'historical system' – whose structural TimeSpace evolved from a European capitalist world economy in the sixteenth century to a global world economy in the twentieth century, and which was reproduced through a cyclical-ideological TimeSpace of distinctive periods that developed through the episodic-geopolitical TimeSpace of events. World-systems integration of time and space is adapted to regional geography as 'historical regions' by Taylor (1988; 1991).

Wallerstein's TimeSpace is not the only use of Braudel's (1980) social times in human geography. Historical geography has developed into a multifarious enterprise and among these different themes there has been an engagement with Braudel's work (Baker, 1984). There is a revived concern for understanding the major patterns of social change in time and space (e.g. Thrift, 1990). Dodgshon (1998) has developed a *geographical perspective on change* wherein he provides an overview of social change in different spaces. He includes a taxonomy of change which compares mechanisms of change (e.g. system feedback), sources of change (e.g. external), products of change (e.g. expansion) and morphologies of change (e.g. non-linear change). The latter encompasses the continuity–discontinuity debate that defines a basic division in the treatment of time as social change: is change a relatively smooth phenomenon or are their major discontinuities in the nature of changes? The former position is represented by theories of progress that see social change as a 'forward march' of humanity; in contrast many geographers identify basic disruptions where there are 'accelerations' in the degree of social change, the 'industrial revolution' being the classic case. With disruptions there is a time divide between patterns of social relations (e.g. preindustrial and industrial society or between medieval 'feudal' Europe and early-modern 'capitalist' Europe). The remainder of this chapter will delve more deeply into these time morphologies.

## MORPHOLOGIES OF TIME

The condition of modernity is generally defined by social pressure for incessant change. Berman (1982: 15) has famously described this condition as 'a maelstrom of perpetual disintegration and renewal, of struggle and contradiction, of ambiguity and anguish'. Change can be both very exciting and very stressful: modern men and women (us) experience modernity as creating new opportunities while simultaneously destroying old cherished ways. The trick is to ensure that we are not mere objects of the vicissitudes of modernity but rather that we become active subjects, participants in the processes of change. Controlling change is the job of *planning*, the archetypal modern activity that focuses upon time.

A planning exercise can be defined as any project that attempts to control social change over a specified time horizon. As modern people we live our lives through many individual and group projects as we plan our short, medium and long-term futures. Modern institutions operate through various planning instruments. For instance, during parts of the twentieth century all the following 'grand planning' activities prospered for at least a period: the urban planning of cities, the corporate planning of firms, the military planning of the cold war military-industrial complexes, the Keynsian[1] economic planning of welfare states, the development planning of third world states and the five-year industrial planning of communist states. The last example reminds us that planning does not always work; in fact, all the surviving examples in the list have been latterly modified to become a much more 'flexible' version of their former selves. Flexible means, of course, much less control on change, an admittance that modern change cannot be simply tamed through the application of planning. Modernity is much too complex to be 'planned'.

In hindsight we can see a common process in action here. Planning is a reaction to a social problem. For instance, urban planning developed in the early twentieth century as a reaction to the legacy of poor living conditions in the Victorian industrial city. The modernist movement in planning set about clearing the 'slums' and relocating people into new, clean, high-rise blocks of housing. All planning provides a solution to its problem as defined at a given time. But modernity is perpetual change so that as soon as grand planning projects begin they progressively get out of date. Planning is condemned to solve yesterday's problems. The classic example is the Soviet Union, which produced the greatest nineteenth-century industrial state in history. The only problem was that they created it in the twentieth century: the modern world had moved on. This is equally true of urban planning wherein high-rise housing became 'slums in the sky' and have had to be abandoned as 'urban solutions'.

There is one example of failed planning that we need to consider further because it is associated with a time-based terminology that continues to distort our thinking. From the 1950s onwards the poorer countries of the world have been encouraged to embark on 'development planning'. Such planning typically involved harnessing investment in order to pass through various stages in a path that culminates in something termed 'development'. In the most famous such model, states were deemed to pass through five stages from 'traditional society' to 'high mass consumption' (Rostow, 1960). Clearly the last term shows that what was happening here is that poor countries were being advised to mimic the growth patterns of rich countries. All countries, it seems, were expected to proceed along identical parallel paths separated only by time (Taylor, 1989). Put another way, poor countries are merely lagging behind; given the right policies, they will soon catch up. Hence the coining of two terms: 'developed countries' for the rich states since they have already reached the goal to be 'developed', and 'developing countries' for the poor countries still on their way to 'development'. Now let's return to the reality that is modernity. In hindsight we can see that economic development in the rich countries is not actually an end-point; economic change in these countries has proceeded in leaps and bounds with new high-tech industries. In contrast, most of the erstwhile 'developing countries' are economically falling further and further behind. What are termed 'developed countries' implying an 'end-state' are developing more and more; what are termed 'developing countries' are, to a large degree, simply not developing, certainly not in the way envisaged by development planning. The terms 'developing countries' and 'developed countries' have become topsy-turvy euphemisms for simply 'poor states' and 'rich states' derived from an optimistic time when it was thought planning could steer modernity over all the world.

Does this all mean that we are condemned to live as objects of modernity with little or no control on our futures, either individually or collectively? Fortunately not. All that planning failures illustrate is that modernity cannot be packaged into controlled timed segments. There are many success stories within modernity and they occur through what has been called *surreptitious modes of change* rather than overt 'top-down' planning (Taylor, 2000). These modes of activity involve a multitude of decisions by ordinary people as part of their everyday lives. Where such a social movement captures an ongoing trend and recreates it as a major form of social change, we have subjects successfully making modernity for their own needs and wants. The rise of the suburb as the dominant urban form is the classic case of a surreptitious mode of change. Ridiculed by architects and planners as 'urban sprawl', ordinary people voted with their feet (actually cars) to create large swathes of single-family dwellings with gardens through the twentieth century (Fishman, 1987; Hall, 1996). The very opposite of state planning, here was a popular and commercially profitable mode of change which – despite the problems it sometimes caused by inscribing a gendered and racialized division of space on the landscape – continues to be popular with many heterosexual nuclear families, and hence property developers, today.

The rise of the suburb was very much a USA-led process of social change and represents an element of what is sometimes called the American hegemonic cycle. This is part of a Wallersteinian cyclical-ideological TimeSpace. In world-systems analysis, the modern world system has developed through three hegemonic cycles each based upon the economic successes of a world hegemonic state. These hegemonic states are defined in terms of their economic world leadership in production, trade and finance that generates concomitant political and cultural leadership as well. The first example of a state achieving this status was the Dutch Republic in the seventeenth century with its mercantile hegemonic cycle. In the nineteenth century, Britain was predominant with its industrial hegemonic cycle and, in the twentieth century, the American hegemonic cycle was based on mass consumption (Taylor, 1996). Each cyclical phase reached its fruition when a basic enabling breakthrough was consolidated by cutting-edge economic activity. For the Dutch it was innovations in shipbuilding that led to them being the leading traders of their era. For the British it was the steam engine that enabled factory textiles to rule the world market. And for the Americans it has been the communications and computer technologies, and their eventual integration, that has enabled the development of a vast advertising industry that produces the necessary demand for consumption.

This cyclical model is directly implicated in the *creation* of modernities (Taylor, 1999). As previously noted, the condition of modernity is experienced as incessant change and the economic upheavals caused by the three hegemons – new mercantilism, new industrialism, new consumerism – intensified the degree of social change. Worlds of new opportunities and dangers were created to which ordinary people were forced to respond. Within each hegemonic state the heightened pressure of change led to new forms of everyday life that has been called *ordinary modernity* (Taylor, 1999). In effect, the traditional household was the invention of the modernist ideal of the home as a haven from the turmoils occurring outside. This is a private world of comfort, an idealized place where the family relaxes away from the stresses of social change.[2]

Under the leadership of American hegemony, everyday life has been centred on historically unprecedented consumption by masses of ordinary people. The suburb is the archetypal landscape of modern consumption. This is our contemporary world of suburban living based upon machines for access (the motor car), machines for domestic work (washing machines, microwaves, vacuum cleaners, dishwashers) and machines for entertainment (TVs, video players, music centres, game players). These large, general and necessary items are supplemented by many individual items of consumption such as furniture, clothing and toys that make each home different and distinctive. This is a world pioneered in the USA in the first half of the twentieth century by such 'household names' as Ford, Hoover and General Electric and diffused to the other parts of the world in the rest of the twentieth century. But this everyday consumer modernity did not arise from nowhere. Before the Americans it was the nineteenth-century British who developed the nuclear family-centred everyday life in their new industrial world. The Victorian home may have lacked the 'modern conveniences' that we expect but it was still, in its own way, an unprecedented zone of comfort. Famously individual in character – Victorian homes were cluttered with family 'knick-knacks' – this was when identifiable modern (comfortable) furniture first becomes widespread. But it was the Dutch in the seventeenth century who invented the modern home itself (Rybczynski, 1986). Before them dwellings were relatively public places where business, entertainment, eating and sleeping were mixed. In seventeenth-century Dutch houses the upstairs became a separate private area for family members and invited friends only. This is the crucial separation of home from paid work and business, a hallmark of modernity. The new Dutch homes were furnished and decorated for and by the family to reflect their individuality, and this included children – the Dutch are said to have invented childhood. Thus our modern everyday life has a time trajectory that goes from the houses of Dutch burghers to contemporary suburbia reflecting the hegemonic cycles of economic change.

Hegemonic cycles, with their associated everyday modernities, are a classic example of a morphology of social time. As a cyclical model of change they counter simple progress models that assume a linear pattern of advancement. However, this is not a case of discontinuity over continuity. The nature of social change in the modern world system is too complex to be captured by such either/or models. Obviously there must be discontinuities in a cyclical model – the worlds of mercantilism, industrialism and consumerism are different – but there can also be continuity. Alongside the differences there are enough similarities for us to identify a generic ordinary modernity of everyday life that has its own trajectory of development culminating in contemporary mass consumption. Social change in our modern world is a complex mix of cycles and trends as social times. Ultimately, however, it may be that it is the trends that are all important to our futures.

## CONCLUSION: TIME TODAY

Contemporary globalization is a classic example of how the concepts of time and space are linked together. The idea of globalization has dominated much thinking in human geography and beyond in the last decade ago or so. It is self-evidently a spatial term since it references and announces a specific geographical scale of activity, the 'global'. But this spatial-scale referens only makes sense in relation also to time. Consider the titles of three classic books on globalization: *Global Formation* (Chase-Dunn, 1989), *Global Shift* (Dicken, 1998) and *Global Transformations* (Held et al., 1999). Each one links global with a particular process of social change: the three different terms describing change reflect alternative social-theoretical bases behind each book's argument. The point is that globalization is studied because it represents an important element of contemporary social change, so important in fact that it is sometimes said to define a new historical era. Globalization is a time-space concept par excellence.

The new communication technologies that make possible instant worldwide connections are the basic enabling mechanism of contemporary globalization. This has created a new relation between time and space: information, knowledge, ideas and instructions can be electronically transmitted instantaneously around the world. It has been said to denote the 'end of geography' (O'Brien, 1991). More realistically it marks another change in the relations between time and space. The most influential writer on this topic is Castells (1996). For him social space is materially produced as a means to facilitate meetings – he calls them 'time-sharing practices' – between social agents. Social space has traditionally been organized so as to bring together people simultaneously so that they can interact as social beings. Such simultaneous practices had always relied upon spatial contiguity. Today there is a global space of flows that enables social practices to occur across large distances: with electronic technology spatial contiguity and temporal simultaneity have been physically separated (Castells, 1996: 411). This does not presage the 'end of geography' but rather points towards exciting new geographies with, for instance, the development of a world city network simultaneously to facilitate global social practices within the new spaces of flows (Beaverstock et al., 2001).

One of the prime characteristics of our 'globalizing' times is that we are very self-conscious of intensive social change. The human geography and social sciences literatures are awash with descriptions of things being new. This is indicated by the many terms that proclaim the passing of a recent past: postcolonialism, postindustrialism, post-Fordism, post-development, post-Marxism, poststructuralism and, of course, postmodernism are the most common. And there are identifications of many associated processes that are supposedly changing our world: restructurings, new orders, new identities, transitions and crises are the most common. The sheer number of 'posts' and related processes indicate that the modern maelstrom of incessant social change remains very much with us. In fact the cacophony of such time-laden concerns brings up the question as to whether contemporary times are indeed a special time of change.

This is where the morphology of social time is crucial. For those who broadly follow a progressive linear social time model, our times are but a stepping-stone to more modern technological breakthroughs leading to a more advanced society. Those for whom cycles are part of the morphology have to ask whether the conditions are right for the creation of a new cycle. Who follows the USA? Given that the last hegemon has led us to mass consumption the question is asked whether this is sustainable – is the earth big enough for ever-growing and never-ending mass consumption? Ultimately time in human geography and the social sciences reduces to a question of social justice across generations (de Shalit, 1995). Taking a progressive position means that we should push on so that as yet-unborn generations can cumulatively reap the benefits of technological advance. On the other hand, if modern consumption is not sustainable, it behoves us to make sure we bequeath to future generations a quality of environment on our planet at least as good as the one we inherited from previous generations.

## Summary

- Social time is a 'time with content', social process.
- Social time is inherently linked to social space or place.
- We live in modern times where change in incessant.
- Geographers have used several time-space concepts to understand modern times: space-time convergence, time geography, space-time compression and TimeSpace.
- Planning is the 'top-down' modern practice for controlling social change.
- Planning is condemned to apply yesterday's solutions to today's problems.
- There are surreptitious modes of change where everyday behaviour creates large-scale historical change.
- Successful surreptitious changes are associated with the hegemonic cycles of the Dutch, British and Americans.
- These changes create time-spaces of ordinary modernity where people find a haven from incessant social change.
- Contemporary globalization is a classic time-space concept based upon new communication technologies.

- Globalization is constituted by new electronic spaces of flows that separate temporal simultaneity from spatial contiguity.
- The bottom line for globalization is intergenerational justice: will we leave the planet in as healthy a physical condition as we inherited?

## Further reading

The best starting point for following up this chapter is Leyshon (1995), whose article provides a comprehensive discussion of the idea of a 'shrinking world'. In terms of communications, the collection of essays edited by Brunn and Leinbach (1991) is useful and includes an update of his ideas by Janelle. The basic statements on time-space compression and TimeSpace are to be found in Harvey's (1990) and Wallerstein's (1991) *The Condition of Postmodernity*, and *Unthinking Social Science*, respectively. Both are difficult reads but well worth the effort. Dodgson's (1998) *Society in Time and Space* provides a valuable recent historical geography contribution to studying social change, one which is well versed in social theory. Finally the links between time and modernity are developed further in Taylor's (1999) *Modernities: A Geohistorical Perspective*.

*Note*: Full details of the above can be found in the references list below.

### NOTES

1    Keynsian economic planning is named after the economist John Maynard Keynes who devised a theory and practice of 'demand management' that dominated economic policy in the mid-twentieth century.
2    Multitudes of modern people aspired to this ideal place of comfort but not all succeeded: havens can also be cages, places of hidden violence (Taylor, 2000). Moreover, as second-wave feminism highlighted, the home can be both a space of isolation and labour for women, as well as a site of resistance.

## References

Baker, A.R.H. (1984) 'Reflections on the relations of historical geography and the *Annales* school of history', in A.R.H. Baker and M. Billinge (eds) *Explorations in Historical Geography*. Cambridge: Cambridge University Press, pp. 1–24.

Beaverstock, J.V., Smith, R.G. and Taylor, P.J. (2001) 'World-city network: a new metageography?' *Annals of the Association of American Geographers*, 90: 123–34.

Berman, M. (1982) *All that is Solid Melts into Air*. New York, NY: Simon & Schuster.

Braudel, F. (1980) *On History*. London: Weidenfeld & Nicholson.

Brunn, S.D. and Leinbach, T.R. (eds) (1991) *Collapsing Space and Time*. London: HarperCollins.

Carlstein, C., Parkes, D. and Thrift, N. (eds) (1978a) *Making Sense of Time*. London: Arnold.

Carlstein, C., Parkes, D. and Thrift, N. (eds) (1978b) *Human Activity and Time Geography*. London: Arnold.

Carlstein, C., Parkes, D. and Thrift, N. (eds) (1978c) *Time and Regional Dynamics*. London: Arnold.

Castells, M. (1996) *The Rise of Network Society*. Oxford: Blackwell.

Chase-Dunn, C. (1989) *Global Formation*. Oxford: Blackwell.

de Shalit, A. (1995) *Why Posterity Matters*. London: Routledge.

Dicken, P. (1998) *Global Shift*. London: Paul Chapman.

Dodgshon, R.A. (1998) *Society in Time and Space*. Cambridge: Cambridge University Press.

Fishman, R. (1987) *Bourgeois Utopias*. New York, NY: Basic Books.

Forer, P. (1978) 'A place for plastic space?' *Progress in Human Geography*, 2: 230–67.

Hägerstrand, T. (1973) 'The domain of human geography', in R.J. Chorley (ed.) *Directions in Geography*. London: Methuen, pp. 67–87.

Hall, P. (1996) *Cities of Tomorrow*. Oxford: Blackwell.

Harvey, D. (1990) *The Condition of Postmodernity*. Oxford: Blackwell.

Held, D., McGrew, A., Goldblatt, D. and Perraton, J. (1999) *Global Transformations*. Cambridge: Polity Press.

Janelle, D. (1969) 'Spatial reorganisation: a model and a concept', *Annals of the Association of American Geographers*, 59: 348–64.

Leyshon, A. (1995) 'Anniliating space? The speed-up of communications', in J. Allen and C. Hamnett (eds) *A Shrinking World?* Oxford: Oxford University Press, pp. 11–54.

O'Brien, R. (1991) *Global Financial Integration: The End of Geography*. London: Pinter.

Pred, A. (1986) *Place, Practice and Structure: Place and Society in Southern Sweden, 1750–1850*. Cambridge: Cambridge University Press.

Rostow, W.W. (1960) *Stages of Economic Growth*. Cambridge: Cambridge University Press.

Rybczynski, W. (1986) *Home: A Short History of an Idea*. London: Penguin Books.

Taylor, P.J. (1988) 'World-systems analysis and regional geography', *The Professional Geographer*, 40: 259–65.

Taylor, P.J. (1989) 'The error of developmentalism in human geography', in D. Gregory and R. Walford (eds) *Horizons in Human Geography*. London: Macmillan, pp. 303–19.

Taylor, P.J. (1991) 'The theory and practice of regions: Europes', *Environment and Planning D: Society and Space*, 9: 183–95.

Taylor, P.J. (1996) *The Way the Modern World Works*. London: Wiley.

Taylor, P.J. (1999) *Modernities: A Geohistorical Perspective*. Cambridge: Polity Press.

Taylor, P.J. (2000) 'Havens and cages: reinventing states and households in the modern world-system', *Journal of World-Systems Research*, 6: 544–62.

Thrift, N. (1990) 'Transport and communication, 1730–1914', in R. Dodgson and R.A. Butlin (eds) *An Historical Geography of England and Wales*. London: Academic Press, pp. 453–86.

Wallerstein, I. (1988) 'The inventions of TimeSpace realities', *Geography*, 73: 7–23.

Wallerstein, I. (1991) *Unthinking Social Science*. Cambridge: Polity Press.

# Place: Connections and Boundaries in an Interdependent World

## Noel Castree

### Definition

Place is among the most complex of geographical ideas. In human geography it has three meanings: a point on the earth's surface; the locus of individual and group identity; and the scale of everyday life. Until recently, all three meanings of place were framed by a 'mosaic' metaphor that implied that different places were discrete and singular. However, in the wake of globalization, it has become necessary for human geographers to rethink their ideas about place. This is not to imply that places are becoming the same, as if globalization is an homogenizing process. Rather, the challenge has been to conceptualize place difference and place interdependence simultaneously. The metaphors of 'switching points' and 'nodes' enable us to see places as at once unique and connected. The chapter shows how these metaphors have been applied to the three definitions of place identified in the chapter.

## INTRODUCTION: THE END OF PLACE? THE END OF GEOGRAPHY?

Geography is concerned to provide accurate, orderly and rational description and interpretation of the variable character of the earth surface (Hartshorne, 1939: viii).

The fundamental fact is that . . . places . . . become diluted and diffused in the . . . [new] logic of a space of flows (Castells, 1996: 12).

Places are not what they used to be. Consider the two quotations above. Writing over six decades ago, Hartshorne, one of the most influential geographers of his generation, famously argued that geography's principal aim was the study of 'areal differentiation'. The world, he argued in *The Nature of Geography* (1939), was a rich and fascinating mosaic of places, and the geographer's task was to describe and explain this 'variable character' in both its human and physical dimensions. Writing on the cusp of a new millennium, the geographer and sociologist Castells sees things very differently. The globalization of production, trade, finance, politics and culture, themselves facilitated by remarkable advances in transport and telecommunications, has made the world a 'global village'. For Castells, globalization thus signals the end of place. In our brave new world, he argues, a 'space of flows' – flows of people, information and goods – is increasingly breaking down the barriers that have hitherto rendered places distinct and different. The contrast between this argument and Hartshorne's is striking. If Castells is right, the twenty-first century arguably entails something Hartshorne could scarcely have anticipated: namely, 'the end of geography' (O'Brien, 1992). In other words, if areal differentiation is diminishing, if places are becoming 'diluted and diffused', geography as a subject arguably loses one of its *raisons d'être*. Globalization, it seems, forments a crisis of disciplinary identity.

Or does it? In this chapter I want to argue that far from signalling the end of place, the global interconnections to which Castells refers have resulted in an exciting and innovative redefinition of what place means. Accordingly, the discipline of geography is still very much about the study of the world's variable character – and thus still very much alive and well. The point, though, as we'll see, is that this variation can no longer be accounted for by treating places as relatively bounded and separate. This 'mosaic view' of the world was already outliving its usefulness in Hartshorne's time. By the 1940s it was becoming clear that places were no longer isolated, a fact that posed a challenge to Hartshorne's idea of 'areal differentiation'. Over 60 years later, places worldwide are, as Castells argues, more intimately interlinked than ever before. However, as we will see in this chapter, contemporary human geographers argue that this does not result in the diminution of place differences. Their challenge is to explain an apparent paradox: how can places remain different at a time when they're more interconnected – indeed interdependent – than ever before? Surely, the globalization of trade, finance and the like to which Castells points signals a more homogeneous world? This paradox, as we shall see, is indeed apparent rather than real: for contemporary geographers have argued that a concept of place fit for our times is one that sees *place differences as both cause and effect of place connections*. Far from heralding the end of place, the argument is that globalization is coincident with new forms of place differentiation. This, if you like, is Harthorne's areal differentiation resurrected but with an important new twist. In the twenty-first century, the geographical study of place cannot afford to remain caught in the conceptual straitjacket of a mosaic view of the world. But neither should it buy into Castells's exaggerated vision of a placeless planet where geographical sameness is replacing geographical difference.

In what follows I want to explain how human geographers have fashioned a concept of place that is appropriate for this era of globalization (see Chapter 10 on place and physical geography). Moreover, I want also to explain why this concept matters – both for geography as a discipline and for people living in the interdependent world geographers study. First, though, we need to look a little more closely at what place means, how geographers have defined it in the past and its importance as a concept to geography as a discipline.

## THE 'PLACE' OF GEOGRAPHY

> . . . the significance of place has been reconstituted rather than undermined (McDowell, 1997: 67).

The term place, as geographer Tim Cresswell (1999: 226) has observed, 'eludes easy definition'. My *Concise Oxford Dictionary* identifies 20 meanings of the term, and this semantic elusiveness is compounded by the fact that human geographers have used it in a variety of ways throughout the discipline's history. John Agnew (1987), writing some years ago, cut through this complexity to identify three principal meanings of the term in geographical discourse. These meanings arguably remain in force today:

1   *Place as location* – a specific point on the earth's surface.
2   *A sense of place* – the subjective feelings people have about places, including the role of place in their individual and group identity.
3   *Place as locale* – a setting and scale for people's daily actions and interactions.

In the following sections of this chapter I want to explore these three meanings of place in more detail. In each case my overarching concern is to explain how contemporary geographers have reckoned with the fact of the increasing interconnections among places while still insisting that places are not somehow becoming more alike (see Figure 9.1).

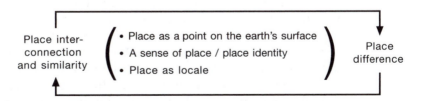

**FIGURE 9.1**   Approaches to place in contemporary human geography

For now, though, I simply want to describe how this triad of approaches to place has emerged, waxed and waned in the years before and since Hartshorne's plenary statement about areal differentiation and the nature of geography (see Figure 9.1). As we'll see in the chronology that follows, the second and third definitions of place emerged to challenge the first in the 1970s, 1980s and 1990s, respectively, since when attempts have been made to synthesize and update them.

### *Beginnings*

Hartshorne used the term place rather imprecisely, often conflating it with the equally complex term region. This fact notwithstanding, it's probably fair to say that Hartshorne viewed place as location – the first and oldest meaning identified by Agnew – and places as distinct points on the earth's surface. Indeed, in the five decades or so that geography had been a university subject in western Europe and North America up to 1939, the normal expectation was that professional geographers would study particular places, in both their human and environmental aspects, in great detail and publish articles and books on them (see Chapters 2 and 4 on the strength of regional geography). Classic examples include *Tableau de la Géographie de la France* (1917) – by French geographer Paul Vidal de la Blache – and H.J. Fleure's *Wales and her People* (1926). For Hartshorne these types of study were what made geography special among academic disciplines. In *The Nature of Geography* (1939) he distinguished among 'systematic', 'chronological' and 'idiographic' subjects. The former take just one main aspect of reality and study it in detail – thus economics studies the economy and chemistry the world's chemical elements and so on. Chronological disciplines study change over time – as history and geology do. However, Hartshorne argued that few disciplines look at how multiple different processes and events come together in the real world in specific places. Geography, he insisted, is precisely this 'synthetic' or integrative discipline. Moreover, because economic, social, political, hydrological, topographic and all manner of other factors never relate in quite the same way in any two places, he argued that geography studies the unique rather than the general. This, for Hartshorne, is what made it an idiographic discipline: it was about accounting for difference rather than sameness.

In truth, Hartshorne exaggerated the importance of place study to geography's disciplinary identity. Others had for decades seen geography less as the study of place and more the study of 'man–land [*sic*] relationships'. Indeed, after Oxford University's first professional geography appointment – Halford Mackinder – had famously defined geography as a 'bridging subject' between the human and natural sciences in 1887, many geographers had devoted their energies to studying these relationships right up to the continental and global scales. Moreover, the 'nature' of the geography Hartshorne sought to define and defend was to change almost as soon as his widely read book was published. There were three reasons why. First, many professional geographers were drafted into the armed forces during the Second World War and soon found that they lacked the technical skills required to undertake military and intelligence activities. The problem was many of the place studies geographers undertook were broad and largely descriptive. Geographers were trained to be jacks of all trades – to know a bit about a lot of things in given places – but masters of none. Secondly, this gave the subject a 'dilettantish image', as historian of geography David Livingstone (1992: 311) has put it, which served it poorly in a postwar educational environment where specialization was the norm. When the Geography Department in America's most prestigious university – Harvard – was closed down in 1951, many geographers keenly felt the need to make the discipline more rigorous and respectable. Finally, the mosaic view of place that seemed common sense to Hartshorne and his predecessors started to look highly unrealistic, both during and after the war. As one geographical critic of the time put it: 'we are no longer dealing with a world of neatly articulated entities . . . Our suspicion . . . [is] that . . . geographers may perhaps be trying to put boundaries that do not exist around areas that do not matter' (Kimble, 1951/1996: 500, 499).

### Dis-placements

Consequently, the concept and study of place fell into disuse for almost three decades after Hartshorne's tome was published. In the immediate postwar era a new generation of geographers instigated what one of them called a 'scientific and quantitative revolution' (Burton, 1963). Learning and applying the tools of mathematics and statistics, this new generation sought to make geography a science. This entailed specialization – including the increasing separation of human and physical geography and the subdivision of each[1] – and the attempt to develop testable theories and laws. Rather than looking for the unique, the different and the particular, geographers sought to mimic the physical sciences by looking for similarity, generality and pattern. In the words of Hartshorne's great rival, the geographer Fredrick Schaefer (1953: 227): '. . . geography has to be conceived as the science concerned with the formulation of the laws governing the spatial distribution of certain features on the surface of the earth'. The keynote titles of the period said it all and were a far cry from the regional monographs of the prewar era: *Theoretical Geography* (Bunge, 1962), *Models in Geography* (Chorley and Haggett, 1967), *Locational*

*Analysis in Human Geography* (Haggett, 1965) and *Explanation in Geography* (Harvey, 1969). To the extent that Hartshorne's vision of place figured at all, it was when events or things in one place were shown to be 'a particular realisation of the laws governing all similar events and things' (Rogers, 1992: 244). So much for place difference and uniqueness. Geography was now to be a 'spatial science', devoted to searching for geographical order at a variety of scales, to measuring numerically both people and things and to testing rigorously hypotheses and models so as to develop generally applicable laws and theories.

Mid-century geography therefore survived quite happily without place as a central, organizing concept. By the early 1970s, however, it started to become clear that scientific geography was not to everyone's liking. Specifically, a cohort of human geographers wondered whether people's activities could and should be studied 'scientifically'. Within a decade this critique of spatial science, as this chapter now goes on to explain, led to what Rogers (1992) has called 'the rediscovery of place'.

### The return of the repressed

This critique and rediscovery came in two phases. To begin with, a set of so-called 'humanistic geographers' argued that spatial science was 'in-human'. By treating people as 'little more than dots on a map or integers in an equation' (Goodwin, 1999: 38), it ignored the subjective, qualitative and emotional aspects of human existence and amounted to a 'Geography without man [*sic*]' (Ley, 1980). Consequently, the attempt to rehumanize human geography took the form of close and careful studies of individual and group 'lifeworlds'. Two classic examples were David Ley's (1974) exploration of gang 'turf' rivalries in poor inner-city neighbourhoods in Philadelphia and Graham Rowles' (1978) detailed analysis of a group of old people's attachment to their home-place. In effect, what Ley, Rowles and other humanistic geographers were doing was resurrecting the importance of place. However, in the humanistic lexicon places were not, *pace* Hartshorne, conceived as objective points on the earth's surface. Rather, the aim was to recover people's varying *sense of place* (the second definition of place identified by Agnew): that is, how different individuals and groups, within and between places, both interpret and develop meaningful attachments to those specific areas where they live out their lives.

This concern with geographical experience was a vital corrective to the passionless, placeless grids of spatial scientific analysis. But it was not the only alternative to scientific human geography. From the early 1970s humanistic geographers were both accompanied and challenged by another group of dissenters from spatial science: Marxist geographers. Led by David Harvey, a former darling of geography's scientific establishment, these politically left-wing geographers argued that spatial science did little to address pressing real-world problems, like poverty, famine and environmental degradation. Moreover, they argued that by hiding behind a mask of 'objectivity' spatial science was dishonest about its own conservative, 'status quo' political commitments. As Harvey made clear in human geography's first overtly Marxist book – *Social Justice and the City* (1973) – a radical geography should be focused on non-trivial issues and should be geared to changing the world rather than simply understanding it. What has all this got to do with place? A good deal as it turns out. Despite their common disdain for spatial science, tensions developed between Marxist and humanistic geography – and it was, among other things, over the question of place. For Harvey and his Marxist colleagues, the humanistic concern for a sense of place was worthy but ultimately problematic, for it tended to treat people and places in isolation and was obsessed with the minutiae of local attachments and local experiences. Against this, the Marxists – pointing to the development of a truly global economy by the early 1970s – argued that places were increasingly not only interconnected but also interdependent. That is, places were not only related to one another but related in ways that meant that what happened in one place could have serious consequences for another place many thousands of miles away. Harvey's (1982) *The Limits to Capital* was a major attempt to explain and criticize the nature and consequences of these global interconnections: namely, those specific to capitalism.

### Overcoming dualisms

This brings me to the second phase in human geography's rediscovery of place. Though the Marxists were right to argue that human geographers needed an objective understanding of what places had in common, they were, by the early 1980s, as guilty as the spatial scientists had been of failing to pay sufficient attention to place difference. They also tended to give far more attention to the global economic and other processes that supposedly 'structured', and even, it was sometimes said, 'determined', the thoughts and actions of people in specific places (Duncan and Ley, 1982). That is to say, the Marxists were preoccupied with interplace connections more than specific place differences. By the same token, though humanistic geographers were right to emphasize the particularity of place experience, their concern with difference and lifeworlds arguably blinded them to the common processes linking places worldwide – 'stretched out' processes that could change the 'objective' nature of place and, thereby, locals' 'subjective' sense of place. Likewise, they tended to overemphasize the degree to which people in place could control their own lives since Marxists like Harvey argued that global systems (like capitalism) constrain people's 'agency' in their home-places. How, then, to connect 'local worlds' with 'global worlds'? This was the challenge taken up by a set of UK and American geographers from the mid-1980s. What inspired these geographers' efforts was a mixture of dramatic real-world changes and new theoretical developments.

During the previous decade, Britain and the USA, like many other countries, had seen their human geography literally remade by the ravages of a sustained economic crisis. The geography of people and places in the two countries was being restructured in the face of global economic competition and neoliberal governments (led by Thatcher and Reagan) intent on creating a new Britain and a new America. But the point, as Doreen Massey (1984) showed in her germinal book, *Spatial Divisions of Labour*, is that the *same* processes of economic competition were having *varying* effects across the face of these and other countries. In other words, the global interconnections that meant that British and American cities and towns could not be analysed in isolation were producing not *geographical similarity* but *geographical difference*. The task of the so-called 'localities projects' which followed Massey's study (and which involved UK human geographers undertaking detailed studies of different British towns and cities) was to explain how global forces could have such variable local effects. Concurrent with the writings of Massey and the localities researchers were those inspired by new theoretical developments from outside geography. In a series of books, the now famous sociologist Anthony Giddens had developed 'structuration theory' in order to overcome the impasse between structural (or determinist) explanations of people's actions and free-will (or voluntarist) explanations. How, Giddens asked, could one combine a focus on 'big social systems' with a focus on individual and group action? In geography the impasse was represented by the Marxist obsession with global socioeconomic processes and the humanistic geographers' concern with locally variable place experiences and actions. The geographers Derek Gregory (1982) and Allan Pred (1986) sought to spatialize Giddens' thinking (and to answer his question) in their innovative books, *Industrial Revolution and Regional Transformation* and *Place, Practice and Structure* – books which used historical examples to show how previously isolated places became embroiled in translocal forces. What Gregory and Pred demonstrated is that social structure and social agency come together differently in different places such that they *mutually* determine one another.

Conducted in the wake of the stand-off between Marxist and humanistic geography, the localities research projects and structuration theory-inspired work of Gregory and Pred sought to find a middle ground between two dualistic and untenable positions: that is, that places are *either* all the same *or* all different and that people in places are *either* free agents – able to develop their own singular attachments to, and practices in, a place – *or* the victims of overwhelming global social forces. The result was a conception of *place as locale* – the third meaning of place identified by Agnew. For Massey, Gregory, Pred and their fellow travellers, a locale was the scale at which people's daily life was typically lived. It was at once the objective arena for everyday action and face-to-face interaction *and* the subjective setting in which people developed and expressed themselves emotionally. It was at once intensely local *and yet* insistently non-local to the extent that 'outside' forces intruded into the objective and subjective aspects of local life in an interdependent world. And every locale was at once unique and particular *and yet* shared features in common with the myriad other locales worldwide to which it was connected (see Figure 9.2).

To summarize, after fading into mid-century obscurity, place is once again 'one of the central terms in . . . geography' (Cresswell, 1999: 226). Over the last decade human geographers have extended and enriched the return to place pioneered by those writing in the 1970s and 1980s. In the remaining sections of the chapter I want to take each of the three approaches to place discussed here and illustrate briefly, using examples, how contemporary geographers have shown that place interconnection and interdependence in the modern world mark not the end but what Neil Smith (1990: 221) called 'the beginning of geography'. In terms of our three definitions of place we can ask three key questions – namely, how can places be unique and yet subject to similar global forces? How is people's sense of place intensely local and yet (implicitly or explicitly) extroverted? And how can human actions be place based, unpredictable and variable and yet

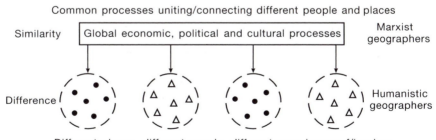

FIGURE 9.2 Marxist and humanistic geographers' approach to place

considerably constrained by extra-local forces hailing from far away? In the last few years human geographers have offered innovative answers to all these questions. It's to these answers that I now turn.

### RETHINKING PLACE AS LOCATION: POROUS PLACES

People and things are increasingly out of place (Clifford, 1988: 6).

I've already called into question the mosaic view of place. Globalization entails the 'stretching' of social relationships across space such that the boundaries between the 'inside' of a place and the 'outside' are rendered porous. Today we must appreciate the openness of places; that is, we need what Massey (1994: 51) calls 'a global sense of the local'. It's not just that today more and more places are interlinked and interdependent. It's also that the *intensity* of these global connections has increased: we live in an age of what Peter Dicken (2000: 316) calls 'deep integration'. In sum, the world is no longer a mosaic of places. At this point is might be tempting to join Castells and declare 'the end of place'. But this would be to confuse the redundancy of *a particular conception of place* with the disappearance of place as such. As I said in the introduction, places are not what they used to be. But places still undoubtedly exist. For instance, Manchester, where I live, is not the same as and remains far distant from, say, Manilla – even though the two cities might be directly connected by relations of finance, trade or immigration. As Massey (1995: 54) puts it, 'we . . . [therefore] need to rethink our idea of places . . .' because 'place has been transformed . . .' (Agnew, 1989: 12).

In metaphorical terms, this rethinking can be evoked as follows. Since the mosaic view conceptualizes places as distinct points in space – which is, today, unrealistic – it is perhaps better to see them as *switching points in a larger global system* or else *nodes in translocal networks* (Crang, 1999) (see Figure 9.3). These metaphors, as I'll now explain, allow us to think of places as inextricably interconnected –

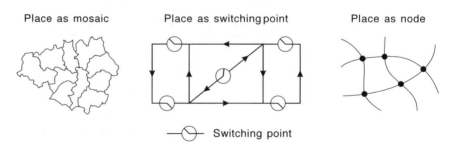

**FIGURE 9.3**  Metaphors for understanding place

indeed interdependent – *and* as different and unique. Let us take each half of this metaphorical equation in turn.

Places in the contemporary world are, clearly, no longer separate. For instance, the bank where I this morning deposited a cheque is but one local fragment of a global financial system, while the apple I just consumed in front of my Japanese computer implicated me in a production network stretching back to an orchard in New Zealand (from whence the apple came). Moreover, with interconnection also comes *interdependence*. For instance, barely a day passes without newspaper reports of job losses and job creation in places as diverse as Chicago, Calcutta or Cairo. Often, though not always, these changing local employment situations can be explained with reference to inter-place competition for investment and markets. For example, if Calcuttan workers can make auto-parts more cheaply than labourers in Chicago, a firm like Ford might favour an Indian auto-parts supplier for its vehicles. In short, what happens *then and there* can have sharp consequences in the *here and now*.

But if places are no longer separate, the more difficult argument to understand is that they somehow remain unique. No two places are quite the same, even in this era of globalization – or so several geographers, disagreeing with Castells, have argued. Notice that I use the word unique and not singular. In Hartshorne's worldview places were *singular*: that is, they were all so obviously or subtlely different from one another as to be absolute one-offs. The same combination of human and environmental factors, the argument went, was never found twice. However, if we see places as *unique* we can argue that they are different *and* that they have something in common in an interdependent world (just as we are all unique as people in terms of looks and personalities and yet share the same biological make-up). This is the argument made by Ron Johnston (1984) in 'The world is our oyster' and more recently by Doreen Massey (1995) in 'The conceptualisation of place'.

The question therefore arises: how can places continue to differ in a world of increasingly intimate global interrelationships? There are five answers. Together these answers explain why the metaphors of switching points and nodes are apt: for both evoke the idea that different places are 'plugged in' to different sets of global relations with different degrees of power over those relations. First, and most obviously, while globalization brings places closer together in terms of the reduced time taken to cross the space between them, the fact of geographical distance still remains. Thus, to return to the example of Manchester and Manilla, while the two cities are *relatively* closer together (see also Chapter 8) their *absolute* locational differences endure. Secondly, globalization has not unfolded across a homo-geneous space. Rather, it has linked places *because* they are different.

For instance, precisely because Boeing is a leading aircraft manu-
facturer, places without the capacity to produce aircraft have imported
Boeing products all the way from its base in Seattle. Thirdly, even
though many places are subject to the same global forces, they react to
and mould them differently. An aerospace company like Boeing, for
example, has a number of choices as to how to respond to foreign
competition. It can close its factories in cities like Seattle altogether,
lay off some but not all workers or retrain these workers and sell new
products to new places. More radically, it could shift production
operations to cheaper or more efficient sites outside the USA. Likewise, McDonald's – sometimes held up as a potent symbol of cultural
globalization and homogenization (Ritzer, 1996) – means different
things in different places. In Moscow it might be a 'trendy' sign of all
that's modern or new, while here in Manchester it's but a familiar and
rather banal marker of consumer culture. Fourthly, even today all or
most social relationships are not global in reach. Many remain insistently local – like the one I enjoy with my family or my local football
club team-mates. Finally, we should not forget that not all places in
the world are equally 'wired in'. Globalization, as Dicken (2000) notes,
can take the form of 'shallow' as well as deep integration. Thus many
places in sub-Saharan Africa, for example, remain partially cut off
from the rest of the world or else subject to very one-sided relationships that exacerbate poverty – the kind of 'difference' that places in
the developing world certainly do *not* want to preserve. For instance,
it may surprise you to discover that Ethiopia – one of the world's
poorest places – continued to produce large quantities of food throughout the horrific famines of the mid-1980s. How and why? Because
wealthy landowners were producing export crops for European and
North American markets rather than food crops for their own people.
For the five reasons mentioned above, it is simply misconceived to
think that globalization equals sameness and homogeneity. On the
contrary, human geographers have shown that the more linked
places become, the more place differences endure and are remade. In
Swyngedouw's (1989) apt neologism, we need to talk less about
globalization and more about an uneven process of 'glocalization'.

## RETHINKING A SENSE OF PLACE: 'GLOCAL' IDENTITIES

> . . . even local identities are completely caught up in a web of global
> interdependence (Mitchell, 2000: 274).

In the previous section we considered places, implicitly, in terms of
their objective properties – that is, as material and physical locations –
and how to conceptualize them. But what of the subjective questions
of how people interpret their home-places and those of others? As

we've seen, humanistic geographers were among the first to take the subjective aspects of place existence seriously. As these geographers were right to argue, the thoughts and feelings that people have towards places are every bit as real and material as the places themselves. Disclosing people's 'sense of place' requires 'empathetic' inquiries into the realms of feelings, emotions and values. Some three decades after the likes of Ley launched this so-called 'hermeneutic approach' in human geography, it's clear that subjective attachments to, or interpretations of place, matter as much as ever. Global interdependency notwithstanding, most people live their lives within just a few square kilometres. Moreover, at certain times of life, people can be *highly* confined to specific places, as with children and many elderly people. So place remains a crucial locus for daily experience. Think about yourself: which places matter to you and why? Your answer will probably involve just a few places, and one of them will almost certainly be your home-place(s). You will have a highly personal sense of place that's bound up with specific events in your life, involving not just your perception of place(s) but your feeling about place(s). So apart from their physical dimensions, there's an imaginative and affective dimension to places too.

How, though, to understand these non-physical realms of thought and feeling? The humanistic desire to disclose people's sense of place will no longer suffice, for two reasons. First, cultural geographers have argued that place is linked to the formation of personal and group *identities* (Keith and Pile, 1993). People have more than just a sense of place: additionally, place is written into their very characters. Think, for example, of how we tend to characterize people – often stereotypically – by their place of origin (e.g. in Britain there are 'Cockneys' and 'Geordies', in North America rural 'rednecks' and an inner-city 'underclass'). And think about how your very sense of self, as a person, is intimately linked to the place you are from. For instance, though I live in south Manchester I'm originally from a town in north Manchester and both my accent and my character still carry the traces, 17 years since I left, of my upbringing. So place runs deep. Secondly, there was an implication in humanistic writing that there was one ultimately 'real' or 'authentic' sense of place for people. The Canadian geographer, Edward Relph (1976), for example, complained about the 'placelessness' of so many modern towns with their highrise towers and bland, serial suburbs. He believed that the spread of faceless modern architecture and planning was 'de-humanizing' place experience such that people's senses of place were being thinned out and rendered uniform. The problems with this kind of argument are manifold. To begin with it's rather conservative in nature, seeing 'outside' influences as a 'threat' to the supposedly 'authentic' nature of places. It's almost as if Relph lamented the fact that places were increasingly interlinked rather than different pieces in a mosaic. As problematically, it underestimates the sheer *variety* of place attachment and identities that people can and do develop in the *same* places. There is ultimately no one sense of place or place identity (think of how a poor immigrant woman in Hackney, London, might view that place as opposed to a wealthy young male professional) but many. Finally, geographers like Relph underestimate how different senses of place and place identity could persist not despite but *because of* 'external' influences hailing from other places.

This last comment brings us to the important insight that different local identities might result from, or be expressed because of, similar global connections. Identities are not natural. They are, rather, socially fabricated over people's life course. People tend, when considering the place element of identity, to conjure up the image of a settled community – literally, a home-place. But in a globalizing world, most places are anything but settled. They are subject to ongoing change, both physically (the factory that shuts down or the new shopping centre that opens) and socially (the foreign immigrants that move in or the older generation who die off) – and much of this change is, as we saw in the previous section, about local changes resulting from global/extra-local processes. So we must recognize that while identities are, today, still formed in places (they are place based) they are not place bound – that is, the result of *purely* local experiences. Rather, locally variable identities partially arise from 'outside' influences, paradoxical though this may seem.

Contemporary human geographers have illustrated this 'glocal' nature of identity in two ways (see Chapter 12 for more on glocalization). First, there are those cases where identities *seem* to be purely local but where human geographers have shown that they are in fact not so. For instance, in mid-2001 a set of serious 'race' riots erupted in the poor, old industrial towns of Bradford, Burnley and Oldham in the north of England. These towns, like so many multicultural places in western Europe, have had large immigrant populations from the Indian subcontinent for over three decades. Yet extreme right-wing political groups – like the National Front – want to expel them, thereby 'purifying' these places and returning them to their purportedly 'true' character as white and English. The irony, of course, is that this attempt to define and defend a 'local' identity from unwanted 'foreign' influences arises precisely in and through the presence of those 'outside' influences! A further irony is that the Indians and Pakistanis being discriminated against consider themselves to be very much *a part of* these three places – and rightly so, having lived there for over two generations. So *seemingly* local identities that attempt to shut out non-local influences – in the three places mentioned, influences of international immigration – are, in the modern world, not straightforwardly local at all (Harvey, 1995).

Secondly, human geographers are also showing that many 'local' identities are overtly and explicitly 'extra-local'. There are two main cases to consider here. The first is where people who are not indigenous to a place characterize it in a way that both reflects their own worldview and which therefore takes on a certain reality – even though it might be a far cry from the local residents' view of that place. The best example here is modern tourism, which serves up the world as a set of idealized places, each with a specific image that is marketed to potential tourists. For example, the Caribbean is usually thought of as a peaceful, paradisical place, full of exotic resorts; what tourists rarely see behind this 'imagined geography' are the slums and poverty that are endemic to most Caribbean towns and cities (Cater, 1995). Secondly, and rather differently, geographers have shown that many place-based identities today are openly 'extroverted' and outward looking – in effect explicitly incorporating 'non-local' influences (unlike the National Front in Bradford, Burnley and Oldham). The best examples come from so-called 'transnational communities' – that is, communities that are spread out among different places but which remain connected. In Vancouver, Canada, for example, there are many Chinese residents who are from Hong Kong and who maintain strong familial and cultural links with this former British colony. So their identity as Vancouverites, living in a western Canadian city abutting the USA, is complemented by their identity as Hong Kong Chinese. Theirs is an avowedly *hybrid* identity, such that even though they live physically in one place their place loyalties are plural and transnational (Mitchell, 1993) (see Figure 9.4).

In sum, in many places in the contemporary world the identities of people who live in those places are rarely local in the 'mosaic' sense of the word. As Massey (1998) insists, we need to look not for the

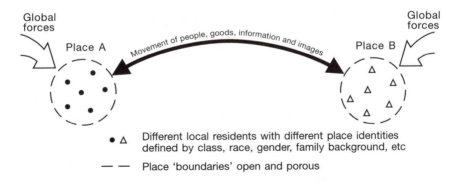

Different local residents with different place identities defined by class, race, gender, family background, etc

Place 'boundaries' open and porous

**FIGURE 9.4** 'Glocal' identities

*roots* of people's identity but the *routes*. That is, we need to trace how 'local' identities are built from the way people internalize a whole array of 'non-local' influences as the latter converge on different places.

## RETHINKING PLACE AS LOCALE: GLOBAL FORCES, LOCAL RESPONSES

Life chances are materially affected by the lottery of location (Crang, 1999: 24).

We live in a highly uneven world. Global interconnection and inter-dependency have been coincident with inequality and uneven development rather than homogeneity. Since the first incursions of Marxism into human geography, geographers have argued that local inequalities are *caused by* global interlinkages, not merely correlated with them. If we take the example of Ethiopian famines cited earlier, it's clear that these local tragedies were a direct outcome of colonial and trade ties to Europe and beyond. But the traffic is not all one way. People acting in places are not simply marionettes whose actions and life chances are dictated by movements of the world economy and global politics. In other words, people acting in place have a degree of 'agency' to control their destinies and those of the places they reside in. So local action cannot only *react to* global pressures but also *act back on them*. Since Gregory, Pred and others, following Giddens, first made this argument in the 1980s, human geographers have not only shown the nature and limits of place-based agency but also how it varies from place to place. This geographically variable interaction between global and international structures and people's place-based agency is the process of what Giddens, as we saw earlier, famously called 'structuration'.

This uneven geography of structuration can be illustrated, in simple terms, by the following interpretation of a recent, little-known but fascinating event: the attempt by the small, central American country Costa Rica to make money by selling off it's 'genetic resources'. Like other central American counties, Costa Rica is relatively poor in global terms and classed as a 'developing country'. Its principal means of income is the export of coffee beans and bananas. However, large, Western transnational pharmaceutical companies have, in recent years, become very interested in tropical countries – like Costa Rica – that are so-called 'genetic hotspots'. The tropics contain the bulk of the world's plant, animal, insect and bacterial species, and it's estimated that some 50% of these species are yet to be discovered. Transnationals like Monsanto, Pfizer and Smith-Kline-Beecham are now actively 'prospecting' for these species, hoping that their physical and genetic properties might some day be usable in the development of pharmaceutical products, like drugs or cosmetics. Among developing countries, Costa Rica has been at the forefront of this 'merchandizing' of currently unowned and undiscovered tropical species and, in 1991, set up an organization – INBio [the National Institute of Biology] – to collect species samples and sell them to interested Western companies. Thus far InBio has made over 3 million US dollars selling Costa Rica's genetic resources.

In this case, the 'structure' that both conditioned the decision to sell Costa Rica's genetic heritage and led to the establishment of INBio was the world economy: an economy in which Costa Rica has become overly reliant on two staple exports, coffee and bananas. The 'agency' at work here, embodied in INBio's everyday operations in the country's capital, San Jose, has yielded Costa Rica 3 million valuable dollars. However, this agency has been unequally distributed within Costa Rica. Historically, Costa Rica was widely populated by indigenous or so-called 'First Nations' peoples. These peoples were displaced during the Spanish conquests of the sixteenth and seventeenth centuries and, today, some 30 000 of them live in small, poor 'native reserves' located in out of the way rural areas. Many of these peoples have a unique knowledge of local environmental resources and, more generally, have legitimate claims to the Costa Rican genetic inheritance being sold off by scientists and bureaucrats at INBio. However, there's little evidence that any of the 3 million dollars earned through INBio has made its way into Costa Rica's native reserves. The country's indigenous peoples are locked in a political structure that offers them little power or opportunity, and their exclusion from INBio's operation illustrates this graphically. On top of this, their physical location in places distant from the centre of political authority in Costa Rica, the capital city, makes it doubly difficult to be heard.

## CONCLUSION: THE MATTER OF PLACE

> . . . the significance of place depends on the issue under consideration and the sets of social relationships that are relevant to the issue (McDowell, 1997: 4).

Place matters and its importance is multifaceted. Some three decades after spatial science reached its zenith, difference is back on the geographical agenda. The discipline is once again concerned with the idiographic, but in a very different and much wider sense than Hartshorne could ever have imagined. Place difference, both objective and subjective, is now understood in terms of uniqueness rather than singularity. We again have a style of human geography that is integrative and synthetic rather than analytical and place-blind. But it must reckon with a world where places are infinitely more complex and changing than they were during geography's first engagements with place in the early twentieth century. In addition, we must also acknowledge that place matters in a very profound and very worldly sense, which is why other subjects – like sociology, communications studies and economics – are now very interested in the difference that place makes. We need to understand the variable nature of places not just out of sheer curiosity (though that's reason enough). More than this, as the bloody struggles over place in Israel, Northern Ireland, the Basque country, Sri Lanka, the former Yugoslavia and elsewhere show so tragically, local attachments and differences remain fundamental aspects of the human condition. In short, the renewed study of place is too important to be left to geographers alone. This is why Massey (1993) argues that geographers need to advocate a 'progressive sense of place' to people in the world at large. What she means is that geographers have a moral obligation to show people that their place-based actions and understandings make no sense without acknowledging all those things impinging on place from the outside. What's 'progressive' about this, for Massey, is that it encourages an openness to the wider world, not a defensive putting up of barriers. We must, she says, live with the incontrovertible fact that the global is *in* the local and vice versa. This is more than a merely academic observation. In a world of place difference, stressing what connects places has real practical and political relevance. It can make all the difference between a world of inward-looking local rivalries and a cosmopolitan world where place differences are respected and place connections celebrated.

## Summary

*   Place is a complex concept with three principal meanings in modern human geography.
*   As the world has changed so too have human geographer's conceptions of place.
*   Human geographers have tried to rethink place in a way that respects place differences while acknowledging heightened place interconnections and interdependencies. That is, places are conceived of as being unique rather than singular.
*   This rethinking has taken human geographers away from older 'mosaic' metaphors of place to newer notions of 'switching points' and 'nodes'.

- Using these notions we can rethink all three definitions of place in order to show how local and non-local events and relations intertwine.
- The importance of a place concept that stresses how 'outside' processes impact on the 'inside' of places is that it challenges the idea that places and the peoples in them can ever thrive by defensively putting up barriers against non-local forces.

## Further reading

A good place to start is with the following entries in *The Dictionary of Human Geography* (Johnston et al., 2000): place, locale, sense of place, placelessness, globalization and boundary. The introduction and Chapter 1 of Hannerz (1997) provide a good general introduction to the meaning of place in the contemporary world. A comprehensive introduction to the place concept in geography is provided by Holloway and Hubbard (2000) in *People and Place*, while Massey (1995) and Allen and Hamnett (1995) offer first-rate general introductions to conceptualizing place in an era of globalization. McDowell's (1997) edited book, *Undoing Place?*, showcases the best recent writing on place in geography and cognate fields. In relation to the three meanings of place explored in this chapter, see the following: on the local, the global, difference and sameness, Crang (1999) and Allen and Hamnett (1995); on 'glocal' identity, Cloke (1999) and Driver (1999); on local action and global processes, Meegan (1995).

*Note*: Full details of the above can be found in the references list below.

### NOTE

1    And this book reflects these enduring divisions, with each key concept given separate treatment by a human and a physical geographer.

## References

Agnew, J. (1987) *Place and Politics*. Boston, MA: Allen & Unwin.

Agnew, J. (1989) 'The devaluation of place in social science', in J. Agnew and J. Duncan (eds) *The Power of Place*. Boston, MA: Allen & Unwin, pp. 9–30.

Allen, J. (1995) 'Global worlds', in J. Allen and D. Massey (eds) *Geographical Worlds*. Oxford: Oxford University Press, pp. 105–42.

Allen, J. and Hamnett, C. (1995) 'Uneven worlds', in J. Allen and C. Hamnett (eds) *A Shrinking World?* Oxford: Oxford University Press, pp. 233–54.

Bunge, W. (1962) *Theoretical Geography*. Lund: Kleerup.

Burton, I. (1963) 'The quantitative revolution and theoretical geography', *The Canadian Geographer*, 7: 151–62.

Castells, M. (1996) *The Rise of the Network Society*. Oxford: Blackwells.

Cater, E. (1995) 'Consuming spaces: global tourism', in J. Allen and C. Hamnett (eds) *A Shrinking World?* Oxford: Oxford University Press, pp. 183–222.

Chorley, R. and Haggett, P. (eds) (1967) *Models in Geography*. London: Methuen.

Clifford, J. (1988) *The Predicament of Culture: Twentieth-century Ethnography, Literature*. Cambridge, MA: Harvard University Press.

Cloke, P. (1999) 'Self-other', in P. Cloke et al. (eds) *Introducing Human Geographies*. London: Arnold, pp. 43–53.

Crang, P. (1999) 'Local-global', in P. Cloke et al. (eds) *Introducing Human Geographies*. London: Arnold, pp. 24–34.

Cresswell, T. (1999) 'Place', in P. Cloke et al. (eds) *Introducing Human Geographies*. London: Arnold, pp. 226–34.

Dicken, P. (2000) 'Globalisation', in R.J. Johnston et al. (eds) *The Dictionary of Human Geography* (4th edn). Oxford: Blackwell, pp. 315–16.

Driver, F. (1999) 'Imaginative geographies', in P. Cloke et al. (eds) *Introducing Human Geographies*. London: Arnold, pp. 209–17.

Duncan, J. and Ley, D. (1982) 'Structural Marxism and human geography', *Annals of the Association of American Geographers*, 72: 30–59.

Goodwin, M. (1999) 'Structure-agency', in P. Cloke et al. (eds) *Introducing Human Geographies*. London: Arnold, pp. 35–42.

Gregory, D. (1982) *Regional Transformation and Industrial Revolution*. London: Macmillan.

Haggett, P. (1965) *Locational Analysis in Human Geography*. London: Edward Arnold.

Hannerz, U. (1997) *Transnational Connections*. London: Routledge.

Hartshorne, R. (1939) *The Nature of Geography*. Lancaster, PA: Association of American Geographers.

Harvey, D. (1969) *Explanation in Geography*. London: Arnold.

Harvey, D. (1973) *Social Justice and the City*. London: Arnold.

Harvey, D. (1982) *The Limits to Capital*. Oxford: Blackwell.

Harvey, D. (1995) 'Militant particularism and global ambition', *Social Text*, 42, 1: 69–98.

Holloway, L. and Hubbard, P. (2000) *People and Place*. Harlow: Prentice Hall.

Johnston, R.J. (1984) 'The world is our oyster', *Transactions, Institute of British Geographers*, 9, 5: 443–59.

Johnston, R.J., Gregory, D., Pratt, G. and Watts, M. (eds) (2000) *The Dictionary of Human Geography* (4th edn). Oxford: Blackwell.

Keith, M. and Pile, S. (eds) (1993) *Place and the Politics of Identity*. London: Routledge.

Kimble, G. (1951/1996) 'The inadequacy of the regional concept', in J. Agnew et al. (eds) *Human Geography: An Essential Anthology*. Oxford: Blackwell, pp. 492–512.

Ley, D. (1974) *The Black Inner City as Frontier Outpost*. Washington, DC: Association of American Geographers.

Ley, D. (1980/1996) 'Geography without man', in J. Agnew et al. (eds) *Human Geography: An Essential Anthology*. Oxford: Blackwell, pp. 192–210.

Livingstone, D. (1992) *The Geographical Tradition*. Oxford: Blackwell.

Massey, D. (1984) *Spatial Division of Labour*. London: Macmillan.

Massey, D. (1993) 'Power geometry and a progressive sense of place', in J. Bird et al. (eds) *Mapping the Futures*. London: Routledge, pp. 62–8.

Massey, D. (1994) *Space, Place and Gender*. Oxford: Polity Press.

Massey, D. (1995) 'The conceptualisation of place', in D. Massey and P. Jess (eds) *A Place in the World?* Oxford: Oxford University Press, pp. 46–79.

Massey, D. (1998) 'The spatial construction of youth cultures', in T. Skelton and G. Valentine (eds) *Cool Places*. London: Routledge, pp. 121–9.

McDowell, L. (ed.) (1997) *Undoing Place?* London: Arnold.

Meegan, R. (1995) 'Local worlds', in J. Allen and D. Massey (eds) *Geographical Worlds*. Oxford: Oxford University Press, pp. 53–104.

Mitchell, D. (2000) *Cultural Geography*. Oxford: Blackwell.

Mitchell, K. (1993) 'Multiculturalism, or the united colors of capitalism?' *Antipode*, 25: 263–94.

O'Brien, R. (1992) *Global Financial Integration: The End of Geography?* London: Pinter.

Pred, A. (1986) *Place, Practice and Structure*. Cambridge: Polity Press.

Relph, E. (1976) *Place and Placelessness*. London: Pion.

Ritzer, G. (1996) *The McDonaldsization of Society*. London: Pine Forge Press.

Rogers, A. (1992) 'Key themes and debates', in A. Rogers et al. (eds) *The Student's Companion to Geography*. Oxford: Blackwell, pp. 233–54.

Rowles, G. (1978) *The Prisoners of Space?* Boulder, CO: Westview Press.

Smith, N. (1990) *Uneven Development* (2nd edn). Oxford: Blackwell.

Staehli, L. (2003) 'Place', in J. Agnew, K. Mitchell and G. Toal (eds) *A Companion to Political Geography*. Oxford: Blackwell, pp. 158–70.

Swyngedouw, E. (1989) 'The heart of a place', *Geografisak Annaler*, B71: 31–42.

# Place: The Management of Sustainable Physical Environments

## Ken Gregory

### Definition

Place has not explicitly been a primary focus for physical geographers, although it has been implicit in much of the development of physical geography for more than a century. The description of places was essential as environments were explored. Such descriptions were then compared, leading to systems of categorization of places so that places could subsequently be evaluated against the background of general models. As physical geography now extends to environmental management, place warrants greater explicit attention by physical geographers in relation to the management of sustainable physical environments.

### INTRODUCTION: PLACE LOCATED

The aphorism 'Geography is about maps, but biography is about chaps' (Bentley, 1905) encapsulates much public perception of geography as the study of places, with the word geographer still connoting someone who not only knows where places are but also what they are like. Paradoxically, physical geographers have given comparatively little explicit attention to place, although it will be argued here that for much of the twentieth century physical geography was *implicitly* concerned with place, a theme now becoming more *explicit*. To the physical geographer, place is the particular part of space occupied by organisms or possessing physical environmental characteristics. Place is associated with a number of related terms including environment, landscape and nature. The range of terms used indicates how place is not exclusive to any one discipline because others, including ecology, geology, other environmental sciences and landscape architecture, also focus upon place, using their own terms and approach. It has been suggested (Rolston III, 1997) that six words model the world we view: nature, environment, wilderness, science, earth, value. In addition to descriptions by physical geographers, our appreciation of the physical character of place is gradually established from a variety of images provided by literature, art, mathematics, science, language and various forms of media. The distinctiveness of places is shown by the fact that, in the Russian language, there are words for unique types of valley, and Finnish and Swedish vocabulary includes words for aspects of winter which do not occur in other countries (Mead and Smeds, 1967). Initially attempts were made to restrict definitions of place to natural conditions, prior to human activity and cultural influence, until it became appreciated that the impact of human activity is such that there are now few if any really natural places, environments or landscapes. Most recently it has been realized that even physical environment is culturally determined: do people from different cultures see physical landscape in the same way?

Just as the definition of place has varied over time, most recently reflecting perceptions of physical environments by particular cultures, so the places studied by physical geographers have changed over time. Places that are the focus of attention for physical geographers are located within the spheres in the envelope from about 200 km above the earth's surface to the centre of the earth. Since 1875, when the Austrian geologist, Suess, invented the terms hydrosphere, lithosphere and biosphere to complement atmosphere, which had been used since the 1700s (Figure 10.1), earth and life scientists have gone

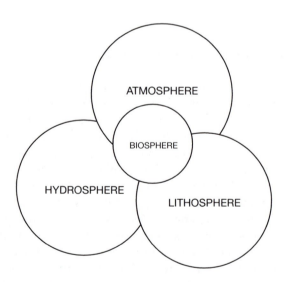

**FIGURE 10.1**   Major earth spheres. The four major spheres recognized by E. Suess in 1875 are depicted as intersecting circles, and other spheres that have been suggested are detailed in Table 10.1

sphere crazy (Huggett, 1995). No one discipline can study all the major spheres now identified (Table 10.1); it is by focusing research upon a particular blend of spheres that the basis is provided for the study of places in a discipline such as physical geography. However, some spheres have not always been studied by physical geographers. Thus the ocean sphere was included in physical geography in the first half of the twentieth century but became the exclusive province of oceanography in the second half, when the physical geographer made increasingly substantial contributions to the terrestrial hydrosphere in the study of hydrology. Whereas other spheres, or combinations of them (Table 10.1), are the province of other disciplines in the way that the biosphere is dominantly the realm of the biological sciences, there is a physical geography perspective which is distinctive in the way in which it links the several spheres together. In focusing upon a particular interaction of spheres the physical geographer has to locate places, to describe them both in terms of their characteristics and their dynamics, to explain them and their development in relation to adjacent places and to evaluate them for particular purposes.

This chapter outlines how the study of places by physical geographers evolved, how it stimulated integrated approaches and how opportunities and challenges are now available for a physical geography perspective.

## PLACE LOCATED AND FRAGMENTED

Since Arnold Guyot (1850) wrote that physical geography 'should compare, it should interpret, it should rise to the how and the wherefore of the phenomena which it describes', it has been appreciated that more than mere description is required. Over the subsequent 150 years, phases of exploration and audit, of classification and of categorization were necessary to establish fundamental knowledge about the physical environment. In the twentieth century, exploration furnished maps that provided basic information and only in the latter half of that century were many of the characteristics of place inventoried in maps or in detailed surveys, greatly accelerated by the advent of remote sensing. Once sufficient information had been amassed, it was necessary to classify it – a process often requiring major schemes of mapping (see Chapter 6), of types of geomorphology, of climate, of soils and of vegetation and, thence, of regions. Significant debates occurred about how classification should be attempted and whether it should be based upon static or dynamic characteristics, for example. It was not until the 1960s, with the arrival of the quantitative revolution, that physical geographers could embark upon the quantitative description of place, greatly enhanced in the late

**TABLE 10.1**   Earth spheres

| Sphere | Interpretation | Source |
|---|---|---|
| Atmosphere | Used since end of seventeenth century for the gaseous envelope of air surrounding the earth up to 200 km | |
| Troposphere | The lowest 12 km of the atmosphere | |
| Geosphere | Used to signify the lithosphere; or the lithosphere + hydrosphere + atmosphere; or any of the terrestrial spheres or shells, e.g.: | |
| | Core, mantle and all layers of the crust | Huggett (1995) |
| | Any of the so-called spheres or layers of the earth | Bates and Jackson (1980) |
| | Zone of interaction on or near the earth's surface of atmosphere, hydrosphere, biosphere, lithosphere, pedosphere and noosphere or anthroposphere | Vink (1983) |
| Geoecosphere | The sphere in which other spheres (biosphere, toposphere, atmosphere, pedosphere and hydrosphere) interact | Huggett (1995) |
| | The landscape sphere | Vink (1983) |
| Pedosphere | Layer of the earth in which soil-forming processes occur | Mattson (1938); Bates and Jackson (1980) |
| | The sphere of the regolith affected by soil-forming processes is the *edaphosphere* and the remainder of the pedosphere of weathered rock and unconsolidated material is the *debrissphere* | Huggett (1995) |
| Cryosphere | The part of the earth's surface that is perennially frozen and includes snow, ice, frozen ground and sea ice | Bates and Jackson (1980) |
| Hydrosphere | Water, both fresh and saline, in liquid, solid or gaseous state close to or on the surface of the earth: 95% in oceans and seas; 2% in glaciers and permanent snow; 2.5% is freshwater | Suess (1875) |

**TABLE 10.1**  Continued

| Sphere | Interpretation | Source |
| --- | --- | --- |
| Toposphere | At the interface of the pedosphere, atmosphere and hydrosphere | Huggett (1995) |
| | *Relief sphere* used for the totality of the earth's topography | Budel (1982) |
| Biosphere | Coined by E. Suess in 1875 but not given strict definition. Now used in at least three ways:<br><br>1  Zone or surface envelope of the earth which is naturally capable of supporting life<br>2  Synonymous with biota as sum of living creatures on the earth<br>3  Where life exists and where its influence extends | Vernadsky (1945); Teilhard de Chardin (1959); Hutchinson (1970); Bates and Jackson (1980); Huggett (1995) |
| Ecosphere | Life and the inorganic environment that sustains it | Cole (1958) |
| Noosphere | The realm of human consciousness in nature or the 'thinking' layer arising from the transformation of the biosphere under the influence of human activity | Developed by Pierre Teilhard de Chardin (1881–1955); Vernadsky (1863–1945) began to use the term in the mid-1930s |
| Lithosphere | The earth's crust and portion of upper mantle | Suess (1875) |

twentieth century with advances in information technology. A methodological change could then occur as the idiographic focus upon the description of the unique character of distinctive places, which had featured in the first half, was succeeded by a more nomothetic, law-giving approach seeking more general explanation and models in the second half of the twentieth century (Gregory, 2000).

As physical geography developed (see also Chapter 6) it became subdivided into branches, broadly corresponding with the spheres that had been identified (Figure 10.1; Table 10.1), with each branch adopting a basic unit of resolution for a particular place in the physical environment. Such units could be the primary basis for characterization and monitoring studies. In respect of the atmosphere, the site conditions were carefully specified for monitoring weather data to ensure that the weather station was typical of the surrounding area. The basic unit for soil study was the vertical section through all the constituent horizons of a soil, recognized as the soil profile since the time of Dokuchaev and other early Russian soil scientists in the late nineteenth century. As the profile is two-dimensional, the pedon was introduced as the smallest unit or volume of soil that represents, or exemplifies, all the horizons of the soil profile: it is a vertical slice of soil profile of sufficient thickness and width to include all the features that characterize each horizon (Wild, 1993), is usually a horizontal, more or less hexagonal area of 1 m$^2$ but may be larger (Bates and Jackson, 1980), and is an integral part of many soil-survey classification systems.

In biogeography basic places were recognized because all organisms live in niches – either a fundamental niche which an individual may occupy in the absence of competition with other species or a realized niche which is the actual niche occupied when competition is in progress (Watts, 1971). The niche has subsequently been defined as the habitat in which the organism lives but also the periods of time during which it occurs and is active, and the resources it obtains there. Other terms have therefore been developed for the organisms, for their place or habitat and for the combination of organisms and habitat. Micro-habitat is a precise location within a habitat where an individual species is normally found; a biotope is the smallest space occupied by a single life-form, as when fungi grow on biotopes found in the hollows of uneven tree trunks. Habitat, a term employed in various ways according to when and in which branch of science it has been used (e.g. Morrison, 1999), can also designate the living place of an organism or a community, applying to a range of scales from the microscale relating to organisms of microscopic or submicroscopic size through to the macroscale at continental or subcontinental scale. In ecology, habitat has come to signify description of where an organism is found, whereas niche is a complete description of how the organism relates to its physical and biological environment. The duality of place implicit in definitions that involve organisms as well as the environment of the organisms was reflected in biogeocoenosis, a Russian term equivalent to the western term ecosystem, and involving both the biocoenosis, a term introduced by Mobius in 1877 for a mixed community of plants and animals, together with its physical environment (ecotope). Biotope was defined as an area of uniform ecology and organic adaptation, although it was subsequently thought of as a habitat of a biocoenosis or a microhabitat within a biocoenosis. In relation to the land surface, the relief sphere or toposphere (Table 10.1), the basic place unit for the geomorphologist is the morphological unit; the undivided flat or slope is the basic unit of relief characterized by Linton (1951) as the electrons and protons of which physical landscapes are built.

The pedon, biotope and morphological unit are all examples of bases for specifying basic characteristics of place which emerged during the course of study of the physical environment, increasingly concentrated upon particular sub-branches of physical geography, including climatology, geomorphology and biogeography with pedology. After the acquisition of information to characterize the physical environment of such basic units, several further requirements had to be fulfilled:

- Parameters were needed to describe the character of place – these were often chosen from the sub-branches of physical geography (associated with one of the spheres of Table 10.1). In climatology mean annual temperature or mean annual precipitation were extensively used but, in other branches of physical geography, there were insufficient basic data available so that mapping schemes devised to characterize place included systems of morphological, geomorphological or vegetational mapping.
- Mapping schemes provided detailed field-survey scale information concerning the spatial units of physical environment necessary for the study of climate, geomorphology (landforms), soils (soil profile) or vegetation (plant community) and had to be related to a range of other scales up to global. This is illustrated by 16 000 different kinds of soil recognized within the USA (Buol, 1999), which had to be arranged hierarchically as part of a soil classification system to the global level. Detailed maps of landforms, local climates, soil series or ecosystems could be amalgamated to larger scales and eventually to the world scale.
- Classification was achieved by reference to some spatial or temporal framework in the way that rivers could be categorized in terms of stream order; and landforms could be dated according to the age where underlying datable material was expected to be contemporaneous in age with the landform.
- Because basic data and many classifications were expressed in terms of single aspects of the environment, of physical form/slope/ geomorphology, of climate parameters, of soil series and of plant or animal communities, ways of integrating the different aspects were needed. The greatest limitation of regional geography was that the so-called integration was left to the reader to accomplish. Hence parameters were sought which described the dynamics or process of place and which reflected the way in which place functioned. With the advent of hydrology this was exemplified by the way in which output parameters of water and sediment delivery reflected the integrated characteristics of the catchment

areas upstream. Four systems (morphological, cascading, process-response and control) incumbent in the systems approach (Chorley and Kennedy, 1971; see also Chapter 14) provided particular ways of characterizing place.

- Place characterized by using single parameters, frequently expressed in terms of average values, was seen to be insufficient in physical geography. The climate at a point location expressed in terms of mean annual temperature or mean annual precipitation did not reflect the extremes of temperature, the incidence of extreme rainfalls or droughts – possibly inducing climatic hazards or the variation from year to year or from decade to decade that was more pertinent to environmental processes and indeed to human activity.

- Not all sciences investigate places to the same level of analysis or resolution. Whereas the basic sciences such as physics, chemistry and biology may investigate sections of the earth spheres at a microscopic or subatomic scale, physical geography and other environmental sciences, conceived as composite sciences (Oster-kamp and Hupp, 1996), adopt a broader scale of resolution. Although the explicit quantitative description of landform development has lagged behind the understanding of processes, recent technical developments in terrain monitoring now apply at the microscale where information can be gathered at the total station level; at the mesoscale aided by global positioning systems (GPS); and at the macroscale facilitated by remote sensing and aided by digital elevation models (Lane et al., 1998).

Place is not exclusive to any one discipline but, as the physical geographer and other scientists initially characterized environment in terms of single characteristics, it was necessary to develop more integrated methods of depiction.

## INTEGRATED PLACES

Advances in the more integrated depiction of place involved more than that of climate, soil, vegetation and surface form and were achieved in three ways: by recognition of integrated units; by the production of system models that depended upon the relationship of basic units; and by integrated classification of places into a hierarchy of levels culminating in world regions or types.

As an example of an *integrated unit*, singularities (short seasonal episodes lasting as little as just a few days, commonly occurring at specific dates of the year) are a way of describing the climate experienced at a place, based upon analysis of meteorological or climatic records over periods of time. Thus, based upon analysis of 50 years of daily weather maps (1898–1947), Lamb (1964) demonstrated the existence of five seasons in the British Isles within which there were 22 singularities, many well known in folklore by terms such as April showers. In some places it is naturally occurring hazards and, often, extreme geophysical conditions threatening life or property that are especially influential. Natural hazards in the hydrosphere, biosphere, lithosphere and atmosphere include the risk of extreme events which differ substantially from their mean values (Alexander, 1999). For southwestern Ontario it was shown (Hewitt and Burton, 1971) that in a 50-year period, there would be 1 severe drought, 2 major windstorms, 5 severe snowstorms, 8 severe hurricanes, 10 severe glaze storms, 16 severe floods, 25 severe hailstorms and 39 tornadoes. It was therefore possible to express the hazardousness of a place as the complex of conditions that define the hazardous part of a region's environment.

In relation to land-surface relief and morphology, places can be characterized as landforms, employing terms such as eskers, limestone pavement or type of river channel pattern. However, more generally, the undivided flat or slope (the morphological unit) is a basic unit of relief (Linton, 1951), having much in common with the *site* originally described as 'an area which appears for all practical purposes to provide throughout its extent similar conditions as to climate, physiography, geology, soil' (Bourne, 1931). Indeed, the site was a primary feature in soil investigations because it was the area over which soil profiles were investigated. Soil profiles and sites were then grouped, or classified, into the fundamental soil-mapping units which might be a series, defined as groups of soils with similar profiles formed on lithologically similar parent materials.

In biogeography, as in ecology, it had been appreciated that a habitat plus the community it contains are a single working system. The term ecosystem was invented by Tansley (1935) for a community of organisms plus its environment as one unit, therefore embodying the community in a place together with the environmental characteristics, relief, soil and rock type (i.e. the habitat) that influence the community. Ecosystems can vary in size from one to thousands of hectares and could be a pond upstream of a debris dam or a large section of the Russian steppe. As examples of integrated descriptions of place, ecosystems represented the more open-system thinking developed in physical geography whereby place in physical geography was described by types of systems. The ecosystem depends upon a dynamic relationship between a community of organisms and its environment.

Other ways of focusing upon dynamic relationships in order to characterize place included the drainage basin, implicitly employed in hydrology since the nineteenth century when the integrity of the

water balance equation was first appreciated (Gregory, 1976a), and
subsequently proposed as the fundamental geomorphic unit (Chorley,
1969). The drainage basin (the area drained by a particular stream or
drainage network and delimited by a watershed) is an integrated unit
that may be described in terms of drainage basin characteristics
including rock type, soil, vegetation and land use, and relief characteristics. It is also a dynamic response unit from which outputs of water,
sediment and solutes reflect the characteristics of the drainage basin
which acts as the transfer function. The drainage basin unit is definable in relation to streams and rivers of a great range of sizes, can be a
basis for spatial variations (Chapter 6) and is often employed in
relation to environmental management (see below).

Just as there are ways of characterizing individual places in physical environments, there are also ways in which adjacent places are
interrelated – expressed in *system models* (see Chapter 14). Such
relationships may arise as drainage basins forming part of a nested
hierarchy or they may be linked by energy flows in the way that
cascading systems have been recognized as structures with output
from one subsystem forming input to the next (Chorley and Kennedy,
1971). Sequences of the character of places can be identified for
climate (climosequences), relief (toposequences), lithology (litho-
sequences), ecology (biosequences) and time (chronosequences). In geomorphology a nine-unit hypothetical land-surface model (Dalrymple
et al., 1967) showed how nine particular slope components could
occur in land-surface slopes anywhere in the world (Figure 10.2a). Each
component was associated with a particular assemblage of processes
so that it was possible to predict how slopes could occur under different morphogenetic conditions. A similar approach was applied to
pedogeomorphic research (Conacher and Dalrymple, 1977), although a
simple five-unit slope may be sufficient (e.g. Birkeland, 1984). Recurrent patterns of spatial variation have included the catena concept
(Milne, 1935), expressing the way in which a topographic sequence of
soils of the same age and usually on the same parent material can
occur in landscape, usually reflecting differences in relief/slope and
drainage (Figure 10.2b) – an arrangement which others have described
as a toposequence (Bates and Jackson, 1980). Most catenas or hillslope
models are thought of in two dimensions but combined with the
drainage basin concept can produce a three-dimensional model of a
small watershed (e.g. Huggett, 1975) embracing the manner in which
water moves through the surface layers and over the surface leading to
the creation of soil profiles, so providing an integrated understanding
of the dynamics of place. Further scope remains to elaborate the
catena idea and, in southeast Australia, sequences of erosion and
deposition over time periods were related to soil profile development
in a series of K cycles (Butler, 1959). Each period of time called a K

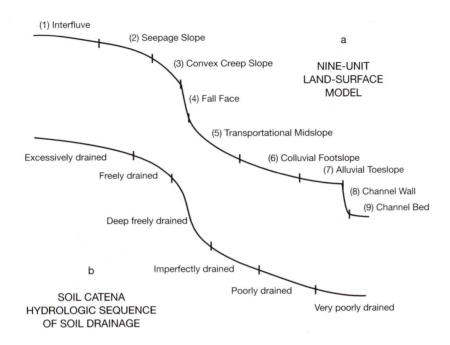

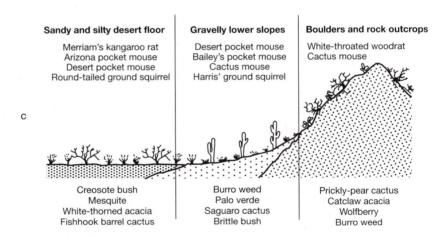

Source: After Vaughan (1978), as developed by Huggett (1995)

**FIGURE 10.2**   Sequences of places in the physical environment: (a) the components of the nine-unit land-surface model (based on Dalrymple et al., 1967); (b) the drainage categories found in a hydrologic sequence of soils which makes up a catena; (c) a particular example showing how communities vary with contrasting types of place in North American deserts

cycle is composed of an unstable phase when erosion or deposition may occur, and a stable phase when soil profile development occurs. Evidence for up to eight K cycles has been found preserved in some Australian soil landscapes. Toposequences of vegetation associations (Figure 10.2c) have been demonstrated showing how slope, soil drainage and vegetation vary empathetically in a repeated sequence across a landscape.

The search for recurrent patterns of place is really a step towards the search for *integrated classifications* of physical environment. As spatial patterns were explored and related to controlling factors such as climate, the characteristics of soil and slope catenas (Ollier, 1976), of hydrological slope models (Kirkby, 1976) or of drainage networks (Gregory, 1976b) were exemplars of the relationships established between geomorphology and climate (Derbyshire, 1976). It was also necessary to relate places at a range of different scales such as that from a point in a river channel to a drainage basin (Figure 10.3). A particular viewpoint can be adopted, and two particular visions, at completely opposite scales, were the global and the local. The global approach started from a world distribution and subdivided it, whereas the local vision described places in detail and amalgamated them to show how they fitted into broader regional, national and even world patterns.

The global vision had been evident in the classification of climates, vegetation, soils and relief. Some interrelationships had been found between such world patterns – for example, one scheme of classification by Köppen attempted to fit climatic values to world vegetation distribution patterns. However such approaches did not really yield integrated approaches to the physical environment so that climatic-based schemes were sought to relate to world distributions in climatic geomorphology. These included the scheme of 9 morphogenetic systems (Peltier, 1950; 1975), each distinguished by a characteristic assemblage of geomorphic processes; 13 morphoclimatic zones (Tricart, 1957) related to climates, geomorphic processes and also to soils and vegetation; and 5, later increased to 8, climato-morphogenetic zones (Budel, 1977), each characterized by particular landscape-forming processes and by relief features related in a distinctive way to past landscape development (Figure 10.4). Although climatic factors are undoubtedly of significance in affecting the global pattern of physical environments, other factors can be equally significant (Twidale and Lageat, 1994); there were dangers in adopting a very simplistic approach – climatic geomorphology was viewed as not new, not well established and premature (Stoddart, 1968) and, hence, climate impacts may have been overestimated (Twidale and Lageat, 1994).

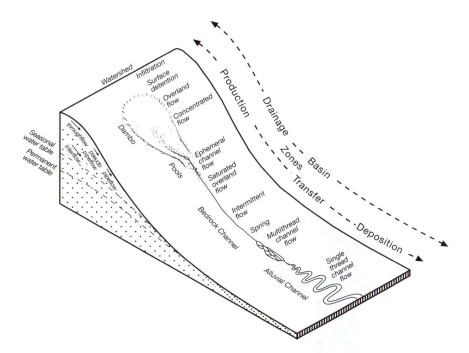

**FIGURE 10.3** Hierarchy of places. The drainage basin is one scale of place, and at the other extreme is study at a point in the river channel. In between these two extremes are major zones such as the production, transfer and deposition zones, within which it is possible to identify segments (bedrock, alluvial or colluvial) of valleys, within each of which there can be different types of reach (e.g. pool riffle), within which there are channel units such as pools and specific aquatic communities. Such a hierarchy of scales of place, rather like a series of Russian dolls, exists in a physical environment which is dynamically changing according to the sequence of hydrologic events. The diagram indicates the way in which the drainage basin network changes dynamically

Perhaps most successful, or at least most industriously pursued, were Russian efforts to develop the energy budget approach of Budyko (1958) which attempted a physico-geographical zonation of the earth and provided the foundation for energy and moisture regimes to be related to vegetation types (Grigoryev, 1961), to genetic soil types (Gerasimov, 1961) and to geographic zonality (Ye Grishankov, 1973). Such schemes could be extended to link with the approach to integrated systems based upon field survey.

A contrasted local vision arose from attempts to describe physical environment in an integrated way at the level of field survey. Russian studies recognized the *urochischa* as a basic physical-geographical unit of landscape with uniform bedrock, hydrological conditions, microclimate, soil and meso-relief (Ye Grishankov, 1973). Reconnaissance investigations in Australia, Africa and New Guinea, aided by

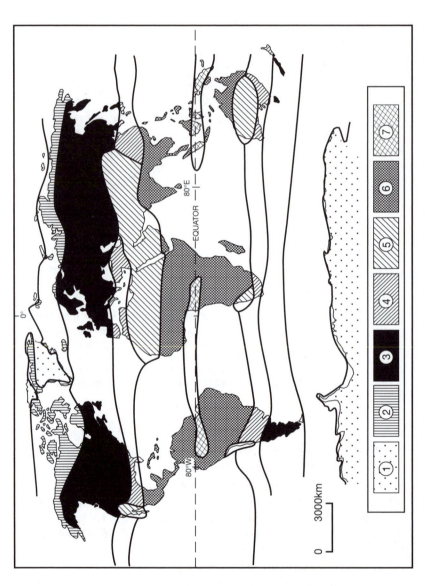

**FIGURE 10.4**  An example of places in a global integrated context: climato-genetic zones (after Budel, 1969). Many attempts have been made to subdivide the earth's surface according to the physical character of places. One integrated approach involved recognition of climato-genetic zones distinguished according to contemporary environmental processes and the past development of the physical environment. The zones were described as: (1) glacier zone; (2) subpolar zone of excessive valley formation; (3) extratropical zone of former valley formation (later called ectropic zone); (4) subtropical zone of mixed relief formation; (5) arid zone; (6) peritropical zone of excessive planation; (7) inner tropical zone of partial planation

the rapidly growing use of air photograph data, involved the recognition of land-systems defined as areas with a recurring pattern of topography, soils and vegetation (Christian and Stewart, 1953). The land-systems approach employed for resource evaluation in particular areas of the world utilized landscape ecology and land evaluation. These two approaches developed from soil surveys sometimes founded upon land capability analysis – a grouping of kinds of soil into classes according to their potential use and the treatments required for their sustained use; and upon suitability analysis, defined as the fitness of a given tract of land for a defined use (see also Chapter 6). Landscape ecology, a term first used by Troll to connote the interaction between landscape and ecology, provides an approach that interprets landscape as supporting interrelated natural and cultural systems (Vink, 1983). The central focus of landscape ecology (which is the study of pattern and process at the landscape scale – Forman, 1995) involves the interrelationship among landscape structure, the spatial patterning of ecosystems and landscape functioning, and the interactions of flows of energy, matter and species within and among component ecosystems (Kupfer, 1995). Whereas landscape ecology focuses on what the systems in the landscape can generally be used for, landscape evaluation is the estimation of the potential of land for particular kinds of use which can include productive uses such as arable farming, livestock production and forestry, together with uses that provide services or other benefits such as water catchment areas, recreation, tourism and wildlife conservation (Dent and Young, 1981). Such approaches have been refined with the advent of information systems (Cocks and Walker, 1987), advanced developments in remote sensing and the development of geographical information systems (e.g. Heywood et al., 1998). The land-system approach provided an applied approach to physical environment, greatly assisted by geographical information systems (GIS) which liberated multi-scale approaches to data acquisition and analysis. Further progress was aided by GPS facilities and by real-time analysis so that the way in which it is now possible to perceive place in physical environment has been completely revolutionized as a result of advances in data acquisition and analysis techniques. Such advances have meant that the initially independent global and local approaches can now be amalgamated by remote sensing and GIS.

## PLACE REDISCOVERED? OPPORTUNITIES AND CHALLENGES

Place has not received such explicit interest from physical geographers as that accorded by human geographers. However, attention has been given implicitly to what place means, how it is described and how it is related at various scales from the local to the global. Inter-relationships between the significance of the physical environment of place (landscape ecology) and the way in which place may be used (landscape evaluation) are just two strands fundamental to the move towards greater understanding of the capacity and potential of environment. However, place is now assuming a greater significance in physical geography in relation to four themes: how do we sustain it, manage it, restore it and design it?

First, how we sustain place requires consideration of *sustainable development*, which originated from conservation ideas in North America and Europe and which came to the fore with the 1992 UN conference in Rio de Janeiro. It has subsequently been thought of as 'development that respects the life quality of future generations and that is accomplished through support for the viability of the Earth's resources and ecosystems' (Saunier, 1999: 587). Instead of thinking about ecosystems (and place) as physical objects, it is possible to visualize them in terms of attributes with value for people as natural assets or 'natural capital' – an approach employed by Haines-Young (2000) to combine the scientific and cultural traditions of landscape ecology in managing landscapes. This challenges the view that the goal of sustainable development was to maintain the quantity and quality of ecological resources at an approximate steady state so as to ensure that they are not depleted at rates exceeding their renewal (Bartell, 1996) – because characterization of environmental management in terms of the 'natural capital' paradigm is fundamentally different. It provides an understanding of how the physical and biological processes associated with landscapes have value in an economic and cultural context, and so it is the study of natural capital from a dynamic, evolutionary and landscape perspective. It is not a steady state that we seek but, rather, a sustainable trajectory for ecosystems and landscapes so that equilibrium models rarely apply with no single sustainable state but a whole set of landscapes that are more or less sustainable (Haines-Young, 2000).

Secondly, in the *management* of place, a more cultural approach to the physical environment is required (Gregory, 2000): two current major approaches are *conservation* and *holistic dynamic management*. Conservation has a long tradition in relation to natural resources, but the conservation of areas or place as nature reserves, sites of special scientific interest (SSSIs) or wilderness areas is more recent and is one approach to management. In the UK regionally important geological and geomorphological sites (RIGSs) were established in 1990 to provide a regional complement to SSSIs. However, it may be more pragmatic to adopt an approach, often styled environmental, which acknowledges the dynamics of environmental systems, requires a sustainable solution and recognizes the totality of all the aspects of the physical environment – to which the term holistic has often been applied. The drainage basin or catchment has been extensively used as a management planning unit, which is capable of being employed for both the physical management of environment and for administrative purposes. However, management has not always covered the entire breadth of environmental characteristics and may not be as holistic as it could or should be so that further progress remains to be made (NRC, 1999).

Thirdly, a range of disciplines is currently engaged with the *restoration* of places because wetlands, prairies, lakes, wildlife habitat and other areas have been damaged by mineral extraction and many other forms of human impact. Restoration ecology is concerned with the restoration of badly damaged ecosystems to some predisturbance condition (Cairns, 1989) although, as we pursue environmental restoration as an addition to conservation (Berger, 1990), the question arises as to what we restore the original 'place' to: the condition that would have developed if disturbance had not occurred or one that appears natural? In which case, what is natural? In river management, restoration is now a major theme and changing public opinion is becoming at least as important as gaining new scientific knowledge (Douglas, 2000). Yi-Fu Tuan wrote his book, *Topophilia* (1977: v); 'out of the need to sort and order in some way the wide variety of attitudes and values relating to man's physical environment'. Awareness of such cultural strands is now more evident in physical geography (Gregory, 2000).

Finally, renewed interest in place by physical geographers and environmental scientists raises the question as to whether physical geographers should be engaged in *environmental design* (Gregory, 2000) and, if so, how? Ian McHarg, in his book *Design with Nature* (1969), proposed ways of approaching natural environment that were of wider importance than landscape architecture itself. When the book was subsequently revised in 1992, McHarg commented that, in 1969, 'scientists had not yet discovered the environment, the mandarins were molecular biologists and physicists concerned with sub-atomic particles' (p. iii) but hoped that his book 'provided a method whereby environmental data could be incorporated into the planning process, by interpreting ecological studies to include the full panoply of the environmental sciences, with the subject of values being crucial to the environmental movement' (p. iv). Although the full range of environmental sciences was embraced, McHarg (1992) identified social systems as being the one significant omission; in 1969 the influence of economics was antithetical to ecology, and other social sciences were then, unlike now, oblivious to the environment. The vision in McHarg's book was characterized by Lewis Mumford (1992: viii) as one of 'organic exuberance and human delight, which ecology and ecological design promise to open up for us. McHarg revives the hope for a better world'.

## CONCLUSION

Physical geographers can now be involved in the design of the physical environment or of place, building upon involvement in the sustainability agenda; greater awareness of different cultural attitudes to place; the need to restore damaged places; and the possibility of using understanding of the dynamics and evolution of place in the physical environment as the basis for sustainable environmental design. As physical geography refreshingly progresses to become more concerned with the holistic design of particular places, we still have to remember that our existing knowledge, models and theories have been built upon the experience of Anglo-American places, of some physical environments more than others, and influenced by dramatic eye-catching examples such as the Grand Canyon, the Amazon rainforest or the Yosemite National Park. In the development of physical geography we can discern threads running through science reflecting the investigation of particular types of place. As we become more holistic, having progressed from an idiographic to a nomothetic approach, are physical geographers sufficiently objective in their attitudes to place so that they can now profit from a postmodern idiographic approach, concerned with the character and design of individual places? Knowing how physical geographers have reacted to place in the past assists in our becoming more holistic in attitude in the future.

## Summary

- Place is central to physical geography but, until recently, it has not been given explicit attention.
- Places studied by physical geographers can be located in relation to environmental spheres of study. Physical geographers focus their research upon a particular combination of these spheres.
- In each branch of physical geography, basic units are required to characterize place. Examples are the morphological unit, the landform, the soil profile and the niche.
- Emphasis upon separate aspects of place, in terms of climate, soil, ecology and geomorphology, meant that more integrated relationships were sought – such as the land system.
- Place now receives more explicit attention by physical geographers in relation to sustainability, management, restoration and the design of physical environments.

## Further reading

Useful background is provided by an up-to-date dictionary, such as the *Encyclopedia of Environmental Science* (Alexander and Fairbridge, 1999). The development of physical geography (including cultural physical geography) is reviewed in Gregory's (2000) *The Changing Nature of Physical Geography*. An integrated approach to physical geography in the context of environmental change can be found in Slaymaker and Spencer's (1998) *Physical Geography and Global Environmental Change*. Lane et al.'s (1998) *Landform Monitoring, Modelling and Analysis* demonstrates how current techniques have advanced the description of landform developments – an example of place in physical geography. McHarg's (1969) *Design with Nature* is a salutory read because it shows how a landscape architect appreciated environmental challenges before they were acknowledged in physical geography. However, Huggett's (1995) *Geoecology: An Evolutionary Approach* and the article by Haines-Young (2000) indicate some ways in which physical geography can progress. An example of how river restoration is being achieved is demonstrated in Brookes and Shields' (1996) *River Channel Restoration*. Finally, Tuan's (1974) *Topophilia* is a typically thought-provoking product of a stimulating writer who began research as a physical geographer and Phillips (2001) makes an intriguing contribution.

*Note*: Full details of the above can be found in the references list below.

## References

Alexander, D.E. (1999) 'Natural hazards', in D.E. Alexander and R.W. Fairbridge (eds) *Encyclopedia of Environmental Science*. Dordrecht: Kluwer Academic, pp. 421–5.

Alexander, D.E. and Fairbridge, R.W. (eds) (1999) *Encyclopedia of Environmental Science*. Dordrecht: Kluwer Academics.

Bartell, S.M. (1996) 'Ecological risk assessment and ecosystem variation', in R.D. Simpson and N.L. Christensen (eds) *Ecosystem Function and Human Activity*. New York, NY: Chapman & Hall, pp. 45–70.

Bates, R.L. and Jackson, J.A. (1980) *Glossary of Geology*. Falls Church, VA: American Geological Institute.

Bentley, E.C. (1905) *Biography for Beginners*. London: T. Werner Laurie.

Berger, J. (ed.) (1990) *Environmental Restoration*. Washington, DC: Island Press.

Birkeland, P.W. (1984) *Pedology, Weathering and Geomorphological Research*. New York, NY: Oxford University Press.

Bourne, R. (1931) *Regional Survey and its Relation to Stocktaking of the Agricultural and Forest Resources of the British Empire*. Oxford Forestry Memoir 13.

Brookes, A. and Shields, F.D. (eds) *River Channel Restoration: Guiding Principles for Sustainable Projects*. Chichester: Wiley.

Budel, J. (1969) 'Das System der klima-genetischen Geomorphologie', *Erdkunde*, 23, 165–82.

Budel, J. (1977) *Klima-Geomorphologie*. Berlin/Stuttgart: Borntraeger.

Budel, J. (1982) *Climatic Geomorphology* (trans. L. Fischer and D. Busche). Princeton, NJ: Princeton University Press.

Budyko, M.I. (1958) *The Heat Balance of the Earth's Surface* (trans. N. Steepanova from the 1956 original). Washington, DC: Weather Bureau.

Buol, S.W. (1999) 'Soil', in D.E. Alexander and R.W. Fairbridge (eds) *Encyclopedia of Environmental Science*. Dordrecht: Kluwer Academic Publishers, pp. 563–4.

Butler, B.E. (1959) *Periodic Phenomena in Landscapes as a Basis for Soil Studies. Soil Publication 14*. Melbourne: CSIRO.

Cairns, J. (1989) 'Restoring damaged ecosystems: is pre-disturbance condition a viable option?' *The Environmental Professional*, 11: 152–9.

Chorley, R.J. (1969) 'The drainage basin as the fundamental geomorphic unit', in R.J. Chorley (ed.) *Water, Earth and Man*. London: Methuen, pp. 59–96.

Chorley, R.J. and Kennedy, B.A. (1971) *Physical Geography: A Systems Approach*. London: Prentice Hall.

Christian, C.S. and Stewart, G.A. (1953) *Survey of the Katherine–Darwin Region 1946. Land Research Series 1*. Melbourne: CSIRO.

Cocks, K.D. and Walker, P.A. (1987) 'Using the Australian Resources Information System to describe extensive regions', *Applied Geography*, 7: 17–27.

Cole, L.C. (1958) 'The ecosphere', *Scientific American*, 198: 83–96.

Connacher, A.J. and Dalrymple, J.B. (1977) 'The nine-unit land-surface model: an approach to pedogeomorphic research', *Geoderma*, 18: 1–154.

Dalrymple, J.B., Conacher, A.J. and Blong, R.J. (1967) 'A nine-unit hypothetical land-surface model', *Zeitschrift für Geomorphologie*, 12: 60–76.

Dent, D. and Young, A. (1981) *Soils and Land Use Planning*. London: Allen & Unwin.

Derbyshire, E. (ed.) (1976) *Geomorphology and Climate*. Chichester: Wiley.

Douglas, I. (2000) 'Fluvial geomorphology and river management', *Australia Geographical Studies*, 38: 253–62.

Forman, R.T.T. (1995) *Land Mosaics: The Ecology of Landscapes and Regions*. New York, NY: Cambridge University Press.

Gerasimov, I.P. (1961) 'The moisture and heat factors of soil formation', *Soviet Geography*, 2: 3–12.

Gregory, K.J. (1976a) 'Changing drainage basins', *Geographical Journal*, 142: 237–47.

Gregory, K.J. (1976b) 'Dainage networks and climate', in E. Derbyshire (ed.) *Geomorphology and Climate*. Chichester: Wiley, pp. 289–318.

Gregory, K.J. (2000) *The Changing Nature of Physical Geography*. London: Arnold.

Grigoryev, A.Z. (1961) 'The heat and moisture regions and geographic zonality', *Soviet Geography*, 2: 3–16.

Guyot, A. (1850) *The Earth and Man: Lectures on Comparative Physical Geography in its Relation to the History of Mankind*. New York, NY: Scribners.

Haines-Young, R. (2000) 'Sustainable development and sustainable landscapes: defining a new paradigm for landscape ecology', *Fennia*, 178: 7–14.

Hewitt, K. and Burton, I. (1971) *The Hazardousness of a Place: A Regional Ecology of Damaging Events*. Toronto: University of Toronto Press.

Heywood, I., Cornelius, S. and Carver, S. (1998) *An Introduction to Geographical Information Systems*. Harlow: Longman.

Huggett, R.J. (1975) 'Soil landscape systems: a model of soil genesis', *Geoderma*, 13: 1–22.

Huggett, R.J. (1980) *Systems Analysis in Geography*. Oxford: Clarendon Press.

Huggett, R.J. (1995) *Geoecology: An Evolutionary Approach*. London: Routledge.

Hutchinson, G.E. (1970) 'The biosphere', *Scientific American*, 223: 45–53.

Kirkby, M.J. (1976) 'Hydrological slope models: the influence of climate', in E. Derbyshire (ed.) *Geomorphology and Climate*. Chichester: Wiley, pp. 247–68.

Kupfer, J.A. (1995) 'Landscape ecology and biogeography', *Progress in Physical Geography*, 19: 18–34.

Lamb, H.H. (1964) *The English Climate*. London.

Lane, S.N., Richards, K.S. and Chandler, J.H. (eds) (1998) *Landform Monitoring, Modelling and Analysis*. Chichester: Wiley.

Linton, D.L. (1951) 'The delimitation of morphological regions', in L.D. Stamp and S.W. Wooldridge (eds) *London Essays in Geography*. London: London School of Economics, pp. 199–218.

Mattson, S. (1938) 'The constitution of the pedosphere', *Annals of the Agricultural College of Sweden*, 5: 261–76.

McHarg, I.L. (1969) *Design with Nature*. New York, NY: Natural History Press.

McHarg, I.L. (1992) *Design with Nature*. Chichester: Wiley.

Mead, W.R. and Smeds, H. (1967) *Winter in Finland*. London: Hugh Evelyn.

Milne, G. (1935) 'Some suggested units of classification and mapping, particularly for east African soils', *Soil Research*, 4: 183–98.

Mobius, K. (1877) *Die Auster und die Austernwirtschaft*. Berlin: Wiegundt, Hampel & Parey.

Morrison, M.L. (1999) 'Habitat and habitat destruction', in D.E. Alexander and R.W. Fairbridge (eds) *Encyclopedia of Environmental Science*. Dordrecht: Kluwer Academic, pp. 308–9.

Mumford, L. (1992) 'Introduction', in I.L. McHarg, *Design with Nature*. Chichester: Wiley, pp. vii–viii.

National Research Council (NRC) (1999) *New Strategies for America's Watersheds*. Washington, DC: National Academy Press.

Ollier, C.D. (1976) 'Catenas in different climates', in E. Derbyshire (ed.) *Geomorphology and Climate*. Chichester: Wiley, pp. 137–70.

Osterkamp, W.R. and Hupp, C.R. (1996) 'The evolution of geomorphology, ecology and other composite sciences', in B.L. Rhoads and C.E. Thorn (eds) *The Scientific Nature of Geomorphology*. Chichester: Wiley, pp. 415–41.

Peltier, L.C. (1950) 'The geographic cycle in periglacial regions as it is related to climatic geomorphology', *Annals of the Association of American Geographers*, 40: 214–36.

Peltier, L.C. (1975) 'The concept of climatic geomorphology', in W.N. Melhorn and R.C. Flemal (eds) *Theories of Landform Development*. Binghamton, NY: State University of New York Press.

Phillips, J.D. (2001) 'Human impacts on the environment: unpredictability and the primacy of place', *Physical Geography*, 22: 321–32.

Rolston III, H. (1997) 'Nature for real: is nature a social construct?' in T.D.J. Chappell (ed.) *The Philosophy of the Environment*. Edinburgh: Edinburgh University Press, pp. 38–64.

Saunier, R.E. (1999) 'Sustainable development, global sustainability', in D.E. Alexander and R.W. Fairbridge (eds) *Encyclopedia of Environmental Science*. Dordrecht: Kluwer Academic, pp. 587–92.

Slaymaker, H.O. and Spencer, T. (1998) *Physical Geography and Global Environmental Change*. Harlow: Longman.

Stoddart, D.R. (1968) 'Climatic geomorphology: review and assessment', *Progress in Geography*, 1: 160–222.

Tansley, A.G. (1935) 'The use and abuse of vegetational concepts and terms', *Ecology*, 16: 284–307.

Teilhard de Chardin, P. (1959) *The Phenomenon of Man*. London: Collins.

Tricart, J. (1957) 'Application du concept de zonalite a la géomorphologie', *Tijdschrift van het Koninklijk Nederlandsch Aarddrijiskundig Geomootschap*, 422–34.

Tuan, Y.-F. (1974) *Topophilia. A study of Environmental Perception, Attitudes and Values*. Englewood Cliffs, NJ: Prentice Hall.

Tuan, Y.-F. (1977) *Space and Place: The Perspective of Experience*. London: Arnold.

Twidale, C.R. and Lageat, Y. (1994) 'Climatic geomorphology: a critique', *Progress in Physical Geography*, 18: 319–34.

Vaughan, T.A. (1978) *Mammalogy* (2nd edn). Philadelphia, PA: W.B. Saunders.

Vernadsky, V.I. (1945) 'The biosphere and the noosphere', *American Scientist*, 33: 1–12.

Vink, A.P.A. (1983) *Landscape Ecology and Land Use* (trans. and edited D.A. Davidson). London: Longman.

Watts, D.A. (1971) *Principles of Biogeography: An Introduction to the Functional Mechanisms of Ecosystems*. London: McGraw-Hill.

Wild, A. (1993) *Soils and the Environment*. Cambridge: Cambridge University Press.

Ye Grishankov, G. (1973) 'The landscape levels of continents and geographic zonality', *Soviet Geography*, 14: 61–77.

# 11

# Scale: Upscaling and Downscaling in Physical Geography

## Tim Burt

## Definition

Physical geographers study the world across a wide range of scales, from the molecular to the global. In any investigation, we will pay attention to some things and ignore others. In terms of spatial scale, the resolution of any study indicates the level at which we focus on a particular item of interest. However, geographers have never confined themselves to one scale alone: on the one hand, they may narrow their perspective in order to focus on the detailed way in which a system operates; on the other hand, they will wish to extrapolate their findings at one scale to the wider area.

## INTRODUCTION: SCALE AND RESOLUTION

All this time the guard was looking at her, first through a telescope, then through a microscope, and then through an opera glass (Lewis Carroll, 1872).

Somewhere between the atom and the universe lies the geographical field of inquiry. The *Penguin Dictionary* defines geography as the science that describes the earth's surface. This clearly provides plenty of scope for the physical geographer and, as Figure 11.1 shows, a scale of interest that ranges over many orders of magnitude (see Chapter 12 on scale and human geography). A number of related issues have always confronted the geographer regardless of the specific topic of interest. This chapter engages with two of these key questions: at what scale should the inquiry be focused? How can findings at one scale be related to another?

The first question relates to the *resolution* of the study. Since one meaning of resolution is breaking into parts, in our context resolution

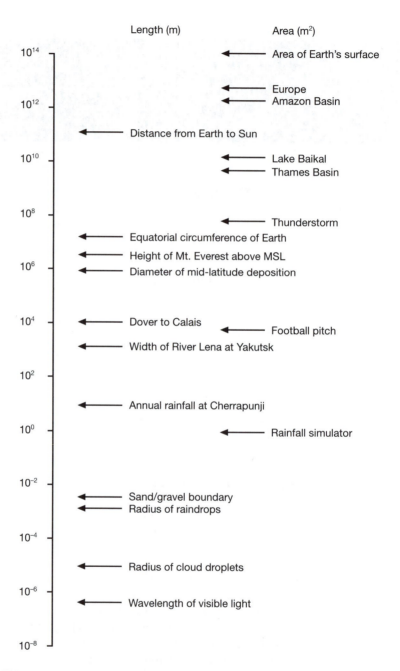

**FIGURE 11.1** Scale in geographical inquiry

indicates the spatial scale at which we observe individual items of interest and, by inference, the scale below which we do not seek to focus our studies. Thus, a study of the particle size of a soil sample might focus on the fractions of sand, silt and clay, but need not extend down to electrochemical bonding mechanisms at the molecular scale. Resolution is a familiar notion in remote-sensing studies where spatial resolution indicates the greatest resolving power of the instrument in question; for example, on early Landsat images, individual pixels (or picture elements) were approximately 80 m × 80 m and features much smaller than this would not be detectable. Spectral resolution indicates both the discreteness of wavebands measured by the satellite's sensor and its ability to measure radiation intensity levels. You could think of this as the numbers of different colours that the sensor can measure. What should we choose – a few primary colours or a larger number of shades? A simple analogy in grain size analysis would be the number of sieves used and thus the number of categories of particle size that can be distinguished.

The second question relates to the need to apply the results of one scale of analysis at different scales. This may involve *upscaling* of results from smaller to larger areas – for example, extending results from small catchment studies to large river basins. It may also in some circumstances involve *downscaling* – for example, applying the results of general circulation models (global scale) to particular regions. It has long been known that generalizations made at one level do not necessarily hold at another, and that conclusions derived at one scale may be invalid at another (Haggett, 1965). For example, consider the recolonization of Krakatau after the 1883 eruption. At one scale, today's plant distribution on the island reflects the local ecological niches – shoreline, slope, wetland, etc. However, at the larger scale, the general biodiversity is dependent on other controlling factors. How far is the island from the mainland (this controls the likelihood of species migration)? What is the biodiversity of the source area (the new community will obviously mirror the biodiversity of its close neighbours)? Clearly, as we change scale, the questions – and the answers – are different.

There is, of course, a strong relationship between the spatial and temporal scales of study. In general, as spatial scale increases, so does the timescale of interest. Thus, Knighton (1984) suggests the approximate linkage between scales in studies of fluvial landforms as shown in Table 11.1. The correlation between spatial and temporal scales of analysis does not always hold but, in general terms, short-term studies tend to focus on process dynamics whereas longer-term studies are more likely to involve statistical analysis of form and structure. In geomorphology this equates to the contrast between functional analysis of dynamic systems as opposed to historical studies of landform evolution. There is also a link between scale and causality, as pointed out in a seminal paper by Schumm and Lichty (1965). At the shortest timescale, processes operate within an essentially fixed environment,

**TABLE 11.1**  Approximate linkage between scales in studies of fluvial landforms

|  | Length scale (m) | Timescale (years) |
| --- | --- | --- |
| Bed forms in sandy channels | 0.1–10 | 0.1–1 |
| Cross-sectional form | 1–100 | 1–50 |
| Bed forms in gravel-bed channels | 1–100 | 5–100 |
| Meander wavelength | 10–500 | 10–1000 |
| Reach gradient | 100–1000 | 50–5000 |
| Long profile analysis | 1000–100 000 | 100–10 000 |

for example, water flow in a channel; form controls process at this scale ('static equilibrium'). However, over the longer term, properties that were fixed at the shorter timescale themselves become variable. Process now controls form and a 'steady-state equilibrium' may be identified. Large events may perturb the system but there is then recovery to a characteristic form. At the longest timescales, even characteristics like the long profile of the river valley will eventually change. A 'dynamic equilibrium' involves progressive evolution of landform as a response to ongoing erosion. This is also the timescale at which major changes in climate can affect landforms at the regional scale – for example, advances and retreats of ice sheets over entire continents. Chorley et al. (1984) provide further discussion of the scale issue in geomorphology, while time is discussed in more detail by Thornes (Chapter 7).

**UPSCALING**

Chorley (1978) has drawn the distinction between functional and realist approaches in geomorphology. Functional theories involve statistical generalizations whereas a focus on process dynamics implies greater realism of understanding. Two questions arise: can the results of small-scale studies, whether functional or realist, be applied to the larger scale? And to what extent are local studies representative of the wider region or merely unique case studies?

*Filtering*

In time-series analysis, techniques are applied to remove small-scale noise (rapid fluctuations in system output) in order to emphasize larger-scale patterns such as periodic cycles, trends and episodes such as major perturbations or threshold changes. Trend-surface analysis has been applied in spatial studies too, generalizing regional patterns from point-scale data. In both cases the filter passes components of certain frequencies while excluding others. Chorley and Haggett (1965) provide a detailed discussion and many examples of trend-surface mapping. One example included there, and discussed in more detail in Chorley et al. (1966), concerns grain size variations of soils within the Breckland, a distinctive region of sandy deposits in East Anglia. A nested sampling design was employed to allow variations at different scales to be identified. Systematic sampling at 2 km intervals indicated a coarsening towards the north east, the presumed source of the wind-blown material. At the smallest scale, a peak in variability at around 8 m spacing showed the effect of sediment sorting under periglacial conditions when classic patterned ground features, polygons and stripes, developed. Residual analysis indicated features at scales of 125–1000 m, possibly related to dune formation during a very dry epoch.

More recent developments in the field of geostatistics include spatial autocorrelation techniques and methods such as kriging, originally developed to map variation in mineral concentrations within ore bodies. Kriging helps identify the scales over which certain processes remain important and how different phenomena show continuity in space. Such methods provide a real improvement on traditional mapping techniques – in the old days, when faced with a scatter of observations in space, you used a mixture of knowledge, judgement and pure guesswork to draw the isolines on your map. Modern geostatistical methods introduce more rigour into such interpolation exercises. A recent example of the use of such techniques in hillslope geomorphology can be found in Burt and Park (1999), who used linear kriging to model the spatial distribution of soil properties across a hillslope. Soil samples were collected from 64 soil pits based mainly on a 25-metre sampling grid. Typical results are shown in Figure 11.2. The map for total exchangeable bases (TEB) shows that there is a reasonably uniform distribution across the slope. This is because cations like calcium, magnesium and potassium are important nutrients for plant growth and their distribution reflects a tight soil–vegetation system. Such nutrients never pass through a soil without biological involvement and they tend therefore to be present in topsoil regardless of location. The map for manganese (Mn) shows a much clearer catenary distribution, with strong leaching in more acid soils upslope and deposition in more neutral footslope soils. This example illustrates the way in which we upscale from point data (soil profile) to the hillslope scale, and additionally shows how the questions asked change subtly as we refocus from vertical processes (e.g. leaching, nutrient cycling) to horizontal processes (subsurface runoff).

a

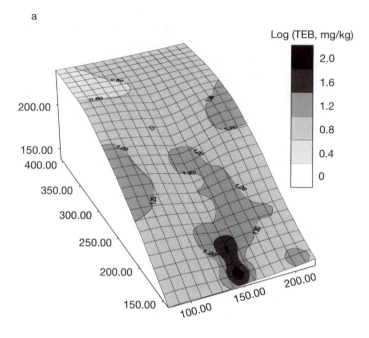

b

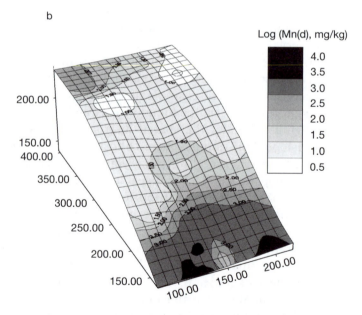

**FIGURE 11.2**    (a) The spatial distribution of total exchangeable bases (TEB); and (b) manganese (Mn) in topsoil (0–10 cm) for a hillslope section at Bicknoller Combe, Quantock Hills, England

### Sampling at the regional scale

In some cases, sampling has been sufficiently widespread that analyses can be conducted over very wide areas. For example, Hewlett et
al. (1977) used data from a large number of stream gauging stations to
categorize the stormflow response of rivers in the eastern USA. It is
critical for river basin managers to understand the flood response of a
river basin following rainfall. The approach differs from trend-surface
analysis in that no use is made of regression-type techniques, but both
methods allow regional responses to be mapped. The flashiness of a
river basin may be judged by calculating the ratio of stormflow
volume $(Q)$ to storm rainfall $(P)$. Figure 11.3 maps the mean stormflow response for all rainfall events over 25 mm. For the eastern USA
as a whole, the mean response is 0.2 – in other words, typically about
20% of rainfall becomes stormflow. However, the stormflow response
varies widely through space, exceeding 0.4 in the Louisville basin
where soils are thin and basin storage low. By contrast, because of
deep, permeable, sandy soils, in the Sand Hills of the Piedmont, the
ratio is as low as 0.04. The approach allows estimates to be made for
ungauged basins, an important benefit since estimating runoff
response in the absence of stream discharge records remains one of the
biggest problems in hydrology. Note that at the regional scale, stormflow response is strongly associated with geology, whereas more
locally (i.e. at the scale of individual fields) spatial variations in
infiltration capacity become more important.

We can develop the idea that different variables are dominant at
different scales by considering the Universal Soil Loss Equation
(USLE), which was develop in the USA to predict soil erosion from
agricultural land. By co-ordinating soil erosion research using standardized plots, more than 8000 plot-years of erosion research data were
compiled from 36 locations in 21 states (Mitchell and Bubenzer, 1980).
While much effort has been devoted more recently to find alternative
methods of modelling soil loss, the USLE remains widely used. It is
essentially a functional approach: process mechanisms are implied but
not explicitly modelled. The generalized form of the equation is as
follows:

$$A = f(R\ K\ L\ S\ C\ P)$$

where $A$ is the soil loss, $R$ is the rainfall erosivity factor, $K$ is the soil
erodibility factor, $L$ is the slope length factor, $S$ is the slope gradient
factor, $C$ is the crop management factor and $P$ is the erosion control
practice factor. As with the map of stormflow response, USLE allows
soil loss at unknown sites to be predicted. In many ways, the approach
is similar to trend-surface analysis, the main difference being the
collation of data from widely separated locations rather than the co-

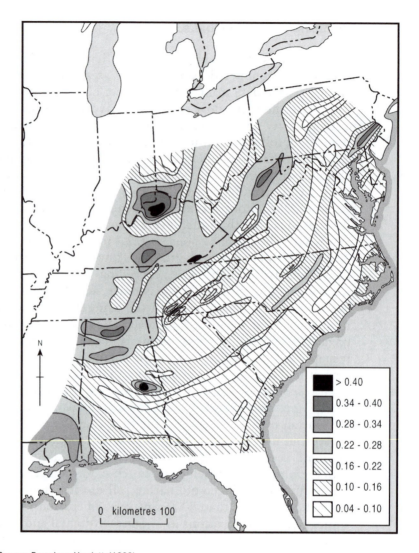

*Source*: Based on Hewlett (1982)

**FIGURE 11.3**   Flood runoff response, expressed as the ratio of stormflow to rainfall, for the southeastern USA

ordinated sampling strategy that usually underpins trend-surface ana-lysis. The difficulty of upscaling plot erosion data to the basin scale is discussed further in the next section.

Both the stormflow analysis and USLE used data collected as part of monitoring programmes. One problem with observations from a single site is knowing how to generalize across space. The examples discussed so far relate to single variables such as soil loss or runoff response. The UK's Environmental Change Network (ECN) was established in 1992 in order to co-ordinate measurements of ecological change, including the major driving variables such as climate and air pollution (Burt, 1994). A wide variety of measurements are made within ECN: physical variables like solar radiation and stream chemistry that need measuring regularly, other characteristics like soil properties that vary more slowly, plus many ecological responses from birds and butterflies to frog spawn. Because of their high cost, such complex and varied monitoring programmes are necessarily few in number, but at least by having a network of stations there is a chance of discriminating regional responses from local effects. It is worth mentioning too that, by maintaining the monitoring programme over several decades, subtle long-term changes can be identified within the noisy short-term record. Once again, changes in time and space go hand in hand.

### Nested experiments

A common problem in geography is how to generalize the results of small-scale experiments at the larger scale. The Breckland example showed the advantage of detailed spatial sampling but this may not always be possible, for reasons of time or the sheer cost of process studies. Of course, results from a single study – in effect, a sample of one – must always be treated with great caution, and this has led to some uncertainty about the wider value of field experiments.

To counter this, a number of studies have adopted a 'nested' approach to field experiments – each one designed to fit inside the next. For example, the newly established CHASM (Catchment Hydrology And Sustainable Management) research initiative in the UK has nested mesoscale (100 km²), miniscale (10 km²) and microscale (1 km²) basins in order to bridge the gap between small-scale studies and the relevant scale for integrated river-basin management. Even smaller 'patch' or plot-scale studies will provide the most detailed data for model calibration and process rate measurements. The whole approach recognizes, as its focus, the need to upscale from detailed local-scale models to larger-scale models with lower spatial resolution but greater areal coverage.

As an example of this sort of approach, Anderson and Burt (1978) presented results from a two-level nested experiment, comparing the hydrological response of a single section of hillslope (3 ha – the same slope as mapped in Figure 11.2) with that of a small first-order catchment (60 ha). This shift of scale is not great enough to lose the dominant control of hillslope processes on the catchment response (Figure 11.4). Even when the shift is larger still, to a 5th order basin (2300 ha), the double-peak hydrographs, characteristic of a delayed throughflow response, remain very obvious (Burt, 1989). It is only in very much larger basins that the detail of the headwater response is lost, subsumed within an aggregate hydrograph that reflects travel time of water through the channel network and the total volume of flood runoff produced, rather than the peak discharge in individual tributary basins. This is illustrated in Figure 11.5 (Hewlett, 1982) which shows the passage of a flood wave down the Savannah River, Georgia, USA. Runoff (discharge per unit area of catchment) at Clayton looks impressive but it is the total discharge at Clyo that is much more likely to be problematic. Plotting simply as discharge rate shows how peak discharges upstream are lost in the volumes of water that create the downstream flood.

When we think about the soil properties shown in Figure 11.2 and tie this in with what we know about the hydrology of the same hillslope (Figure 11.4), we begin to see further links between local processes and patterns at the landscape scale. Ecologists have been much interested recently in 'patch dynamics'. One way of viewing an

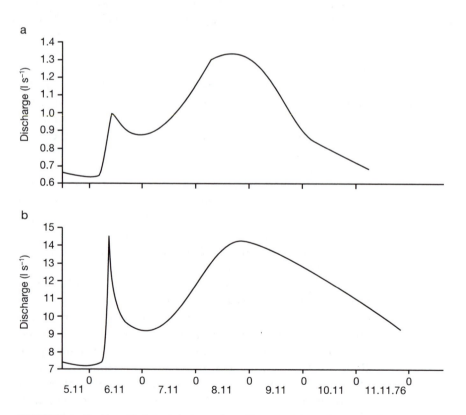

**FIGURE 11.4**  Double-peaked storm hydrographs at Bicknoller Combe, Quantock Hills, England, for (a) a hillslope section; and (b) the entire catchment area

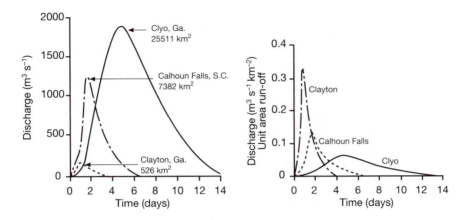

*Source*: Based on Hewlett (1982)

**FIGURE 11.5**  Passage of a flood wave along the Savannah River shown both as discharge and unit area runoff

area is as a mosaic of patches, each patch with its individual assemblage of plants and processes. This is a largely 'vertical' view of ecosystems in that most of the important transfers of energy and matter move up and down, from atmosphere to root zone or vice versa. However, the distribution of patches in space is often orderly rather than random – when we upscale we see that patches are functionally related to one another. Thus, on the hillslope already considered, we see the following toposequence: podzolic soils and heathland vegetation on the interfluve, merging downslope to brown earth soils with grass and bracken – a typical soil catena, in other words (Gerrard, 1981). The important point is that pattern is related to process, with the distribution of soil and vegetation being intimately related to the hydrological and biogeochemical processes operating across the slope. When we add a third dimension, we find that some footslopes are wetter than others because water converges into hillslope hollows (Anderson and Burt, 1978). Soil saturation can allow different processes to occur: soils become anoxic and reduction processes like denitrification take over. In terms of stream-water quality, these footslope wetlands can provide a buffer between the upslope regions (where leaching is another of our vertical processes) and the stream (Cirmo and McDonnell, 1997). For example, acid waters draining from upslope may become more neutral during their residence within the saturated zone downslope making the stream less vulnerable to acid pollution. The distribution of manganese reflects this, with leaching from acid soils upslope and deposition at the base of the slope where the pH is greater. In the same way, nitrate draining from farmland may be denitrified within a riparian buffer zone, thereby

helping to keep river water below the legal limit for nitrate concentration (Haycock et al., 1997; Burt et al., 1999). Given the importance
of such abrupt changes in biogeochemical environment, it is important that any field measurement programme reflects the landscape
scale, but focuses particularly on the near-stream zone (Cirmo and
McDonnell, 1997). At the landscape scale, it is not so much the
individual patch dynamics that are important, more the way in which
patches link together.

### Rainfall–runoff modelling

Discussion of the different controls on the flood hydrograph at
increasing scale leads neatly into a brief consideration of rainfall–
runoff modelling. There has been a great deal of effort devoted to this
task over recent decades, aided by a dramatic increase in computer
power. As a basic premise, model structure has tended to become
more detailed in space and time as the scale of interest *de*creases. At
the hillslope scale, models are likely to incorporate equations describing all the important surface and subsurface processes. Such models
are also usually fully distributed, dividing up the catchment area into
a large number of elements or grid squares and solving the equations
for the state variable at each point. As the scale of interest increases,
hydrological models are likely to be more conceptual in structure,
formulated on the basis of a simple arrangement of a relatively small
number of components, each of which is a simplified representation of
one process element in the system being modelled. These conceptual
models tend also to be 'lumped', treating the catchment as a single
unit, with state variables that represent averages over the entire area
(Beven, 2000). Empirical or black-box models have also proven reliable
for making predictions at the catchment scale despite their lack of
theoretical structure. Rather, they rely on establishing a statistical
correspondence between input and output using data already collected. By definition, their resolution precludes any interest in within-
catchment variation in runoff processes. Beven (2000) provides an
extensive and thorough review of rainfall–runoff modelling while
several different models are described in Anderson and Burt (1985).
Beven concludes that physically-based distributed hydrological
models are the state of the art but, despite much effort and great
investment in computing resources, such models have often been
restricted to research programmes with simpler models continuing to
be preferred (both in terms of cost and accuracy) for practical purposes.
Of course, hydrology straddles the divide between science and engineering. Geographers have been much involved over the years in the
study of process hydrology (including modelling) but from the modelling point of view it is notable that, as scale increases, the spatial unit

of interest gets larger and the process description gets simpler. One of the great challenges in hydrological research therefore is how to upscale the results from the small-scale to the large-scale. No doubt, as we have already seen, as size of drainage basin increases, the important variables will change too.

### Soil erosion and sediment delivery

A major problem arises from the difficulty of upscaling soil erosion plot results to the level of sediment delivery at the basin scale (Walling, 1983). USLE measurements were obtained from the use of standardized erosion plots (22 m long), but the delivery of eroded sediment to the river channel involves consideration of the whole landscape. An isolated plot may provide useful information on erosion itself, but its very isolation precludes a complete appreciation of the erosion problem. Real slopes will behave differently from a 22 m plot: for example, longer flow paths and convergence of water into hollows can both encourage gullies to develop, something ignored in USLE. There will also be deposition of eroded soil, for example, on shallow footslopes or behind hedges; and once in the channel, further storage is possible, in-channel or on floodplains. Typically, only a fraction of the eroded material makes it to the basin outlet – but what fraction exactly? There have been two research groups working on these over-lapping but essentially different problems: those studying soil erosion *per se* and those studying sediment delivery systems. Not surprisingly their conclusions have differed somewhat, especially when it comes to assessing the significance of soil loss. Catchment-scale processes can buffer the erosion process so that the pattern of in-field soil erosion can look very different from the pattern of sediment yield at the basin outlet. This difference between the field and landscape response is the reason why a vigorous debate arose following the publication of Trimble (1999; Table 11.2). His study of the Coon Creek basin, Wisconsin, showed that erosion rates in the period 1975–99 were much less than in the 1930s (as little as 6% if today's rate is compared with the *maximum* rate from the 1920s and 1930s), and he argued that soil loss is no longer a major concern. Moreover, because of lags within the system, sediment yield remained constant throughout the 140-year period of study; in other words, the dramatic soil erosion of the early twentieth century did not produce a signal at the basin outlet. A number of soil erosion experts condemned Trimble for daring to say that the soil erosion threat was less than (they) generally estimated. However, his data certainly do suggest a large reduction in 'upland' (i.e. field) sources. More importantly from our point of view, the com-plexity of the sediment system shows that a local-scale viewpoint is not enough if we are to extrapolate to larger basins.

**TABLE 11.2**  A sediment budget for Coon Creek, Wisconsin

|  | 1853–1938 | 1938–75 | 1975–93 |
|---|---|---|---|
| *Sources* | | | |
| Net upland sheet and rill erosion | 326 | 114 | 76 |
| Upland gullies | 73 | 64 | 19 |
| Tributaries | 42 | 35 | 9 |
| Upper main valley | 0 | 27 | 13 |
| *Sinks* | | | |
| Upland valleys | 38 | 38 | 0 |
| Tributary valleys | 87 | 0 | 25 |
| Upper main valley | 71 | 27 | 4 |
| Lower main valley | 209 | 139 | 51 |
| Total sources | 441 | 240 | 117 |
| Total *soil erosion* sources | 399 | 178 | 95 |
| Net export to upper main valley | 316 | 175 | 79 |
| Net export to lower main valley | 245 | 175 | 88 |
| Sediment yield to Mississippi | 38 | 36 | 37 |

*Source*: **Based on Trimble (1999)**

**DOWNSCALING**

Despite an obvious geographical need to generalize over wide areas, the focus of analysis may move in the opposite direction. Such reductionism has been the hallmark of science, in some cases quite literally using the microscope to examine underlying structures and processes. Thus, the basis of Anderson and Burt's (1978) study was a detailed examination of soil moisture movement in soil, in order to better understand the way in which delayed throughflow hydrographs were generated. Downscaling has been an important activity in some areas of modelling too, notably in relation to general circulation models (GCMs) of global climate. It has been necessary to develop regional-scale models in order to show how forecast changes in global climate will manifest themselves at, say, the scale of North America or Europe. The resolution of GCMs is of the order of 500 × 500 km. Processes operating at finer spatial scales, such as cloud formation and precipitation, are 'parameterized', estimated by empirical or conceptually based relationships. Processes at the land surface operate at this scale too and are similarly parameterized (Arnell, 2002). Clearly, treating a large area of land as a single column of homogeneous soil and vegetation with a uniform climate is very unrealistic. Models with improved spatial resolution are therefore used to help translate the results of GCM simulations to the regional scale. One important study region has been the Amazon basin where the interaction of climate change and the impact of deforestation may have a significant impact on local rainfall and runoff. Arnell (2002) notes that, although different models vary, a complete removal of Amazonian rainforest could reduce evaporation by up to 20% and thereby decrease rainfall by up to 30% across the basin as a whole (Nobre et al., 1991). Field studies reduce the spatial scale of investigation much further, of course, by several orders of magnitude; plot studies in the Amazon basin (Shuttleworth, 1988) have measured interception and evaporation rates to provide a basis for forecasting the effect of forest canopy removal.

## ANALYSIS AND SYNTHESIS

There has always been a tension in physical geography between the need to understand the detail of process mechanisms set against the need to understand how broader, complex systems operate. Analysis encourages use of the microscope while synthesis requires the wide-angle lens! For much of the first half of the twentieth century, geography was dominated by regional studies (see Chapter 1). At the largest scale, Herbertson (1905) divided the earth into major natural regions, mainly on the basis of climate. Small areas were expected to show the same individuality. Thus, the focus was on areal differentiation, on the varying character of the earth's surface. However, by the 1950s, geographers had become disillusioned with the regional paradigm. There was a major reorientation in the nature of geographical research, involving the strengthening of systematic studies, attempts to develop law, theories and models, and application of mathematical and statistical procedures to facilitate the search for generalizations (Johnston, 1983). One almost inevitable outcome, at least in the short term, was a focus on small-scale process studies (Anderson and Burt, 1990).

Inevitably there has been reaction to this focus and, in recent years, physical geographers have raised their sights somewhat, whether in looking once again at large-scale landforms (Sugden et al., 1997) or the impact of climate change and human impact on the world's major biomes (Goudie, 2000). The difference today is that knowledge of process mechanisms is fundamental, providing the basis for large-scale synthesis via modelling studies. However, as we have seen, the important variables change as we increase our scale of interest, and we must learn how to restructure our models as we move to increasingly larger scales, a point discussed by Beven (2000) in relation to hydrological modelling, for example. It may be that scale and complexity go hand in hand but the challenge is to find order and pattern in larger as well as smaller systems. Noble (1999: 297) nicely confronts this issue in relation to the ecological complexity of landscapes: 'We live in landscapes; we manage landscapes. We often describe the environment around us in terms of landscapes. Yet landscapes have long been a scientific blind spot.'

We cannot just add up the individual parts of a landscape to tell us what we have – the whole is, in that sense, greater than the sum of the parts because of functional linkage between different elements of the system. Nevertheless, because complex, large-scale systems do function, we have the opportunity to link pattern and process at that scale. Thus, for physical geographers, whatever their specialist interest, the challenge remains to understand this scale linkage, bridging the gap between studies of process dynamics and the application of such knowledge to larger areas.

## CONCLUSION

The scale of interest in physical geography is extremely wide, from molecular to global. All scientists aim to make generalized statements about the phenomena they study and, in so doing, there are choices to be made – what to focus on and what to ignore. For the physical geographer, this choice also involves a decision about the resolution of study – at what scale to focus the investigation. Whatever scale is selected, geographers have nevertheless always realized that results obtained at one scale are not enough. On the one hand, they will wish to delve beneath their current level of interest to understand more about the process mechanics of the system being studied. And at the same time, they will want to demonstrate the relevance of their work at the larger scale and to use results from one study to speculate about other places.

## Summary

- When we approach any topic, we must define the scale at which we want to focus our attention, but we must also consider what other scales may be relevant to our work.
- Upscaling can be achieved in a number of ways: through the use of 'nested' field experiments, by applying appropriate statistical methods and through the use of computer simulation models.
- Downscaling is traditionally concerned with a reduction of our scale of interest in order to learn more about the dynamics of the system being studied. But much work today is focused at the global scale and the results invariably need translating to the regional scale – a rather different type of downscaling.

- In practice, physical geographers work across a range of scales. As the scale changes, so do the questions being asked and the results obtained. Factors important at one scale may be less relevant at a different scale, so a flexible and perceptive approach is needed.

## Further reading

For those interested in computer modelling, Beven's (2000) *Rainfall-Runoff Modelling* is a purpose-built and up-to-date analysis of hydrological modelling and contains much of interest on the scale issue. Brunsden and Thornes' (1979) 'Landscape sensitivity and change' is an important paper that introduces the concept of landscape sensitivity and demonstrates how scale and location are important factors in landscape change. Despite its age, Chorley and Kennedy's (1972) *Physical Geography: A Systems Approach* is still a very relevant and readable introduction to systems analysis in physical geography. Material covering a wide variety of scales and applications is included. *The Geographer's Art* (Hagget, 1990) is not a physical geography text but it is easy reading and very stimulating for a geographer of any hue! Most of the case studies involve the scale issue. *Fluvial Forms and Processes* (Knighton, 1998) provides an accessible guide to fluvial geomorphology, the different scales of interest involved and how these couple together. Schumm and Lichty's (1965) 'Time, space and causality' is a classic paper in geomorphology, explaining how causal variables differ with changing scale of interest, in both space and time. Not an easy read but a 'must' nevertheless!

*Note*: Full details of the above can be found in the references list below.

## References

Anderson, M.G. and Burt, T.P. (1978) 'The role of topography in controlling throughflow generation', *Earth Surface Processes*, 3: 331–4.

Anderson, M.G. and Burt, T.P. (1985) 'Modelling strategies', in M.G. Anderson and T.P. Burt (eds) *Hydrological Forecasting*. Chichester: Wiley, pp. 1–13.

Anderson, M.G. and Burt, T.P. (1990) 'Geomorphological techniques. Part one. Introduction', in A.S. Goudie (ed.) *Geomorphological Techniques* (2nd edn). London: Unwin Hyman, pp. 1–29.

Arnell, N. (2002) *Hydrology and Global Environment Change*. London: Prentice Hall.

Beven, K.J. (2000) *Rainfall–Runoff Modelling*. Chichester: Wiley.

Brunsden, D. and Thornes, J.B. (1979) 'Landscape sensitivity and change', *Transactions, Institute of British Geographers*, 4: 463–84.

Burt, T.P. (1989) 'Storm runoff generation in small catchments in relation to the flood response of large basins', in K.J. Beven and P.A. Carling (eds) *Floods*. Chichester: Wiley, pp. 11–36.

Burt, T.P. (1994) 'Long-term study of the natural environment: perceptive science or mindless monitoring?' *Progress in Physical Geography*, 18: 475–96.

Burt, T.P., Matchett, L.S., Goulding, K.W.T., Webster, C.P. and Haycock, N.E. (1999) 'Denitrification in riparian buffer zones: the role of floodplain sediments', *Hydrological Processes*, 13: 1451–63.

Burt, T.P. and Park S.J. (1999) 'The distribution of solute processes on an acid hillslope and the delivery of solutes to a stream. I. Exchangeable bases', *Earth Surface Processes and Landforms*, 24: 781–97.

Chorley, R.J. (1978) 'Bases for theory in geomorphology', in C. Embleton et al. (eds) *Geomorphology: Present Problems and Future Prospects*. Oxford: Oxford University Press, pp. 1–13.

Chorley, R.J. and Haggett, P. (1965) 'Trend surface mapping in geographical research', *Transactions, Institute of British Geographers*, 37: 47–67.

Chorley, R.J. and Kennedy, B.A. (1972) *Physical Geography: A Systems Approach*. London: Prentice Hall.

Chorley, R.J., Schumm, S.A. and Sugden, D.E. (1984) *Geomorphology*. London: Methuen.

Chorley, R.J., Stoddart, D.R., Haggett, P. and Slaymaker, H.O. (1966) 'Regional and local components in the areal distribution of surface sand facies in the Breckland, eastern England', *Journal of Sedimentary Petrology*, 36: 209–20.

Cirmo, C.P. and McDonnell, J.J. (1997) 'Linking the hydrologic and biogeochemical controls of nitrogen transport in near-stream zones of temperate forested catchments: a review', *Journal of Hydrology*, 199: 88–120.

Gerrard, A.J. (1981) *Soils and Landforms*. London: George Allen & Unwin.

Goudie, A.S. (2000) *The Human Impact on the Natural Environment* (5th edn). Oxford: Blackwell.

Haggett, P. (1965) *Locational Analysis in Human Geography*. London: Edward Arnold.

Haggett, P. (1990) *The Geographer's Art*. Oxford: Blackwell.

Haycock, N.E., Burt, T.P., Goulding, K.W.T. and Pinay, G. (1997) *Buffer Zones: Their Processes and Potential in Water Protection*. Harpenden: Quest Environmental.

Herbertson, A.J. (1905) 'The major natural regions', *The Geographical Journal*, 25: 300–10.

Hewlett, J.D. (1982) *Principles of Forest Hydrology*. Athens, GA: University of Georgia Press.

Hewlett, J.D., Cunningham, G.B. and Troendle, C.A. (1977) 'Predicting stormflow and peakflow from small basins in humid areas by the R-index method', *Water Resources Bulletin*, 13: 231–53.

Johnston, R.J. (1983) *Geography and Geographers* (2nd edn). London: Edward Arnold

Knighton A.D. (1984) Fluvial Forms and Processes. London: Edward Arnold.

Knighton, A.D. (1998) *Fluvial Forms and Processes* (2nd edn). London: Edward Arnold.

Mitchell, J.K. and Bubenzer, G.D. (1980) 'Soil loss estimation', in M.J. Kirkby and R.P.C. Morgan (eds) *Soil Erosion*. Chichester: Wiley, pp. 17–62.

Noble, I.R. (1999) 'Effect of landscape fragmentation, disturbance, and succession on ecosystem function', in J.D. Tenhunen and P. Kabat (eds) *Integrating Hydrology, Ecosystem Dynamics and Biogeochemistry in Complex Landscapes*. Chichester: Wiley, pp. 297–312.

Nobre, C., Sellers, P.J. and Shukla, J. (1991) 'Amazonian deforestation and regional climate change', *Journal of Climatology*, 10: 957–88.

Schumm, S.A. and Lichty, R.W. (1965) 'Time, space and causality', *American Journal of Science*, 263: 110–19.

Shuttleworth, W.J. (1988) 'Evaporation from Amazonian rainforest', *Philosophical Transactions of the Royal Society of London*, B233: 321–46.

Sugden, D.E., Summerfield, M.A. and Burt, T.P. (1997) 'Editorial: linking short-term geomorphic processes to landscape evolution', *Earth Surface Processes and Landforms* (special issue), 22: 193–4.

Trimble, S.W. (1999) 'Decreased rates of alluvial sediment storage in the Coon Creek Basin, Wisconsin, 1975–1993', *Science*, 285: 1245–7.

Walling, D.E. (1983) 'The sediment delivery problem', *Journal of Hydrology*, 65: 209–37.

# 12 Scale: The Local and the Global[1]

**Andrew Herod**

## Definition

Within human geography scale is typically seen in one of two ways: either as a real material thing which actually exists and is the result of political struggle and/or social processes, or as a way of framing our understanding of the world.

## INTRODUCTION

It is argued by many that contemporary economic, political, cultural and social processes, such as globalization, are rescaling people's everyday lives across the planet in complex and contradictory ways. Thus we have seen the creation of supra-national political bodies such as the European Union at the same time that we have witnessed the devolution of political power from the nation-state to regional political bodies, such as the new Scottish Parliament and Welsh Assembly. Equally, we appear to be witnessing an increased homogenization and 'Americanization' of global culture while, simultaneously, we have also seen the growth of localist tendencies among those who have sought to defend traditional ways of life in many parts of the world, with the famous example of the French farmer who attacked a McDonald's fast-food restaurant in 1999 being a case in point.[2] Such examples of contemporary economic, political and cultural forces resulting in an apparent simultaneous globalization and localization of everyday life, together with myriad others like them, have raised important conceptual questions about this process of rescaling of people's lives and, particularly, about the relationship between what are often taken as the two extremes of our scaled lives – namely, the 'global' and the 'local'. For instance, what does it really mean when we say that what started as a 'local' family business has now grown to become a 'global' transnational corporation? What exactly is the relationship between 'global' climate change and 'local' weather patterns? How is a 'global' language such as English 'localized' in different parts of the world, so that British English, American English, Australian English, Indian English, Nigerian English and Singaporean English appear as quite distinct? What does it mean to talk about a war on 'global' terrorism if acts of violence occur in quite 'local' settings – particular streets in New York City, Ramallah, Belfast or wherever?

In this chapter, then, I explore some of the issues related to how we use the concepts of the 'global' and the 'local' to make sense of the world around us. Specifically, I want to discuss three aspects of the local and the global, namely: (1) their ontological status;[3] (2) how the relationship between the global and the local has often been conceptualized within geographic writing; and (3) how the use of different metaphors concerning scales such as the global and the local can shape the ways in which we understand the scaled relationships between different places.

## THE ONTOLOGICAL STATUS OF THE GLOBAL AND THE LOCAL

Although scale has long been considered one of geography's core concepts, until the 1980s it had largely been a taken-for-granted concept used for imposing organizational order on the world. Thus, while geographers – both physical and human – had frequently employed scales such as the 'regional' or the 'national' as frames for their research projects, looking at particular issues from a 'regional scale' or a 'national scale', they had spent very little time theorizing the nature of scale itself. However, following the publication of two articles by Peter Taylor (1981; 1982) and of Neil Smith's (1984) book, *Uneven Development*, the issue of what came to be known as the 'politics of scale' was hotly debated within human geography during the late 1980s and continues to be so today, particularly as it relates to processes of globalization (for an extended discussion of this debate, see Herod, 2001: esp. pp. 37–46).

Although there were a number of issues at stake, one key concern was scale's ontological status, particularly whether scale is simply a mental device for categorizing and ordering the world or whether scales really exist as material social products. This debate reflected diverse epistemologies (theories of knowledge) upon which different geographers drew for understanding the world, particularly those epistemologies which were idealist and those which were materialist in their origins.[4] Hence, while geographers such as John Fraser Hart (1982) drew upon Immanuel Kant's idealist philosophy to suggest that scales were no more than handy conceptual mechanisms for ordering the world, others drew upon Marxist ideas of materialism to argue that scales were real social products (that is to say, that scales really exist in the world) and that, as such, there was a politics to their construction. Neil Smith, in particular, argued that the production of scales emerged out of contradictions within capital.[5] By way of example, he examined how restructuring in the US economy during the 1980s was leading to a rescaling of its industrial landscape (Smith and Dennis, 1987; Smith, 1988). He also argued that the scales at which social life is organized are produced as a result of political struggles, making his point through an analysis of how the anti-gentrification movement in New York City involved efforts by activists to expand the scale of their endeavours so that isolated actions in different neighbourhoods could be united into a citywide movement (Smith, 1989; 1993). Although much of this debate in the 1980s about scale concerned whether economic regions really exist or not and was tied up with efforts to develop a 'reconstructed regional geography' which focused upon processes of regional formation (for a discussion of this debate, see Pudup, 1988), it has also affected how we conceptualize the 'global' and the 'local'.

For those who draw their inspiration from Kantian idealism, therefore, the local and the global are seen as part of a pre-existing conceptual matrix of scales within which social life is lived. As such, they are simply mental devices for circumscribing and ordering processes and practices so that these may be distinguished and separated from one another – a particular process or set of social practices can thus be considered to be 'local' whereas others are considered to be 'global' in scope. For idealists, the 'global' is usually defined by the geologically given limits of the Earth, whereas the 'local' is seen as a spatial resolution useful for comprehending processes and practices which occur at geographical ranges smaller than the 'regional' scale, which in turn is seen to be anything which is smaller than the 'national' scale (which, for its part, is seen as the next smallest scale after the 'global' scale). For materialists, on the other hand, the key aspect of geographical scale is to understand that scales are socially produced through processes of struggle and compromise. Hence, for instance, the 'national' scale is not simply a scale which exists in a logical hierarchy between the global and the regional but, instead, is a scale that had to be actively created through economic and political processes which consolidated into larger nation-states the various duchies, principalities and fiefdoms that had been the major political units (at least in Europe) until the Middle Ages. This was a process that was formally recognized by the 1648 Treaty of Westphalia (which legally established the notion of the territorial integrity of each nation-state, even those which had been defeated in war, and the primacy of the nation-state over its citizenry) and was not completed in some European countries such as Germany and Italy until the late nineteenth century. In the case of the global and the local, then, a materialist would maintain that both these scales are actively created through the practices of various social actors. Scales such as the global do not simply exist, waiting to be utilized, but they must instead be brought into being. Hence, transnational corporations do not simply adapt their activities to a premade global scale defined by the Earth's geologically given limits but must, instead, actively build their own global scale of operation. They must, in effect, *become* 'global'.

The notion of 'becoming' and the focus on the politics of producing scales have been central to materialist arguments concerning the global scale. Thus there has been much attention paid to how transnational corporations have 'gone global', how institutions of governance have 'become' supra-national and how labour unions have sought to 'globalize' their operations to match those of an increasingly 'globalized' capital. In such an approach, the various scales at which social actors and processes operate cannot be conceived of as separate from the actors and processes that create them. In making such an argument, though, some materialists have tended to assume that, while the global is actively forged through social practices, the local is somehow a more 'natural', less socially produced 'default' scale (i.e. that all social actors start as inherently 'local' actors and subsequently *become* regional, national and/or global). However, such an approach is problematic for it privileges the local scale over all others, viewing it as a kind of foundation upon which all other scales are built. Consequently, other commentators have argued that the local, too, is 'produced' and is no more a natural scale than is the global – social actors must work to become local in the same way that they must work to become global (or regional or national).

Drawing upon the notion of becoming, therefore, means that the key issue for those who argue that scale is socially produced and is not simply a mental device for ordering the world is to examine *how* social actors make themselves global and/or local – that is to say, how they embed themselves locally or extend themselves globally. Instead of viewing the local as the default or starting scale at which all social actors operate and from which some break free to become regional, national or global in their reach, it is important to recognize that social actors may have to work just as hard to become 'local' as they have to work to become 'global'. For example, some manufacturers may rely upon suppliers located thousands of miles away, whereas others may rely upon suppliers located within the same community as themselves. However, before they can use such local suppliers, these latter manufacturers may have to 'localize' themselves by developing business linkages with such suppliers, by training the community's workforce to operate particular types of machinery used in the plant, by establishing credit with the community's banks and other financial institutions, and by building trust with politicians who represent the particular community in different levels of government (see Cox and Mair, 1988, for examples of how such 'locality dependence' comes about). Through such activities, they actively become local in much the same way that other manufacturers may become global through establishing business relationships with suppliers, financial institutions, workforces and politicians in communities located throughout the world.

A useful attempt to think about such social practices theoretically has been Kevin Cox's (1998a: 2) distinction between what he calls *spaces of dependence* ('those more-or-less localized social relations upon which we depend for the realization of essential interests . . . for which there are no substitutes elsewhere [and which] define place-specific conditions for our material well being and our sense of significance') and *spaces of engagement* (the spaces in and through which social actors construct associations with other actors located elsewhere). In making this distinction, Cox suggests that it is important to understand the production of scale as emerging from the interactions that link one particular actor's spaces of dependence with those of other actors as part of a strategy of engagement with them. Certainly, some writers (e.g. Jones, 1998; Judd, 1998) have questioned what exactly Cox means by terms such as 'localized social relations', while others (e.g. Herod, 2001) have suggested that 'spaces of dependence' may not necessarily be only 'local'. A transnational corporation, for instance, may be dependent upon resources found in several different places across the globe, such that its space of dependence may be argued to be 'global'. Nevertheless, Cox's approach does allow us to think about how different social actors may be dependent on particular spaces yet may seek to engage with other social actors operating within their own, quite different, spaces of dependence.

In such an approach, the production of scale is conceptualized as emerging out of the ways in which actors build spaces of engagement to link the various spaces upon which they, or those with whom they must deal, are dependent. Hence, for Cox (1998a: 20), moving from the local to the global scale 'is not a movement from one discrete arena to another' but a process of developing networks of associations that allow actors to shift between various spaces of engagement. Scale is thus seen in terms of a process rather than in terms of a fixed entity. In other words, the global and the local are not static 'arenas' within which social life plays out but are constantly remade by social actions. This allows us to consider not only how a firm or political organization might attempt to 'go global' to engage with actors or opportunities not present within its own local spaces of dependence, but also how a particular social actor such as a transnational corporation may attempt to 'go local' through tailoring its products and operations so as to appeal to consumer tastes in different places or to reflect particular communities' cultural values.

Having considered, therefore, the idealist/materialist debate concerning scale's ontological status, in the next section I want to highlight how discourses pertaining to scale (especially the global and the local) have been an important aspect of human geography during recent years.[6] Regardless of how scale is thought of ontologically, it is important to understand that the ways in which the global and the local – and especially the relationship between them – are presented rhetorically can fundamentally shape how we conceptualize the world and its social processes.

## DISCOURSES OF THE GLOBAL AND THE LOCAL

As I alluded to above, within debates over processes of globalization the local and the global have frequently been thought of as the two ends of the scalar spectrum, with the local being understood through its contrast with, and status as 'Other' to, the global, and vice versa.[7] Within this binary way of thinking, the global and the local have often been associated, respectively, with other sets of binaries – for instance, there exists a correspondence in much Western thinking between the global and the abstract, and between the local and the concrete, such that global activities are often perceived as somehow more abstract and less concrete than are local ones.

Within such binary thinking, Gibson-Graham (2002) has identified at least six ways in which the relationship between the local and the global is often viewed.[8] These six ways are as follows:

1    The global and the local are seen not as things in and of themselves, but are viewed instead as interpretive frames for analysing situations. For example, when considering processes of economic restructuring, what is seen from a 'global perspective' (perhaps a worldwide economic slowdown) may appear different from what is seen from a 'local perspective' in particular places (some places may actually be experiencing economic expansion during such a global economic slowdown).

2    The global and the local each derive meaning from what they are not. Much like our conception of what a slave is only makes sense if we can contrast that with a conception of what constitutes a free person (and vice versa), the global and the local only make sense when contrasted with each other. Thus, drawing on Dirlik (1999: 4), Gibson-Graham (2002) suggests that in such a representation the global is 'something more than the national or

regional . . . anything other than the local'. In turn, the local is seen as the opposite of the global. This view, however, represents an important semantic shift because, whereas once the local 'derived its meaning from its contradiction to the national', now anything other than the global is often seen to be 'local' – for instance, in much rhetoric about globalization, nations and even entire extra-national regions, such as the European Union, are frequently referred to as 'local actors' within the process of globalization.

3   Whereas many writers have viewed spatial scales in terms of a hierarchy of fixed and separate arenas within which social life occurs (what could be called the 'process X occurs within global space but process Y occurs within local spaces' approach), some, such as French social theorist Bruno Latour, have tended to see the world as constituted through a series of networks which link different places. In such an approach, both the local and the global 'offer points of view on networks that are by nature neither local or global, but are more or less long and more or less connected' (Latour, 1993: 122). Thus, much like it would be impossible to describe a spider's web in hierarchical spatial fashion – where does one part of the web end and another begin? – in a view such as Latour's it is impossible to distinguish where the local ends and the global (or other scales) begins. Instead, such a view sees the global and local as simply different 'takes' on the same universe of networks, connections, abstractness and concreteness. The global and the local are not so much opposite ends of a scalar spectrum but are a terminology for contrasting shorter and less connected networks with longer and more connected networks.

4   The global *is* local. In this perspective, Gibson-Graham argues, the global does not really exist, and if you scratch anything 'global' you will find localness. In such a view multinational firms, for instance, are actually 'multilocal' rather than 'global'.

5   The local *is* global, and place is a 'particular moment' in spatialized networks of social relations. Much like the tips of an octopus's legs might touch particular locations on the sea floor while the body of the octopus floats above it, in such an approach the local appears as the location where global forces 'touch down' on to the Earth's surface. In turn, the local is not a place but is an entry point to the world of global flows which encircle the planet.

6   The global and local are not locations but processes. Put another way, globalization and localization produce all spaces as hybrids, as 'glocal' sites of both differentiation and integration (Dirlik, 1999: 20).[9] Thus, the local and the global are not fixed entities but

are always in the process of being remade. Hence, local initiatives can be broadcast to the world and adopted in multiple places across space, while global processes always involve localization. For instance, in the process of globalizing itself McDonald's tailors its products to particular local tastes, serving beer in France, pineapple fritters in Hawaii and vegetarian 'hamburgers' in India.

In reviewing these different ways of articulating the relationship between the local and the global, Gibson-Graham (2002) suggests that the history of this binary has been one in which the power of the global has usually been assumed to be greater than that of the local. Part of this results from the widely held view in Western thought that greater size and extensiveness imply domination and superior power, such that the local is often represented as 'small and relatively powerless, defined and confined by the global' (Gibson-Graham, 2002: 27). In such a representation 'the global is a force, the local is its field of play . . . the global is penetrating, the local penetrated and trans- formed' (p. 27). Thus the global is conceived of as 'synonymous with abstract space, the frictionless movement of money and commodities, the expansiveness and inventiveness of capitalism and the market. But its Other, localism, is coded as place, community, defensiveness, bounded identity, in situ labor, noncapitalism, the traditional' (p. 27). Hence, Gibson-Graham (2002: 33) argues, what emerges from the above review 'is an overriding sense that [power] is either already distributed and possessed or able to be mobilized more successfully by "the global" ' than by the local.

Paradoxically, such a vaunting of the power of the global has frequently been engaged in both by neoliberals and by many Marx- ists.[10] Thus neoliberals frequently use the rhetoric of globalization to undermine local opposition to their agenda. For example, such rhet- oric is evident in claims that national welfare systems and local union work rules 'must' be given up 'due to the imperatives of global capitalism'. For their part, many on the political left appear to have given up any hope of challenging capitalism globally and prefer instead to 'think globally' but to 'act locally' (for more on this argument, see Herod, 2001: 128). However, Gibson-Graham argues that such a discursive reification of the global and diminution of the power of the local restricts the possibilities for progressive political action, because it suggests that capital, which is usually seen as being global in its operation, can always outmanoeuvre its opponents who, as Other to capital, consequently are usually viewed, by association, as non-global or local in their operation. Instead, Gibson-Graham has sought to deconstruct the local–global binary to suggest ways in which the local can be thought of as enabling political struggle and as a powerful basis for challenging (global) capital (along these lines, for an example of how a strike in a single community in 1998 brought to a virtual standstill in North America the operations of automobile giant General Motors, see Herod, 2000). Relatedly, others (e.g. Wills, 1998; Herod, 2001; esp. pp. 128–60 and 197–221) have shown, through examination of international trade-union activities, that it is empirically incorrect to assume that workers and others opposed to global capital have not themselves been able to act globally, thus shaping the political and economic geographies of 'the global'. Put another way, critics of the view which holds that the local is inherently weaker than the global as a scale of political action and that the global is a scale capable of construction only by capital, have argued both that organizing locally can indeed be effective in particular circumstances and that we should not think of the global as a scale at which only capital can operate effectively. Nevertheless, the ability of neoliberal ideologues discursively to represent the global as more powerful than the local can secure them significant advantages, for it may encourage their opponents to believe that organizing locally is doomed to failure. Likewise, corporate executives' success in representing the global as capital's domain may encourage labour unions and others not to attempt to organize globally because they may feel they have no hope of succeeding on 'capital's terrain'. Clearly, then, how the terms 'local' and 'global' are deployed discursively can be very important for the politics of political struggle.

Having explored how the global and the local have frequently been represented as part of a binary of scale, in what follows I want to examine how different metaphors of scale have been used in geography. As will become clear, this is important because how we conceptualize the ways in which the world is scaled will shape how we engage with that world.

## METAPHORS OF SCALE

Following from the above, it is clear that within human geography during the past decade or so a great deal of attention has been paid to the politics of discourse concerning scale – that is to say, how the local and the global (and other scales) are talked about and represented. Centred upon the work of writers such as Jacques Derrida, Michel Foucault, Bruno Latour and others, much of this 'postpositivist' concern with language has focused upon the metaphors used to describe and make sense of the world, for metaphors can be powerful shapers of how we understand things.[11] When thinking about the use of metaphor, however, it is important to recognize that the choice of one metaphor over another is usually not made on the basis of which is empirically a 'more accurate' representation of something but, rather, on the basis of how someone is attempting to understand a particular phenomenon. Thus, for example, the Victorians used the metaphor of the steam engine to describe the functioning of the human body – muscles were described as the body's pistons, food as its fuel, lungs as the body's boilers, etc. – because that was a technology with which they were very familiar. Today, the body is often described using metaphors from the world of computing – for example, the brain is seen as the 'central processing unit'. In thinking about the metaphors used to describe the rescaling of contemporary social life, therefore, it is important to realize that changing the metaphors we use to describe the world does not change the way the world actually is, but it does change the ways in which we engage with the world. Hence, no one would (I hope) suggest that switching from thinking of the body in language redolent of the age of the steam engine to thinking of it using the language of computers actually changes the way bodies function, but such a switch in language does clearly alter fundamentally the ways in which we think about how they work.

In the case of metaphors used within geography to describe the relationship between the global and the local, several have been fairly commonplace. One of those used most frequently has been that of scale as a hierarchical ladder, such that one 'climbs' up the scalar hierarchy from local to regional to national to global, or down it from global to national to regional to local. In such a metaphor, the various scales are considered to be like the rungs on the ladder and there is a strict progression between them (see Figure 12.1). In using such a metaphor, the global – as the highest rung on the ladder – is seen to be 'above' the local and all other scales. At the same time, each scale is

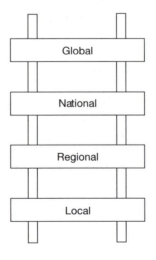

**FIGURE 12.1**  Scale as a ladder

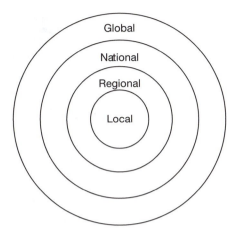

**FIGURE 12.2** Scale as concentric circles

seen to be distinct from every other scale. Clearly, then, the use of such a metaphor leaves us with a particular way of conceptualizing the scalar relationships between places. However, if we choose a different popular metaphor to conceptualize scalar relationships – that of scales represented as a series of ever-larger concentric circles – we get an understanding which is in some ways similar to, and in some ways quite different from, that derived from the ladder metaphor (see Figure 12.2). Thus, in this second metaphor the local is conceived of as a relatively small circle, with the regional as a larger circle encompassing it, while the national and the global scales are still larger circles encompassing the local and the regional. In some ways, this second metaphor has similarities to the first – scales are still seen as being quite separate entities, for example (e.g. the circle representing the local is quite distinct from that representing the global). Yet there are also some distinct differences. Whereas in the ladder metaphor the global was seen as being 'above' other scales, this is not the case with the circle metaphor. Instead, the global is seen to encompass all other scales but is not necessarily seen as being 'above' them. In these metaphors, then, we have two varied understandings of the relationship between the global and the local.

Of course, the metaphors of the ladder and of concentric circles are not the only metaphors we could use to think about the relationship between the local and the global. Frequently, scales are also talked about as being part of a 'nested hierarchy' which can be thought of in terms similar to that of Russian Matryoshka ('nesting') dolls (see Figure 12.3). In such a representation, each doll (i.e. each scale) is separate and distinct and can be considered on its own. However, the piece as a whole is only complete and can only be comprehended in its totality with each doll/scale sitting inside the one that is immediately

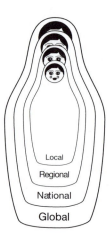

**FIGURE 12.3**  Scale as Matryoshka (nesting) dolls

bigger than itself, such that the dolls and scales fit together in one and only one way (that is to say, a larger doll/scale simply will not fit inside a smaller one). If we think about scale in this way, we have a situation in which there is no scale which is 'above' any other in the vertical sense that the ladder metaphor suggests. Likewise, the global scale (the outside doll) is viewed as being 'larger' than all the other 'smaller' scales, such that the global can contain other scales but this does not work the other way round (i.e. the local cannot contain the global). On the other hand, the Matryoshka doll metaphor implies much more forcefully than does either the circle or the ladder metaphor that there is a *nested* hierarchy of scales, with each scale fitting neatly together to provide a coherent whole.

Still another metaphor for thinking about scale, one popularized by French social theorist Bruno Latour, is that of a world of places which are 'networked' together. Thus, he argues (1996: 370), the world's complexity cannot be captured by 'notions of levels, layers, territories, [and] spheres', and should not be thought of as being made up of discreet levels (i.e. scales) of bounded spaces which fit together neatly. Rather than portraying scales as capable of somehow being stacked one above the other (as in the ladder metaphor), placed within one another (as in the circle metaphor), or fitted together like Matryoshka dolls, Latour maintains that we need to think about the world as being 'fibrous, thread-like, wiry, stringy, ropy, [and] capillary'. Clearly, using such a metaphor gives us yet another way of thinking about the scaled relationships between places. Thus, drawing upon Latour's imagery, we might think of scale as more akin to a set of earthworm burrows or tree roots which are intertwined through different strata of soil (see Figures 12.4 and 12.5). Such metaphors leave us with an image of scale in which the global and local, together with other scales, are not

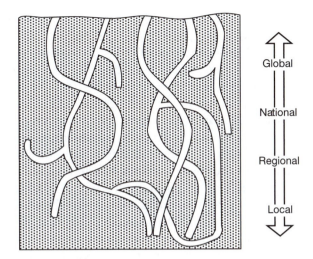

**FIGURE 12.4**  Scale as earthworm burrows

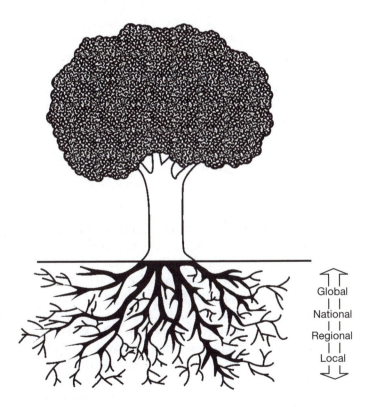

**FIGURE 12.5**  Scale as tree roots

separate from one another but are connected together in a single whole. Moreover, although it is possible to recognize that different scales clearly exist (much like it is possible to think of earthworm burrows or tree roots penetrating different strata of the soil, with some going deeper than others), it is difficult to determine exactly where one scale ends and another begins.[12] This is a quite different way of conceptualizing scale from the other metaphors discussed above. Hence, if we adopt Latour's representation then we are no longer really talking of scale in terms of bounded spaces or in terms of the metrics of the clearly defined hierarchies of Euclidean space that were at the heart of the ladder, circles and Matryoshka doll metaphors.[13] Such a different metaphor dramatically changes the way we think about scale and the relationship between the global and the local, leading us to question what exactly it would mean to talk of a 'larger' scale if we use the metaphor of earthworm burrows, or whether it makes sense to talk any more of the global scale as 'encompassing' the local if both are seen to be 'located' at the respective ends of such burrows.

Again, the significance of using different metaphors to talk about the scaled relationships between places is not to suggest that these different metaphors represent empirically different situations or that one is necessarily a better representation of the world and all its complexities than is another. Rather, such an appreciation of metaphor is important because it suggests that how we talk about scale impacts upon the ways in which we engage socially and politically with our scaled world, and that this, in turn, may impact on how we conduct our social, economic and political praxis and so make landscapes.

## CONCLUSION

In sum, there are clearly a number of important issues concerning how we think about geographical scales in general, and the local and the global in particular. These issues go not only to the heart of what scale actually is (a real social product or a taxonomic mental device) but also highlight the fact that how we conceptualize the relationship between the local and the global scale will, in large measure, determine how we understand the processes, both social and natural, which structure human and physical landscapes. Furthermore, in this age of globalization and anti-globalization protests, it is apparent that the ability to encourage people to think one way rather than another about the relationship between the local and the global can have immense political consequences.

# Summary

- There has been a debate in geography about whether scale is a real thing made through political and economic processes or is merely a mental device for imposing order on the world.
- There are six different ways in which geographers have thought about the relationship between the global and the local:

  1   the global and the local are not actually things but ways of framing situations;
  2   the global and the local each derive meaning from what they are not;
  3   the global and the local offer different points of view on social networks;
  4   the global is local; scratch anything global and you find the local; multinational firms are actually multilocal rather than global;
  5   the local is global; the local is only where global processes 'touch down' on the Earth's surface; and
  6   all spaces are hybrids of the global and the local; they are glocal.

- Typically, in Western thought, the global has been thought of as more powerful and active than the local; the local is seen as small and relatively powerless. However, the local can serve as a powerful scale of political organization; the global is not a scale just controlled by capital, and those who challenge capital can also organize globally.
- Five different metaphors have been popular in representing how the world is scaled. These are: a ladder, concentric circles, Matryoshka (nesting) dolls, earthworm burrows and tree roots.

## Further reading

Smith (1992) is an important early piece which examines how scale is produced out of political conflict between different social actors. Smith argues, among other things, that much of the postmodern theorizing in the social sciences during the late 1980s/early 1990s had focused upon issues of difference, using a spatialized language of 'subject positionality', 'location', etc., yet had failed to think seriously about the production of material spaces and scales. One of the first collections of papers to deal explicitly with the scale question can be found in a special issue of *Political Geography* (1997), Volume 16, issue 2. The papers focus upon a number of different empirical examples, including immigration politics in the European Union, labour union politics in the US docking industry, Italian electoral politics and the politics of the US anti-nuclear movement. Cox's work has developed the conceptualization of scale. His paper in *Political Geography* (1998a) lays out a conceptual argument for understanding politics as concerned with how social actors negotiate the connections between their various spaces of dependence, in the process producing spaces of engagement. His edited book, *Spaces of Globalization* (1998b), contains nine chapters which examine how the local as a scale of political and economic organization is still important under processes of globalization. The chapters also challenge a number of popularly held views about globalization, particularly that it is solely capital which has engaged in processes of globalization. Marston (2000) argues that much of the writing during the 1980s and 1990s concerning the politics of scale focused almost exclusively on what Marxists would call the 'sphere of production'. Marston argues for a broader consideration of the politics of scale in what Marxists would call the 'sphere of consumption' and the 'sphere of social reproduction'. Gibson-Graham (2002) presents six prevalent ways of thinking about the global and the local; she offers both intellectual and political strategies for undermining the presumed power of the global and liberating the transformative potential of the local. Herod and Wright's (2002a) *Geographies of Power: Placing Scale* contains 11 essays which explore three aspects of the politics of scale, namely, theorizing scale, rhetorics of scale and scalar praxis.

*Note*: Full details of the above can be found in the references list below.

## NOTES

1   Parts of this chapter draw upon Herod and Wright (2002b).
2   In July 1999, José Bové, a French sheep farmer, together with nine members of the Peasants' Confederation, attacked a half-built McDonald's restaurant in Millau, southwestern France. The protest was over a number of things, including tariffs placed by the US government upon French exports such as Roquefort cheese and the general 'Americanization' of French culture that McDonald's has come to represent. In 2000 Bové was convicted and sentenced to 90 days in gaol, though he did not begin serving his sentence until 2002.
3   Ontology refers to the science of what exists and the nature of being. It is a term often used in conjunction with 'epistemology', which refers to theories of knowledge (e.g. positivism, realism, Marxism, feminism) that allow us to make sense of the world.
4   Idealism and materialism are two of the most significant ontologies to have shaped Western thought during the past two hundred years. Exemplified by the writings of Immanuel Kant, idealism assumes that any pattern we see in the world is a result of how our mind categorizes and orders spaces. Indeed, Kant viewed space and time as simply mental frameworks for structuring existence, rather than observable realities. Materialism, on the other hand, is exemplified by the writings of people such as Karl Marx, who argued that there was indeed a real, material order to the world, one which was imposed by the workings of economic and political processes (such as those under capitalism).
5   Smith (1990: 97–154) argued that there were two contradictory tendencies within capital, those towards spatial differentiation and those towards spatial equalization. Thus, he argued that capital had to be sufficiently embedded in space so that accumulation could occur (a process which led to the differentiation of the economic landscape as some places received more investment than others) yet also remain sufficiently mobile that it could take advantage of opportunities for investment that opened up elsewhere (a process which led capital to try to equalize the rate of profit across the landscape). The geographical resolution of these opposing tendencies

was seen in the landscape through the production of scales at various spatial resolutions (the urban scale, the regional scale, the national scale and the global scale).

6    The term 'discourse' relates to how particular issues are represented through language.

7    One could just as easily argue for other scalar extremes, such as the body on the one hand and the extra-global on the other, given that our planet is now surrounded by myriad artificial satellites and that a geopolitics of space appears now to be being constructed (e.g. through some nations' efforts to lay claim to the resources of various parts of the Earth's upper atmosphere – for commercial and defensive purposes – and beyond). However, it is the local and the global that appear to have caught the scalar imagination, at least with regard to understanding processes of globalization.

8    The following draws extensively on Gibson-Graham's (2002) argument. For ease in reading, I have not included quotation marks except for quotations of authors cited directly by Gibson-Graham.

9    'Glocal' is a combination of the two terms 'global' and 'local'. It expresses the tension between the two scalar extremes, such as when social actors like transnational corporations have a global outlook but must tailor their products or practices to local situations and conditions.

10   Neoliberalism is the ideology of free trade. It is the belief that government should intervene in the market as little as possible.

11   Positivism was a popular epistemology within human geography in the 1950s and 1960s. It argued that the world was knowable in all its facets and that the application of 'science' would reveal this world to us. Positivists believe that the world can be understood 'objectively'. Postpositivism, on the other hand, suggests that it is never possible to know the world in its entirety and that our understandings of things will always be partial. Furthermore, many postpositivists argue that there are multiple truths to the world, and that different representations of the world are, to all intents and purposes, different truths about that world. Consequently, they argue that language is an important arena of political struggle, for how we represent things will shape how we understand them.

12   In Figures 12.4 and 12.5 the 'global' has been represented as being closest to the surface of the Earth, whereas the 'local' scale is seen as being deeper in the soil. Such a view could be taken to indicate that the 'local' scale is somehow more embedded and runs deeper in people's everyday lives than do the other scales. However, both these figures could just as easily have been drawn to indicate the 'global' scale stretching deepest into the soil, with the 'local' scale closest to the surface of the Earth (i.e. with the arrows to the side of each diagram flipped the other way up). In such a situation, it could be argued that the deeper 'global' scale represents a more extensive spread of worm burrows or tree roots (which can be taken as surrogates for social and political relations), whereas where the burrows or roots come together and break the surface of the ground could represent the 'local', the point where global social relationships and processes become visible to the eye. Clearly, the metaphor of the worm burrows or of the tree roots can, in fact, express two quite different situations, neither of which could be taken to be more 'accurate' in some absolute empirical sense. Such is the power of metaphor.

13   The term 'Euclidean space' refers to the belief that space is absolute and that it can be carved up into smaller and smaller parts of separate, self-contained units. This is in contrast with views which see space as relational – that is to say, space as

produced by social processes which link different actors and which, therefore, cannot be carved up neatly in to infinitely divisible units.

## References

Brenner, N. (1998) 'Between fixity and motion: accumulation, territorial organization and the historical geography of spatial scales', *Environment and Planning D: Society and Space*, 16: 459–81.

Cox, K.R. (1998a) 'Spaces of dependence, spaces of engagement and the politics of scale, or: looking for local politics', *Political Geography*, 17(1): 1–23.

Cox, K.R. (ed.) (1998b) *Spaces of Globalization: Reasserting the Power of the Local*. New York: Guilford.

Cox, K.R. and Mair, A. (1988) 'Locality and community in the politics of local economic development', *Annals of the Association of American Geographers*, 78: 307–25.

Dirlik, A. (1999) 'Place-based imagination: globalism and the politics of place.' Unpublished manuscript, Department of History, Duke University, Durham, NC.

Gibson-Graham, J.K. (2002) 'Beyond global vs. local: economic politics outside the binary frame', in A. Herod and M.W. Wright (eds) *Geographies of Power: Placing Scale*. Oxford: Blackwell, pp. 25–60.

Hart, J.F. (1982) 'The highest form of the geographer's art', *Annals of the Association of American Geographers*, 72: 1–29.

Herod, A. (2000) 'Implications of just-in-time production for union strategy: lessons from the 1998 General Motors–United Auto workers' dispute', *Annals of the Association of American Geographers*, 90: 521–47.

Herod, A. (2001) *Labor Geographies: Workers and the Landscapes of Capitalism*. New York, NY: Guilford.

Herod, A. and Wright, M.W. (eds) (2002a) *Geographies of Power: Placing Scale*. Oxford: Blackwell.

Herod, A. and Wright, M.W. (2002b) 'Placing scale: an introduction', in A. Herod and M.W. Wright (eds) *Geographies of Power: Placing Scale*. Oxford: Blackwell, pp. 1–14.

Jones, K.T. (1998) 'Scale as epistemology', *Political Geography*, 17: 25–8.

Judd, D.R. (1998) 'The case of the missing scales: a commentary on Cox', *Political Geography*, 17: 29–34.

Latour, B. (1993) *We Have Never Been Modern* (trans. by C. Porter). Cambridge, MA: Harvard University Press.

Latour, B. (1996) 'On actor-network theory: a few clarifications', *Soziale Welt*, 47: 369–81.

Marston, S.A. (2000) 'The social construction of scale', *Progress in Human Geography*, 24(2): 219–42.

*Political Geography* (1997) Special Issue on 'The political geography of scale', *Political Geography*, 16(2).

Pudup, M.B. (1988) 'Arguments within regional geography', *Progress in Human Geography*, 12: 369–90.

Smith, N. (1988) 'The region is dead! Long live the region!' *Political Geography Quarterly*, 7(2): 141–52.

Smith, N. (1989) 'Rents, riots and redskins', *Portable Lower East Side*, 6: 1–36.

Smith, N. (1990) *Uneven Development: Nature, Capital and the Production of Space* (2nd edn; originally published in 1984). Oxford: Blackwell.

Smith, N. (1993) 'Homeless/global: scaling places', in J. Bird et al. (eds) *Mapping the Futures: Local Culture, Global Change*. London: Routledge, pp. 87–119.

Smith, N. and Dennis, W. (1987) 'The restructuring of geographical scale: coalescence and fragmentation of the northern core region', *Economic Geography*, 63: 160–82.

Taylor, P.J. (1981) 'Geographical scales within the world-economy approach', *Review*, 5: 3–11.

Taylor, P.J. (1982) 'A materialist framework for political geography', *Transactions, Institute of British Geographers*, 7: 15–34.

Wills, J. (1998) 'Taking on the CosmoCorps? Experiments in transnational labor organization', *Economic Geography*, 74: 111–30.

# Social Formations: Thinking About Society, Identity, Power and Resistance

## Cindi Katz

### Definition

Society is used to describe social relations at a range of spatial scales and even to conceptualize stretched-out social networks that are not place based. Societies are often produced through a sense of sameness or shared identification. Identity is a complex and contested term that concerns how we understand and construct who we are. Identities are relational – that is, the Self is always defined in terms of what it is not which, in theoretical terms, is called 'the Other'. In other words, social formations – selves and others – are produced through social relations of identification and difference. Questions of identification and difference necessarily entail issues of power and resistance. In these arenas, power is often conceptualized as 'disciplinary power', a diffuse and unlocatable force that permeates all levels of society and that is reproduced in indirect and often erratic ways through multiple mediatory networks. In this way, all individuals are in a position of simultaneously exercising and being subject to power. Nevertheless, it is important to remember that, under contemporary conditions, the most formidable sources of power remain capitalist production, patriarchy and racism. Resistance places the emphasis on the way individuals and groups find ways of reworking their situation in the face of power and oppression from these and other sources.

#### INTRODUCTION

Society is a deceptively simple term. Common usage defines society as a readily identifiable social group, the nature of which depends upon the temporal and geographical scale at which we address the question. In a historical sense, society is defined by time periods. Thus we refer to ancient society, medieval society, contemporary society and so forth. Some social scientists also continue to speak of societies in terms of a sense of development or progress in which they refer to complex and simple societies, the status of which depends upon such things as geographic location, level of integration with other social groups and political economic processes, characteristic relations of production and their level of development, and internal coherence (see Chapter 8 for a critique of this mode of thinking). In a geographic sense we refer to broad social formations that encompass geographical or territorial divisions. When we talk about urban society, rural society, Texan society, Hungarian society, European society, etc., we are imputing social relations at a host of different scales from the local, the state and regional scales, up to the national or even trans-national scale. The term society is also used to refer to those who inhabit common spaces either by choice or coercion for various periods of time, such as prison societies, college societies or workplace societies. At the same time we also use society to describe social relations without propinquity – in other words, people can develop and maintain social ties and networks that have no geographical basis. Think, for example, of so-called cybersocieties – online communities of interest – or campaign groups and professional societies (the Society for the Prevention of Cruelty to Animals or the Society for Chartered Accountants) that may only get together face to face occasionally. And then, of course, we also use society as an all-encompassing term to describe the human race.

My brief introductory reflection on all the ways that we use society in everyday life illustrates how little thought is often given to the sorts of relations that might make these groups coherent or recognizable as societies, their internal differences or the webs of connection that bind them to other social formations. In this chapter I want to explore the notion of society in more detail before considering ideas about identity and difference, power and resistance on which it rests. In doing so, I draw out the ways in which societies, identities and power constitute, are constituted within and reconstitute space and place.

## SOCIETY: SAME AND OTHER

Demarcating a given social group as a society presumes a sense of sameness – a shared sense of identification, interests or belonging at a particular geographic scale. To be American, Norwegian or Thai is to be part of a nation. Membership of a particular society in this way helps to construct our identity even if what that means is quite individualistic, shifts over time and is frequently hyphenated as in

Norwegian-Americans. Likewise, to be a member of voluntary or formally constructed societies such as a philatelic society, the Royal Geographical Society or an association of race-car drivers is not only to mark one's identity by choice but to seek out others who do the same so that specialized interests may be shared. Such societies offer arenas – where particular normative patterns of behaviour, practices and relations are produced – that enable people simultaneously to share common interests and construct their identities in particular ways that make them philatelists, geographers or race-car drivers. So, for example, if you regularly attend the Royal Geographical Society meetings, take part in email discussion lists run by its members or receive its newsletter, you may start to build up a particular shared knowledge or history with other members of the society, to use a shared language or vocabulary, or even to start to dress like other geographers. In turn this means you might begin to feel part of this society, that you belong, that you identify as a geographer while at the same time other people may start to recognize from the way you talk, dress and the things you do, etc., that you are a geographer. In other words, there is a mutually constitutive relationship between the making of particular identities and the social groups that render them meaningful and set the norms and parameters of their production and reproduction.

Of course, however, even though societies are linked by common concerns or a shared sense of sameness, this does not necessarily mean that all their members share the same geographical space/ territory, meet each other face to face, know one another or would even like or actually identify with each other if they did encounter each other. Rather, societies might be conceptualized as 'imagined communities', to use Benedict Anderson's phrase. Anderson (1983: 15) suggests that nations are '[I]magined because the members of even the smallest nation will never know their fellow-members, meet them, or even hear of them, yet in the minds they each carry the image of their communion'. He further suggests that nations are imagined communities because fellow citizens usually have a deep sense of comradeship or identity with each other even though, in practice, there may be differences, exploitation or inequality between them. In this way, the concept of imagined communities acknowledges the way members of societies identify and reproduce a sense of sameness with each other while, at the same time, recognizing the limitations of this identification. Anderson's work is problematic, however, in so far as he places far too much emphasis on the *idea* of the nation than on the historical geographies of nationhood. Nations – and membership in them, as with other communities – are produced and reproduced through a host of uneven political, economic, sociocultural and environmental relations. These relations – which can be violent, exclusionary and ideological – are often mobilized to invoke, and sometimes even coerce, a sense of nation and belonging that is then naturalized by the very same social relations (cf. Marston, 1990).

Recognizing that social relationships are defined and worked out in and through place, space and nature, geographers more than other social scientists are interested in the ways societies and geographies are mutually constitutive (e.g. Massey, 1984; Pred, 1985; Lefebvre, 1991; Harvey, 1992). To this end, research addresses how particular geographies sustain or weaken social relations and interactions, and how these social relations and power struggles work, in turn, to make and remake their geographies. Geographers use the term spatiality to denote this understanding of the intertwining of the social and the spatial which makes clear that society and space are simultaneously produced (Keith and Pile, 1993).

Societies, however, are not only produced through processes that foster a sense of sameness but also those that emphasize differentiation from others. Each society tends to define itself against other social formations. The construction of any particular notion of society also rests on making boundaries – however porous – between those who belong, those who are the same; and those who do not, those who are different. To be human is not to be a Martian or plant. Asian society is not European society, rural society is not urban society. Yet each of these societies is unintelligible without the other. In other words, all societies are relational in that they are always constructed and understood in terms of their sameness to, and difference from, others. This is perhaps best understood by looking at the work of the Palestinian literary theorist, Edward Said. In his book *Orientialism*, Said (1978) reflects on the way that the Orient or the East has been an object of fascination to the West for centuries. Drawing on the evidence of paintings, photographs and writings (novels, poetry, etc., as well as academic work) he highlights how Western visitors have represented the Orient as exotic, mysterious, decadent, corrupt, barbarous and so on. In doing so, he argues that the traditions of thought and imagery give the Orient a reality – in particular, the discourse of the mystic Orient – which is, in fact, a European invention, a product of the European imagination. Moreover, he argues that, by containing and representing the Orient through this dominant framework, European culture has gained in strength and identity, setting itself off against the Orient in a series of asymmetrical relationships in which it is always seen in a favourable light (i.e. as civilized and ordered compared to the barbarism and corruption of the East, etc.) (cf. Clifford, 1988). Thus Said argues that '[B]y dramatizing the distance and difference between what is close to and what is far away' (p. 55) imaginative geographies not only produce images of the Other but also of the Self. I will return to the notion of Self and Other in the following section when I focus more specifically on the question of identity.

## IDENTITY

In recent years the notion of identity, and with it identity politics, has become quite popular within the social sciences. In this work, identities such as gender, 'race' and sexuality are understood to be social constructions – a product of the way we understand and interpret our bodies and subjectivities – rather than natural or biological essences (for further discussion of essentialism and social constructionism, see WGSG, 1997; Blunt and Wills, 2001; Valentine, 2001). By conceptualizing identity in this way theorists have problematized understandings of subjectivity as fixed, rooted in place and ahistorical. Rather, identity is regarded not only as changeable over both time and space but also as potentially voluntaristic in that individuals may work out new forms of identification and differentation within and against the social relations of their everyday lives.

Further, this understanding of identity as a social construction has led to a greater recognition of the fact that people are multiply positioned in the world. We are all simultaneously classed, raced, gendered, have a sexuality and so on. For some theorists this recognition led to notions of fluidity or mobility wherein identity appeared almost as a sort of Brownian motion in which people were imagined to shuttle among their multiple identifications depending upon their circumstances and whim. For instance, someone who is simultaneously Welsh, British, working class and a young man might in these terms be understood to identify as Welsh when he is watching the national football team, British when he is travelling on holiday, working class when he is at home in his neighbourhood community or as a teenager when he is out clubbing with his friends. Other writers, however, have pointed out that singular understandings of social identification along the lines of, say, class or gender are inadequate to explain and understand the complex power dynamics and injustices of social life. For example, black feminists have argued that it does not mean the same thing to be a black woman as it does to be a white woman or a black man, identities are not merely something one opts into or out of at different moments, nor are they additive; rather, being black changes what it means to be a woman and being a woman changes what it means to be black, and each of these are altered by class position and so on (cf. Mohanty, 1988).

In a similar vein, David Harvey (1993) has argued that the simple recourse to identity in a 'vulgar' way – as if it can be compartmentalized into separable dimensions – is a dead end politically. The ideas associated with what might be thought of as 'vulgar' identity politics have all too often led to debates about the relative importance of particular identities and forms of oppression rather than a critical examination of the dialectical relations of power that work within and across particular identifications. Not only can such narrow-gauge politics lead to the cul-de-sac of competitive victimization but also the results of such politics can be deadly. David Harvey, for example, often draws upon the example of the 1991 fire at Imperial Foods, a chicken processing plant in Hamlet, North Carolina, to ask why there was so little outcry – let alone political mobilization – when 25 people were killed and another 56 seriously injured in a fire whose deadliness was directly attributable to illegal and grossly negligent working conditions. He argues that the stubborn focus on 'difference' of so much poststructuralist theory, and the ways this has been problematically adopted in identity politics, seems to have evacuated the grounds of class politics that might have redressed the abhorrent working conditions at Imperial Foods and in the US poultry industry more broadly. Far from rejecting notions of difference and situatedness, however, Harvey, following Marx and Hegel, suggests that a dialectical understanding of difference and identity – one that understands them as power relations between oppressor and oppressed, exploiter and exploited, *and* at the same time recognizes that every particular identity embraces otherness or difference within it – could invigorate class politics in important ways. Such a politics – attentive to the ways that multiple forms of oppression combine and work against one another in the workplace – might enable the development of alliances and solidarities between different social groups rather than eclipsing or undermining such alliances. Harvey (1993: 47) seeks to avoid the erosions and political paralysis that he sees in part as the result of the fragmentation of 'progressive politics around special issues and the rise of the so-called new social movements focusing on gender, race, ethnicity, ecology, multiculturalism'. Such fragmentation, he argued, obscured more universal notions of social justice that are important to political mobilization.

Other writers have argued that focusing on hierarchies of oppression – assuming that particular social relations matter more than others – has instigated what has threatened to become a new form of essentialism, valorizing particular social locations, such as class or gender, at the heart of multiple oppressions rather than theorizing carefully which differences matter under what conditions. More productively, still others have developed the notion of intersectionality as a means of understanding multiple identifications (Crenshaw, 1995). Rather than a celebration of difference or a compounding of oppressions, Crenshaw's analysis carefully examined how particular social relations of oppression and exploitation associated with different identifications intersected under specific circumstances and might in the process be compounded, work at cross purposes, be contradicted or otherwise altered (cf. Bondi, 1993).

These debates aside, geographers have particularly sought to think about identities in terms of how they are formed, maintained and contested in and through spatial relations, among other things. Space works in a number of ways to maintain the various inequalities, uneven power relations and social injustices associated with contemporary forms of domination, oppression and exploitation, notably capitalism, sexism, racism, imperialism and heterosexism (cf. Keith and Pile, 1993; *The Professional Geographer*, 2002). Questions of identity are often at the nexus of exclusionary practices in space. Sibley (1995) has used psychoanalytic understandings of Self and Other to theorize the tendency of powerful or hegemonic social groups to purify and dominate space. He defines the purification of space as the rejection of difference and the securing of boundaries to maintain homogeneity. Such practices, which are rooted in uneven social relations of production and reproduction, take many forms. For instance, through various modes of residential segregation that result from covenants, bank-lending policies, gating or informal means such as the results of income disparities or certain cultural practices, some groups may be excluded from particular residential neighbourhoods. Smith's (1987) work on the UK housing market and Davis's (1990) study of Los Angeles both provide examples of different ways that residential neighbourhoods become segregated by race and class.

Exclusionary practices are also evident in non-residential spaces. Public environments, such as particular streets or parks, may have normative codes of behaviour or be policed by public or private security services (see Fyfe and Bannister, 1998; Katz, 2001b) in ways which serve to exclude groups or individuals who are regarded as undesirable Others. For example, Valentine (1996) has shown how young people often have their access to spaces such as shopping malls restricted by the police and public security guards because they are regarded as disorderly or disruptive in public space. Likewise, Parr's (1997) research shows how people with mental ill-health can be denied the freedom enjoyed by other citizens to use and occupy everyday spaces such as city centres; while Don Mitchell (1997) has shown how anti-homelessness laws are being used to clean up US parks and streets.

Just as we can imagine spaces associated with, excluding or embracing particular identities, we can also see identity as spatially formed and enacted. The idea of diasporic identity is one such idea (cf. Anderson, 1991; Gilroy, 1993). Diaspora literally refers to the dispersal or scattering of a population from an original homeland. But it is now used more loosely to capture the complex sense of belonging that people can have over several places, all of which they might think of as home (Clifford, 1994). Paul Gilroy (1993), for instance, speaks of the Black Atlantic to frame a diasporic African population that shares cultural and political-economic histories that are drawn on differently across specific geographies. Kay Anderson (1991) and Peter Kwong (1987) have examined the Chinese diaspora in the production of Vancouver's and New York City's Chinatowns. But as these works and others make clear, diasporic identities are not homogenized across space and time. Rather, they are altered by particularities of place and the unevenness of the ways the social relations of production and reproduction are played out in different locations and at different scales. In other words, if spatiality confers a certain notion of fluidity and the social constructedness of identity against earlier ideas of fixed identity rooted in a given essence, it also makes impossible notions of identity as infinitely malleable. In the simultaneous working out of spatial and social life, particular identifications inevitably rub up against and confront others, calling into question which differences matter under what circumstances (Mitchell, K., 1997a). For example, in Katharyne Mitchell's (1997b) work on Chinese immigrants to Vancouver, British Columbia, the question of class insistently disturbs the construction of homogenized identities of race or nation, while the differences produced around national identity often undermine potential class solidarities among the bourgeoisie of the area.

Questions of identity should, then, raise questions of politics. Why are particular identifications mobilized, under what circumstances and by whom? When identity is understood in more fixed terms such as class, gender or race it provides a more transparent basis for political organizing. However, given the by now well recognized truism that all political actors are multiply positioned, attempts to organize around more singular notions of identity may founder or produce problems of their own, on the one hand by eclipsing important differences among those mobilized, and on the other by failing to recognize potential solidarities people identified in one way might have with others who are differently identified. Effective politics rarely arise from pigeon holes. The challenge for political organizing, then, is to recognize which differences matter at particular historical and geographical conjunctures and to develop the sort of politics that can work the intersections of several modes of identification such as class, race, sexual orientation and gender (cf. Haraway, 1985; Bondi, 1993; Crenshaw, 1995).

### POWER AND DIFFERENCE

As these discussions of society and identity suggest, one of the ways that *power* operates and is felt is through the production and reproduction of *difference*. The French theorist Foucault (1977) has been particularly influential in shaping the way geographers think about power. Foucault argues that power is not something that one person or group of people holds over another or others, or that operates in a unidimensional way. It is not something that can be won or lost. Rather, he conceptualizes power as a diffuse and unlocatable force that permeates all levels of society and that is reproduced in an indirect and often erratic way through multiple mediatory networks. In this way all individuals are simultaneously in a position of exercising power while also being subject to it. Disciplinary power, as Foucault called it, is witnessed in its diffuse, capillary effects as people conform to norms.

This way of thinking about power is perhaps best understood through an example. Taken at face value prisons might be conceptualized as institutions where power is unidirectional – in other words, where the warders have power over the inmates. However, studies of prison life demonstrate that, in practice, power is much more fluid and diffuse. For example, within prisons there is often an informal economy with inmates trading commodities they have bought or own (phone cards, food, personal possessions, etc.) for commodities that are scarce and therefore precious, such as cigarettes, or those that are illegal, such goods that have been smuggled into the prison, like drugs and alcohol. Inmates who build up stockpiles of these valuable commodities – who are often known as barons – can exert power over others: bullying and blackmailing inmates who want these commodities into paying extortionate amounts for them or working for them within the prison. Indeed, sometimes the prison authorities allow such illegal trading and the bullying and violence that goes with it to take place because they may receive overtime payments if they have to carry out extra duties that result from illegal activities; or because the inmates may use their contacts outside the institution to threaten and intimidate the officers' families. Likewise, different barons within the prison will also try to advance their own power within the institution by keeping the influence of others in check. In this way, prison officers can benefit from these struggles because order is indirectly maintained in the prison. In other words, power is not something that officers just have and hold over inmates; rather, from this brief example it is apparent that officers and inmates are linked together through various webs of power that involve intimate alliances and collusions, such that prison societies encompass their own version of play of dominations and oppressions that characterizes other societies (Valentine, 1999; see also Sharp et al., 2000). Nevertheless, it is important to remember that prisons work institutionally in quite oppressive ways, reflecting and reinforcing all the inequalities produced by capitalism, racism and patriarchy and increasingly warehousing those who have been marginalized by these forces from participating in the social and political-economic life of the larger community (Parenti, 1999).

While it can be illuminating to imagine power in the way Foucault suggested, and indeed he was most informative in demonstrating how the diffuse effects of power were expressed and internalized by people and institutions, it remains important to name names so people can – as film-maker Spike Lee exhorts – 'fight the power'. Under contemporary conditions the most formidable sources of political-economic power remain capitalist production, patriarchy and racism. Each of these social relations of domination interacts – often, but not always, reinforcing each other and producing difference. The differences associated with capitalist relations of domination are those of class and, as discussed below, the geographical differences produced in the course of uneven development, including the differences of nation associated with imperialism. With patriarchal relations of domination the key differences are those of gender and sexuality. Finally, racist relations of domination produce differences around race and ethnicity. It is important for social theory to take into account each of these relations of domination and the kinds of differences they produce, not only to account for the ways these relations converge and diverge and reveal the ways they work with and against one another under particular circumstances, but also because each provides a broad and important arena for resistance and opposition.

In a particularly geographical vein, theories of uneven development have made clear that capitalist accumulation works in and through the production of difference in space – not only the development of certain areas and resources at the expense of others but also the deliberate development of particular places and regions, while others are *underdeveloped* (Harvey, 1982; Smith, 1984). The production of difference works at different scales. Indeed, geographic scale is a means of organizing spatial difference (Smith, 1992; Swyngedouw, 1997; Marston, 2000; see also Chapter 12). For instance, at the global scale the underdevelopment of the global south is bound directly with the development of the northern industrialized nations, many of which are former imperial powers. At the national and regional scales the processes of industrial development and decline are best understood within the framework of capitalist uneven development. Finally, at the urban scale gentrification is predicated on long-term patterns of investment and disinvestment in urban neighbourhoods. The spatiality of uneven development is witnessed in the coexistence of environments of value whose landscapes reflect wealth, investment and development and deteriorating natural and built environments that, under particular political economic circumstances, become targets for reinvestment and development. These produced differences, which reflect the peculiar dynamics of capitalist accumulation, are often naturalized by observers and analysts. For example, difference is explained as a result of such things as environmental conditions, social inadequacies, historical accident or explained away with recourse to an evolutionary understanding of development that imagines various places or regions at different stages in the process.

The differences produced by patriarchal relations of power are spatialized as well. One of the most significant forms of spatial difference associated with patriarchy is that between private and public space. As feminist and other theorists have made clear, this commonly naturalized distinction is a political artifact and not so easily maintained; the two spaces and the spheres they represent blur and intersect. Nevertheless, maintaining the distinction in theory and practice has worked to limit women's place and sphere of influence to the home and thereby diminish their exercise of power. Racism produces difference spatially as well through, among other things, formal and informal patterns of residential segregation, controls on international migration and the making and maintenance of racialized environments through such things as fear and violence, that produce places that are simultaneously exclusionary and inclusive depending upon one's racialization and experience (Anderson, 1991; Pred, 2000; *The Professional Geographer*, 2002).

## RESISTANCE

In the last decades of the twentieth century, geographers, like other social scientists, addressed the production of difference and its spatial consequences, complicating the assumed relationships between place and identity and making clear that all places were simultaneously unique and differentiated. The production of difference has been looked at with reference to the increasingly porous boundaries of the nation-state with the globalization of capitalist production and cultural forms and practices, as well as with reference to the production of difference through and in place. To this end, geographers have examined the contemporary romance with place as an often reactionary means to create a bulwark against the homogenizing effects of globalization (see also Chapter 9). Research on the geographies of difference demonstrates not only the ways that power works to produce difference but also how deployments of difference in the production of space, place and nature can counter and redirect power. Thus geographies of difference often suggest geographies of resistance.

If part of the operation of society is to set and maintain appropriate norms of behaviour and the social institutions that uphold them, and these norms, values and social relations generally – and not coincidently – work to ensure the endurance and reproduction of the society, we can expect *resistance* to the inequalities and uneven power relations fostered by them. Many geographers have looked at the ways that social power is produced and reinforced spatially and, more recently, some have addressed the numerous means by which uneven power relations are contested and resisted. In various ways this work examines how hegemony is spatially secured – and thus might be interrupted, compromised and undone on spatial as well as social and political grounds.

While geographers have until recently been more interested in analysing the operations of power in and through space, scholars in cultural studies and other fields have made the material social practices of resistance their focus since the 1970s (Hall and Jefferson, 1976; Willis, 1977; Scott, 1990). The burgeoning interest in resistance during this period seems to have led to its dilution as a political practice and analytic category. In all too many texts it seemed that any 'independent initiative', to use Gramsci's (1971) phrase, no matter how small, was understood as resistance to the social relations of domination. Meanwhile, in this same period, capitalist production has gone global, the disparities between rich and poor nations and between rich and poor people within nations have increased almost everywhere; violence has spread and intensified at all scales from the personal and domestic to the national and global; and indicators of poverty remain stubbornly high and a large and growing number of young people in rich as well as poor countries seem likely to be denied the promises of modernity. For these reasons it is important to parse resistance to develop more subtle and workable distinctions to understand its effects.

In my research on social reproduction and global economic restructuring in rural Sudan, I have distinguished among resilience, reworking and resistance as ways of understanding the material social practices of everyday life through which people confronted and coped with shifting relations of domination associated with, among other things, global economic restructuring, civil war and the local intrusion of a state-sponsored agricultural development project (Katz, 2001a; 2003). In the village where I worked, deforestation and pasture deterioration along with the restrictions on arable land associated with the agricultural project and its effects compromised residents' ability to stay in the village and continue working the mix of agriculture, animal husbandry and forestry that had long sustained their village. I anticipated that as young adults came of age they would have to leave for urban areas, where they would be ill-prepared to find sustaining work. Instead, the local population adjusted to these rather dramatic shifts by radically expanding the terrain of their work. In less than two decades the area they drew on to cut wood, produce charcoal, graze animals and cultivate was a startling 1600 times larger than what it took before (Katz, 2003). Through this spatial strategy of resilience, young people were able to remain in the village even if their work trajectories included occasional stints in nearby towns. In another vein, as adults in the village witnessed these changes and the ways their children were potentially deskilled in the process, they effected a series of self-help initiatives designed to increase school attendance and, in particular, to educate girls. Through the installation of stand-pipes that saved much of the time girls spent fetching water, the construction of separate classrooms for girls with female teachers hired to teach them and, finally, the construction of a secondary school in the village, school attendance was increased dramatically. These efforts of reworking demonstrate the flexibility of people's responses to the shifting conditions of their everyday lives. Finally, there were some events in the village and more broadly that called forth practices that might be understood as resistance in that they were consciously directed at altering a condition that people recognized as oppressive. For instance, the Ministry of Agriculture initially required all tenants in the agricultural project to cultivate cotton and groundnuts exclusively, forbidding the cultivation of the staple dietary crop, sorghum. During the first decade of the project's operation the price of sorghum increased 2000% while cotton and groundnut prices remained relatively stagnant. Tenants organized and sent representatives to the Ministry of Agriculture to fight for the right to cultivate sorghum on project lands. They eventually won the right to devote their groundnut allotment to sorghum. As this brief summary might suggest, distinguishing among strategies that enable people to survive under difficult circumstances, those that rework particular conditions that compromise the conditions of their existence and those that resist relations of power that are exploitative or oppressive to them, can render clearer the political effects of varied material social practices exercised in the face of power. More nuanced understandings of these effects can be of use in the development of political strategies to redress social injustices and economic exploitation.

## CONCLUSION

This chapter has provided an overview of several interrelated concepts: society, identity, difference, power and resistance. It presented society as a social convention that demarcates people who share particular interests and/or identifications. Membership in any particular society thus simultaneously confers and rests upon social identity.

Identity is likewise understood to be a social construction, not resting upon some sort of biological essence or any sort of 'natural' distinction. Rather, identity reveals such invocations of nature as social to their core, raising the question of why certain identities are made manifest while others are obscured. Identity, I have argued, is more productively understood in a dialectical manner. It produces sameness (identity) out of difference. Thus the construct of identity is simultaneously the construction of difference. One of the most breathtaking and influential dialectical theorizations of difference was Edward Said's *Orientalism* (1978). In this brilliant book, Said makes clear that the 'Orient' was not only a product of the Western imagination but also that it was through producing it as such that the 'West' made itself intelligible to itself.

If any notion of society embraces particular formations of identity and difference, it is in their production, reproduction and alteration that power and resistance are worked out. I have suggested that power operates through the production of particular forms of difference and that, for geographers, it is particularly important to understand how power is spatialized. While power may be reinforced spatially, it can also be redressed through spatialized strategies. I draw on my own research in Sudan to argue that contemporary understandings of resistance have become sloppy and offer a means of parsing responses to various social relations of domination and exploitation as resilience, reworking and, finally, resistance.

## Summary

- All societies are relational in that they are produced through uneven social relations that create and maintain particular identifications and formations of difference. These identifications and differences become part of people's collective imaginings and understandings of themselves and others.
- Identities are social constructions rather than natural or biological essences. This understanding of identity problematizes subjectivity as fixed and ahistorical.
- A recognition that we are multiply positioned has led to a debate about hierarchies of oppression (which forms of oppression are most important) and to diverse ways of theorizing multiple oppressions.
- Power is theorized as a diffuse and unlocatable force that permeates all levels of society and that is reproduced in an indirect and often erratic way through multiple mediatory networks. The limitations of this Foucauldian idea are highlighted by analysing how particular relations of domination, such as those

associated with capitalism, patriarchy and racism, work to produce difference and inequality.

- Resistance is a broad category that, until recently, has been used by social scientists to describe any form of independent initiative. It is now being unpacked to emphasize such elements as the way people are *resilient* in the face of power and *rework* situations of oppression.

## Further reading

Said's classic book, *Orientalism* (1978), is a good place to start when thinking about questions of sameness and difference. *Place and the Politics of Identity*, edited by Keith and Pile (1993), is a collection of essays that illustrates and debates the ways that identities are spatially as well socially constituted. There are too many books and special issues of journals that deal with specific forms of identification such as gender, 'race' or sexuality to list, but see, for example, Bell and Valentine (1995); WGSG (1997); Dwyer (1999): *The Professional Geographer* (2002). Sharp et al.'s (2000) *Entanglements of Power* provides a collection of essays, drawing on a diverse range of case studies, that explore the ways practices of domination and resistance cannot be separated and are integral to all workings of power. In a similar collection of essays (*Geographies of Resistance*), Pile and Keith (1997) focus more closely on the concept of resistance, while Katz's (2003) *Disintegrating Developments* unpacks this term into different components. Most of these terms are also considered in general human geography textbooks such as Blunt and Wills (2000) and Valentine (2001).

*Note*: Full details of the above can be found in the references list below.

## References

Anderson, B. (1983) *Imagined Communities: Reflections on the Origin and Spread of Nationalism*. London: Verso.

Anderson, K. (1991) *Vancouver's Chinatown: Racial Discourse in Canada 1875–1980*. Montreal: McGill-Queens University Press.

Bell, D. and Valentine, G. (eds) (1995) *Mapping Desire*. London: Routledge.

Blunt, A. and Wills, J. (2000) *Dissident Geographies*. Harlow: Pearson.

Bondi, L. (1993) 'Locating identity politics', in M. Keith and S. Pile (eds) *Place and the Politics of Identity*. London and New York, NY: Routledge, pp. 84–101.

Clifford, J. (1988) 'Introduction: partial truths', in J. Clifford and G.E. Marcus (eds) *Writing Culture: The Poetics of Ethnography*. Berkeley, CA: University of California Press, pp. 1–26.

Clifford, J. (1994) 'Diasporas', *Cultural Anthropology*, 9: 302–28.

Crenshaw, K.W. (1995) 'Mapping the margins: intersectionality, identity politics, and violence against women of color', in K. Crenshaw et al. (eds) *Critical Race Theory: The Key Writings that Formed the Movement*. New York, NY: New Press, pp. 357–83.

Davis, M. (1990) *City of Quartz: Excavating the Future in Los Angeles.* London: Verso.

Dwyer, C. (1999) 'Contradictions of community: questions of identity for young British Muslim women', *Environment and Planning A*, 31: 53–68.

Foucault, M. (1977) *Discipline and Punish: The Birth of the Prison.* London: Penguin Books.

Fyfe, N. and Bannister, J. (1998) 'The eyes upon the street: closed circuit television surveillance and the city', in N. Fyfe (ed.) *Images of the Street.* London: Routledge, pp. 254–67.

Gilroy, P. (1993) *The Black Atlantic: Modernity and Double Consciousness.* Cambridge, MA: Harvard University Press.

Gramsci, A. (1971) *Selections from the Prison Notebooks* (ed. and trans. by Q. Hoare and G.N. Smith). New York, NY: International Publishers.

Hall, S. and Jefferson, T. (eds) (1976) *Resistance through Rituals: Youth Subcultures in Post-war Britain.* London: Unwin Hyman.

Haraway, D. (1985) 'A manifesto for cyborgs: science, technology and socialist feminism in the 1980s', *Socialist Review*, 80: 65–107.

Harvey, D. (1982) *Limits to Capital.* Oxford: Blackwell.

Harvey, D. (1992) 'Postmodern morality plays', *Antipode*, 24: 300–26.

Harvey, D. (1993) 'Class relations, social justice and the politics of difference', in M. Keith and S. Pile (eds) *Place and the Politics of Identity.* London: Routledge, pp. 41–66.

Katz, C. (1993) 'All the world is staged: intellectuals and the projects of ethnography', *Environment and Planning D: Society and Space*, 10: 495–510.

Katz, C. (2001a) 'On the grounds of globalization: a topography for feminist political engagement', *Signs: Journal of Women in Culture and Society*, 26: 1213–34.

Katz, C. (2001b) 'Hiding the target: social reproduction in the privatized urban environment', in C. Minca (ed.) *Postmodern Geography: Theory and Practice.* Oxford: Blackwell, pp. 93–110.

Katz, C. (2003) *Disintegrating Developments: Global Economic Restructuring and Children's Everyday Lives.* Minneapolis, MN: University of Minnesota Press.

Keith, M. and Pile, S. (1993) *Place and the Politics of Identity.* London and New York, NY: Routledge.

Kwong, P. (1987) *The New Chinatown.* New York, NY: Hill & Wang.

Lefebvre, H. (1991) *The Production of Space.* Oxford: Blackwell.

Marston, S.A. (1990) 'Who are "the people"?: Gender, citizenship, and the making of the American nation', *Environment and Planning D: Society and Space*, 8: 449–58.

Marston, S.A. (2000) 'The social construction of scale', *Progress in Human Geography*, 24: 219–42.

Massey, D. (1984) *Spatial Divisions of Labour: Social Structures and the Geography of Production.* London: Macmillan.

Mitchell, D. (1997) 'The annihilation of space by law: the roots and implications of anti-homeless laws in the United States', *Antipode*, 29: 303–35.

Mitchell, K. (1997a) Different diasporas and the hype of hybridity', *Environment and Planning D: Society and Space*, 15: 533–53.

Mitchell, K. (1997b) 'Conflicting geographies of democracy and the public sphere in Vancouver, BC', *Transactions, Institute of British Geographers*, 22: 162–79.

Mohanty, C.T. (1988) 'Feminist encounters: locating the politics of experience', *Copyright*, 1: 30–44.

Parenti, C. (1999) *Lockdown America: Police and Prisons in the Age of Crisis.* London and New York, NY: Verso.

Parr, H. (1997) 'Mental health, public space and the city: questions of individual and collective access', *Environment and Planning D: Society and Space*, 15: 435–54.

Pile, S. and Keith, M. (eds) (1997) *Geographies of Resistance.* London: Routledge.

Pred, A. (1985) 'The social becomes the spatial, the spatial becomes the social: enclosure, social change and the becoming of places in the Swedish province of Skane', in D. Gregory and J. Urry (eds) *Social Relations and Spatial Structures.* London: Macmillan, pp. 337–65.

Pred, A. (2000) *Even in Sweden: Racisms, Racialized Spaces, and the Popular Geographical Imagination.* Berkeley and Los Angeles, CA: University of California Press.

*The Professional Geographer* (2002) Volume 54(1). Focus section on race, racism and geography.

Said, E. (1978) *Orientalism: Western Conceptions of the Orient.* Harmondsworth: Penguin Books.

Scott, J.C. (1990) *Domination and the Arts of Resistance: The Hidden Transcript.* New Haven, CT: Yale University Press.

Sharp, J.P., Routledge, P., Philo, C. and Paddison, R. (eds) (2000) *Entanglements of Power: Geographies of Domination and Resistance.* London: Routledge.

Sibley, D. (1995) *Geographies of Exclusion.* London: Routledge.

Smith, N. (1984) (2nd edn 1990) *Uneven Development: Nature, Capital and the Production of Space.* Oxford: Blackwell.

Smith, N. (1992) 'Geography, difference and the politics of scale', in J. Doherty et al. (eds) *Postmodernism and the Social Sciences.* London: Macmillan, pp. 57–79.

Smith, S.J. (1987) 'Residential segregation: a geography of English racism?' in P. Jackson (ed.) *Race and Racism: Essays in Social Geography.* London: Allen & Unwin, pp. 25–49.

Swyngedouw, E. (1997) 'Neither global nor local: "glocalization" and the politics of scale', in K. Cox (ed.), *Spaces of Globalization: Reasserting the Power of the Local.* New York, NY: Guilford, pp. 137–66.

Valentine, G. (1996) 'Children should be seen and not heard? The role of children in public space', *Urban Geography*, 17: 205–20.

Valentine, G. (2001) *Social Geographies: Space and Society.* Harlow: Pearson.

Valentine, G. and Longstaff, B. (1998) 'Doing porridge: food and social relations in a male prison', *Journal of Material Culture*, 3: 131–52.

Willis, P. (1977) *Learning to Labor: How Working Class Kids get Working Class Jobs.* New York, NY: Columbia University Press.

Women and Geography Study Group (WGSG) (1997) *Feminist Geographies: Explorations in Diversity and Difference.* London: Longman.

# 14 Physical Systems: Environmental Systems and Cycles

**Barbara A. Kennedy**

## Definition

Systems are sets of objects and their attributes – for example, a drainage basin is a system of hillslopes of varying angles, soil depth, vegetation cover and of channels of different dimensions. The basin is an open system through which both energy and mass move: most systems in physical geography are open, although the earth itself exchanges so little mass with space that it may be considered an isolated system. Cycles are constantly repeating series of events or changes – for example, high and low tides and the seasons. They may have return periods of millions of years or days.

## INTRODUCTION

Some of the most fundamental aspects of the physical world are obviously cyclical: the sequence of day and night; the phases of the moon; the shifting of the stars and constellations of the night sky; and, in much of the world, the movement of the overhead sun with the associated seasonal shifts in temperature, precipitation, vegetation and animal life. Until the shocking events in France in 1789, such processes were considered 'revolutionary' in that they were repeated with each 'revolution' of the globe or the planetary system (Cohen, 1985). Nowadays, we tend to take many of the basic cyclic phenomena for granted: the widespread availability of artificial lighting, for instance, makes us less concerned with the nature of the day/night cycle and yet day length remains a very important control of human psychology as well as plant and animal physiology.

Each of the main cyclic phenomena is linked to a particular subset of physical entities which we may term 'systems': the solar system, the earth–moon system, the earth itself (whether viewed as 'Gaia' –

Lovelock, 2000a – or not) and the biosphere. We can and do identify smaller components as discrete systems – drainage basins; ecosystems – although these may prove difficult to define uniquely and securely. A system is, simply, as defined by Hall and Fagen (1956) 'a set of objects, together with their attributes' (see Huggett, 1980). It is, then, only necessary to specify *which* objects will form the set and, usually, state how the set will be distinguished from all others – i.e. by the drawing of a boundary. Once identified, there are three possible kinds of systems:

1   *The closed.* A set of objects exchanging neither mass nor energy with its surroundings. Few, if any, natural systems of interest to geographers fall in this category.
2   *The isolated.* A set of objects exchanging energy, but not mass, with its surroundings. The earth itself approximates to this condition, as do the other planets, considered as entities.
3   *The open.* A set of objects exchanging both energy and mass with its surroundings. Virtually all systems in physical geography are open in this sense.

A biologist, Ludwig von Bertalanffy, suggested in 1950 that the open systems of both physics and biology had many essential features in common. He went on to advocate the development of what he termed general systems theory or GST, which had five main aims (1956: 2). To:

1   encourage the general tendency towards integration in the natural and social sciences;
2   centre this integration on a general theory of systems;
3   show how this theory might give important assistance to the development of exact theory in the non-physical sciences;
4   develop unifying principles running 'vertically' through each of the individual science; and
5   encourage the integration of material in teaching about science.

Of these, the fourth notion is of particular significance since it emphasizes the hierarchical and 'nested' nature of systems of any kind (the most obvious would be cell, organ, organism, population). A key tenet of GST is the importance of identifying the levels within each hierarchy and the appropriate control mechanisms between and within the levels (see the recent discussion, with reference to bio-diversity, by Willis and Whittaker, 2002).

Von Bertalanffy's ideas were taken up by economists such as Kenneth Boulding (1956) and by the geomorphologist, Arthur Strahler, in an extremely influential paper which called for 'an analysis of geomorphic processes in terms of clearly defined open systems' (1952: 923). This idea was expanded and introduced into British geomorphology by Strahler's pupil, Richard Chorley (first by his 1960 paper on 'Geomorphology and general systems theory' and then in a more extensive discussion across the whole of physical geography: Chorley and Kennedy, 1971). The ideas formulated by von Bertalanffy have, therefore, been around for over half a century and the term 'system' is wellnigh ubiquitous: its most recent manifestation is in the guise of earth surface systems (cf. Allen, 1997).

In order to understand the systems we identify and, in particular, in order to be able to evaluate the flux of energy and mass across their boundaries and the ways in which this drives their internal operations, it is absolutely crucial to be able to define those boundaries. Unfortunately, this proves wellnigh impossible. Even for the earth itself, the precise point at which the exosphere (the outermost layer of the atmosphere) passes into interstellar space is selected arbitrarily. The fact that the earth's systems do not reveal themselves as neat, discrete sets of objects makes for very real difficulties when we wish to evaluate the ways in which they operate: it may sometimes seem that uncertainties are so small as to be irrelevant. But the more we think we understand about the way the world works, the clearer it becomes that tiny variations in energy or mass can produce dramatic shifts in the ways systems operate. Such variations are thought to drive chaotic behaviour (see Gleick, 1998) and also to impel sharp changes across thresholds (Schumm, 1980; see also Chapter 7). So the absence of 'true' system boundaries suggests we should be extremely cautious both in our assumptions that the key elements of the physical world have been firmly and finally identified and in our views of causality. For example, if we work with one hillside, we may see the onset of gullying in terms of a particular, deterministic process such as overgrazing. However, if we look at the larger unit of a drainage basin, it may be that what we are seeing is an integral and essentially random mechanism of slope development. In the former view, the gully is, preventable; in the latter, inevitable.

That understood, it is the cyclic processes – within the limits of the fuzzy recognition of system boundaries – which are crucial to the operations of the physical world. These extend from the astronomical, to the planetary-based geological, to the biogeochemical cycles combining biotic and abiotic factors, down to the 'purely' biological cycles of flora and fauna (see Chapter 10 for more on these environmental spheres).

## ASTRONOMICAL CYCLES

The variable and cyclic character of solar activity characterized by 'sunspots' has been recognized for centuries, and there has been a persistent feeling that these rather regular 11 and 22-year cycles must have a reflection in some aspects of the earth's climate. This has proved very elusive to uphold. However, from the mid-nineteenth century onwards, there has developed a more general and convincing vision of links between the earth's status as a planetary body and its climate. There is also, of course, the early-recognized and still crucial earth–moon coupling which drives the tidal system.

### *Croll–Milankovitch cycles*

Very shortly after the recognition – by the Swiss, J.R.L. Agassiz – of the Ice Age in the 1830s, it was suggested that variations between more and less widespread ice cover might be related to the cyclic changes in the earth's astronomical relationship to the sun. The ideas were worked out by the American, James Croll, in the 1860s and then more formally extended and refined by the Serbian physicist, Milutin Milankovitch, in the 1920s (Roberts, 1998). There are three key elements of the earth's relationship to the sun: its orbit, which varies in its departure from the circular, becoming more or less elliptical on a 100 000-year cycle; the time of perihelion (or the moment when the earth is closest to the sun) which varies on a cycle of 21 000 years, known as the precession of the equinoxes, caused by a slight wobble in the earth's axis; and, finally, shifts in the angle between the plane of the earth's orbit and that of its own rotational equator ('the obliquity of the ecliptic') – this has a 41 000-year cycle (see Figure 14.1).

As with all cyclic phenomena, these three astronomical factors are superimposed, so that – at times – all of them act in concert, while at others their effects come close to cancelling each other out. It follows that calculating their impact is by no means straightforward (Figure 14.1b). However, since an influential study showed an apparently close fit between the three Milankovitch forcing functions and the record of global temperature obtained from deep ocean cores (Hays et al., 1976), there has been a general consensus that, although the scale of variation in solar radiation receipt attributable to these fluctuations *may* be small (perhaps ±10% for the Northern Hemisphere: Allen, 1997), they are the key to much of the cold–warm cycles of the latter Pleistocene (*c.* 700 000 years). What is far from clear is either how far back this link extends – calculations have been made back as far as 2.5 billion years ago (Allen, 1997: 81; Crowley, 2002) – or, precisely, how the radiation variation is converted into climatic shifts. It is interesting to note that these mechanisms should operate on our sister

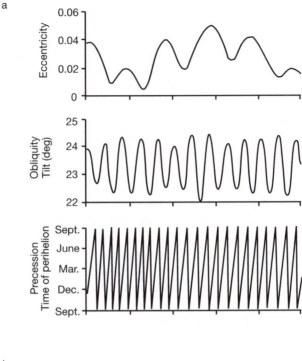

*Source*: From Allen (1997: Figs 2.23 and 2.24)

**FIGURE 14.1**   Milankovitch cycles for the earth for the past 50 000 years: (a) the frequencies of the three orbital elements; (b) the combined effect of the superimposed cycles on solar insolation

planets, most notably Mars: will the apparent banding of Mars's polar ice caps ever be related to *its* astronomical cycles? Such a link would prove very interesting confirmation of the terrestrial significance of Milankovitch cycles (see Smith et al., 2001).

### Solar emissions

The so-called solar constant, or the rate at which solar radiation is received at the outer edge of the earth's atmosphere when the earth is at its mean distance from the sun (Atkinson, 2000), is about 1400 W m$^{-2}$. It is not, in fact, constant, even in the short term; and it is estimated (Allen, 1997: 79) that there has been a 30–40% increase in overall solar luminosity in the c. 4.5 billion years since the sun and earth were formed. More recently (see Kerr, 2001) a range of evidence has come in which links shorter-term fluctuation in solar emission with a cyclic period of c. 1500 years to substantial climate shifts, at least in the Holocene. Again, how such shifts can influence climate is uncertain: one suggestion is that they create changed circulation in the stratosphere, with a knock-on effect to the troposphere and even to the oceans. What is evident is the *very* close fit obtained for solar activity and evidence of cold conditions.

At what are termed 'sub-Milankovitch frequencies' – which includes all the present Holocene interglacial (Roberts, 1998) – the role of cyclic solar output variations seems set to be given increased attention and accorded increased significance. Even the 11 and 22-year sunspot cycles are now considered serious contenders for climatic links, probably through alterations in the emission spectra.

### The earth–moon system

A fundamental and extremely influential astronomical cycle involves the relative location of the earth and moon vis-à-vis the sun: the gravitational pull exerted on the oceans – which nearly balances the centrifugal force created by the earth's rotation – causes two maxima and two minima every 24 hours in the level of the ocean (there are also virtually imperceptible effects on the solid earth and the atmosphere). The precise nature of the tides experienced in any location can be dramatically influenced by coastal topography and by the reflection and amplification of the tidal bulge (Pugh, 2000). The impact of the tidal cycle on coasts and on coastal ecosystems is profound. In human terms, areas of substantial tidal range are at risk from flooding if the high 'spring' tides of the monthly cycle coincide with other, non-periodic phenomena. The severe flooding of the Thames Estuary and large sections of the Kent and Essex coasts in the spring of 1953 was created by a combination of very high spring tides, high water levels with spring runoff in the Thames and a deep depression so located in the North Sea that strong on shore winds increased the height of the flood tide and also backed up river water in the estuaries.

At all scales of time and space, our planetary circumstances create superimposed cycles which have and have had major impacts on the distribution of energy, in particular.

## THE GEOLOGICAL CYCLE

There is, in truth, only one 'geological cycle': that which sees the creation and destruction and recreation of both rocks and relief. As far as we can judge, it has been in ceaseless operation since the origin of our planet about 4.5 billion years ago. It was first identified by the Scottish doctor, James Hutton, in 1785 (published 1788). In a famous phrase, he concluded that 'the result, then, of our present enquiry, is that we see no vestige of a beginning, no prospect of an end' (to the cycle of erosion, lithification and uplift and, hence, to the lifespan of the earth).

Hutton's views had a mixed reception (see Davies, 1969), partly because he was felt to be calling not for an indefinite, but rather an *infinite*, length of the earth's history. He was also limited in his ability to describe the processes whereby the sediments delivered to the oceans are translated into first rocks and, finally, new land. Vindication of Hutton's vision of the ceaseless and fundamental operation of the geologic cycle came, finally, in the twentieth century. First, the discovery of radioactivity was applied, by Rutherford, to the question of the planet's age and, ultimately, settled to the generally accepted figure of 4.5 billion years in the 1940s (see Burchfield, 1990). Secondly, the confirmation of continental plate movements (first proposed by the German, Alfred Wegener, in 1915) in the 1960s gave an apparent machinery whereby the continental crust could be both created and destroyed. This process – while fundamental – does not seem to have reworked all parts of the globe: as Van der Lee (2001) points out, not only are there rocks in northern Canada which appear to be over 4 billion years old, but there is also a huge region ('Laurentia') underlying much of central North America where virtually all rocks are older than 2.5 billion years and where there has been, effectively, no influence by plate tectonics in that timespan.

The nature of the operation of the geological cycle, then, varies in speed and character over time and space. Areas such as Laurentia experience minimal alterations in mass; other old 'shield areas', such as western Australia, have similar passive characteristics. But where plates collide – as in the northern Indian subcontinent or New Zealand – the violent and rapid uplift of rocks is accompanied by equally violent and rapid processes of subaerial denudation – glacial, fluvial and mass movements – with consequences not only for the landscapes but also the atmosphere as carbon dioxide is sequestered and released in the weathering process. Even where plates merely move past each other, the resulting pattern of earthquakes – as on the west coast of North America – produces substantial increase in the rate of mass movements.

The depositional processes themselves are frequently cyclic, on a huge range of temporal scales. The great Carboniferous deposits which have become the global coal measures show particularly fine examples of the regular sequences of beds known collectively as cyclothems (see Woodcock and Strachan, 2000); on shorter timescales, the annual cycle of deposition in many lakes – with one peak runoff period, succeeded by still-water conditions – gives rise to the bands termed varves; and every glacial outlet stream experiences a diurnal cycle of flow and sediment movement, with peak rates in the late afternoon and very low or no flow in the early morning.

We have every reason to suppose that the geologic cycle has operated for virtually all our planet's history. Its speed and character have, however, been substantially varied by the interaction with biota and the resulting biogeochemical cycles.

## BIOGEOCHEMICAL CYCLES

There is debate about the oldest fossils although there is general agreement about the 3-billion-year-old stromatolites; less clear is the evidence for microlife (and similar debates cloud the discussion of 'life forms' in Martian meteorites). What is certain is that the evidence for both the early evolution and the environmental range of the bacteria, in particular, implicates biological processes in the recycling of earth materials for by far the greater part of the planet's existence (see Gross, 2001). Whether these processes have resulted in the emergence of the entity termed 'Gaia' (Lovelock, 2000a; 2000b), which serves to regulate the cycling of material in such a way as to maintain the planetary surface within a narrow range of conditions favourable to life, remains debatable. But Wegener's views – and Hutton's – were equally debated when first proposed. (Lovelock (2000b: 15) now defines Gaia as follows: '[it] is best thought of as a superorganism. These are bounded systems made up partly from living organisms and partly from nonliving structural material. A bee's nest is a superorganism and like the superorganism, Gaia, it has the capacity to regulate its temperature'.)

As noted in the introduction to this chapter, the earth is a more or less isolated system. This situation demands the recycling of elements and compounds – especially once life forms enter the picture – if there is to be continuing response to the input of energy: a very different situation exists on the moon. The earth's surface shows virtually nothing of its early history. In contrast, the moon's face exhibits evidence for aeons of bombardment by meteorites in its suites of craters. Without the continuous internal energy supply to drive geophysical processes and without life, the moon is truly dead. No cycles.

On the earth, in contrast, there are cycles for all the chemical elements and also for the most vital compound – water or $H_2O$ – which can, broadly, be considered to give our planet its key characteristics. Several features of the biogeochemical cycles are significant: first, where are the principal reservoirs? Secondly, how readily can the elements or compounds move between reservoirs? Thirdly, how does the operation of one cycle mesh with those of other elements? In the case of the hydrologic cycle, the principal reservoir is the oceans (97%): of the 3% of freshwater, the majority is held as ice in glaciers, ice caps, ice sheets and permafrost. Fortunately, water moves very freely in and out of the atmospheric portion of the cycle and requires relatively limited amounts of energy to change its state from liquid to gas or vice versa. So there is about 1000 mm of water precipitated, on average, over the earth's surface each year and it is this precipitation which not only drives the key abiotic forces of weathering and erosion (by ice and running water) but also provides the crucial input in to biological processes (most protoplasm is largely $H_2O$). The water cycle intersects with just about all other biogeochemical cycles: one obvious link is with the movement of oxygen. If water flow stagnates, the supply of oxygen is cut off and, ultimately, biochemical reactions switch from oxidizing to reducing: the 'bad eggs' smell of mud in a stagnant pond represents the replacement of oxidizing breakdowns of organic material by the reducing processes that create hydrogen sulphide as a by-product.

One particular subset of multiple, linked biogeochemical cycles is that linked to fire in natural (and managed) ecosystems. In nature, forest and scrub and heath communities frequently experience a regular cycle of fire, driven by the build-up of flammable litter. Such communities rely on the fire cycle both for regeneration of key plant species (many of which have evolved a requirement for seeds to be opened by fire) and for the recycling of nutrients. Human firing of natural communities has mimicked the natural process.

The interlinkage of biogeochemical cycles is extraordinarily complex. It seems that the productivity of ecosystems may often hinge on a surfeit or lack of trace elements required in minute quantities. The German chemist, Justin Liebig, proposed in the nineteenth century that the productivity of ecosystems depended upon the nutrient in shortest supply. Liebig's Law is now extended to state that ecosystem productivity is governed by the most crucial bottleneck in biogeochemical cycles (see Huggett, 1995).

## BIOLOGICAL CYCLES

In many parts of the world, the abundance of plants and animals demonstrates a cyclical pattern. Obvious examples are the annual sequence of plants – both growing and flowering – and the appearance and disappearance of migrants such as swallows in Great Britain. The regularity or otherwise of these annual events has been a source of comment for centuries: the great eighteenth-century English naturalist, Gilbert White, was particularly perplexed by the behaviour of the swallows of Selborne (White, 1789). A careful study of the timing of key annual events in the life cycles of both plants and animals – the subject of phenology – is providing increasingly important evidence for the impact of climatic change (Peñuelas and Fillella, 2001).

But more dramatic are the longer and more mysterious cycles which are evidenced by some plants – the flowering of the 'century plant' or *Agave* every 100 years – and many animals, from desert locusts to lemmings to lynx. Alongside these we should also place the infectious organisms which give rise to the cyclical outbreaks of diseases such as measles (Haggett, 1990). In some cases, these outbreaks can be thought of as very similar to the cycles of fire: a measles epidemic requires a sufficient build-up of 'susceptibles' before it can take off. In the case of locusts, there seems to be a fairly direct linkage to the cyclical abundance of food plants: as food resources shrink, so locusts become crowded together and, ultimately, undergo the dramatic physiological changes which characterize the swarming individual (Evans, 1970). But in the case of the 5–7-year cycles of northern small mammals – the voles, mice and lemmings of Charles Elton's classic study (1942) – the link to environmental cycles and, in particular, to the abundance and nature of food supplies has proved very difficult to establish (see Norrdahl, 1995, for a detailed overview).

Where there is evidence for even longer cycles of abundance – as in the case of the predatory starfish *Acanthaster planckii* which rose to notoriety in the 1960s with the discovery that huge local concentrations on Australia's Great Barrier Reef were killing the corals – a common difficulty is that, without good long-term observations or well documented proxy data, an outbreak (or disappearance) which is entirely natural may not only be ascribed to human activity but also, possibly, seriously misguided attempts may be made to rectify the situation. The proposal made in the 1960s to dynamite 'starfish breaks' in the Great Barrier Reef would almost certainly have proved far more calamitous than the *A. planckii* themselves. (This intervention would not only have destroyed parts of the reef but also altered sediment and nutrient flows and done little if anything to halt the movement of either adult starfish or their larvae.)

Finally, whole communities frequently exhibit fairly regular cycles of abundance of species, termed taxon cycles (Brown and Lomolino, 1998). Once again, it is unusual to have direct observations which cover long enough periods to be sure of the timescales involved, and the cycles are often inferred by comparing ecosystems or communities which are spatially separate and which are presumed to represent temporally distinct stages.

No biological cycles are clockwork in their operations. The longer the periodicity, the more mysterious they seem to human onlookers with imperfect records. The fossil record may be used, of course, to pick up what may be very long-term cycles, notably those of major extinctions (Sepkoski and Raup, 1986): but the reality of the cycles, is also highly debated (see Lawton and May, 1995).

**CONCLUSION**

Many features of the physical world – and many of its systems – exhibit some form of cyclic behaviour. Such periodicities are not by any means always easy to identify. Where – as with orbital patterns – there are several cyclical components with different wavelengths superimposed, it is no trivial matter to detect patterns, even if the individual causal mechanisms are well established. Where – as with ecosystems – we have only fluctuations and a plethora of potential causal agencies, it may be near impossible to account for the cycles – or even be sure that the fluctuations are, indeed, cyclic.

## Summary

- Closed, isolated and open systems need clear boundaries to be defined: this is difficult.
- Caution is therefore needed in identifying both key components and key processes in systems.
- Astronomical cycles influence terrestrial processes on all scales between ice ages and tides.
- The geological cycle is at least 4 billion years old. All depositional processes tend to be cyclic.
- The biogeochemical cycles may be over 3 billion years old.
- Lovelock's Gaia concept sees the earth and its biogeochemical processes as a super-organism.
- Liebig's Law of the Minimum sees ecosystem productivity controlled by limiting nutrient availability and, hence, cycling.
- Biological cycles at all scales are identified with apparent extinctions and plagues.
- In any system where the cyclic mechanisms are not well understood, there is a need for records long enough to separate true cycles from individual events.

## Further reading

For a comprehensive and well illustrated coverage of the basic systems and cycles in physical geography, Briggs et al.'s (1997) *Fundamentals of the Physical Environment* is to be recommended. The rather laboured attempt by Chorley and Kennedy (1971) in *Physical Geography: A Systems Approach* to fit a mass of previous work into the systems framework is probably only of historical interest nowadays: Huggett's (1980) *Systems Analysis in Geography* is an easier read. Ideas of scientific change are well described in Cohen's (1985) *Revolution in Science* and – for geomorphology – in Davies's (1969) *The Earth in Decay*. More recent views which will probably colour our ideas of earth systems and cycles are those in *Chaos: The Amazing Science of the Unpredictable* (Gleick, 1998), *Life on the Edge: Amazing Creatures Thriving in Extreme Environments* (Gross, 2001) and Lovelock's recent books (2000a; 2000b). Finally, the journal *Science* is published weekly and provides a range of accessible updates on all the topics covered in this chapter.

*Note*: Full details of the above can be found in the references list below.

## References

Allen, P.A. (1997) *Earth Surface Processes*. Oxford: Blackwell.

Atkinson, B.W. (2000) 'Solar constant', in D.S.G. Thomas and A. Goudie (eds) *The Dictionary of Physical Geography*. Oxford: Blackwell, p. 448.

Boulding, K. (1956) 'General systems theory – the skeleton of science', *General Systems Yearbook*, 1: 11–17

Briggs, D., Smithson, P., Addison, K. and Atkinson, K. (1997) *Fundamentals of the Physical Environment* (2nd edn). London: Routledge.

Brown, J.H. and Lomolino, M.V. (1998) *Biogeography* (2nd edn). Sunderland, MA: Sinauer Associates.

Burchfield, J.D. (1990) *Lord Kelvin and the Age of the Earth*. London: University of Chicago Press.

Chorley, R.J. (1960) *Geomorphology and General Systems Theory*, US Geological Survey Professional Paper 500–B.

Chorley, R.J. and Kennedy, B.A. (1971) *Physical Geography: A Systems Approach*. London: Prentice Hall.

Cohen, I.B. (1985) *Revolution in Science*. London: Belknap Press.

Crowley, T.J. (2002) 'Cycles, cycles everywhere', *Science*, 295: 1473–74.

Davies, G.L. (1969) *The Earth in Decay*. London: Macdonald Technical and Scientific.

Elton, C.S. (1942) *Voles, Mice and Lemmings: Problems in Population Dynamics*. London: Oxford University Press.

Evans, H.E. (1970) *Life on a Little Known Planet: A Journey in the Insect World*. London: Andre Deutsch.

Gleick, J. (1998) *Chaos: The Amazing Science of the Unpredictable*. London: Vintage.

Gross, M. (2001) *Life on the Edge: Amazing Creatures Thriving in Extreme Environments*. New York, NY: Plenum Press.

Haggett, P. (1990) *The Geographer's Art*. Oxford: Blackwell.

Hall, A.D. and Fagen, R.E. (1956) 'Definition of system', *General Systems Yearbook*, 1: 18–28.

Hays, J.D., Imbrie, J. and Shackleton, N.J. (1976) 'Variations in the earth's orbit: pacemaker of the ice ages', *Science*, 194: 1121–32.

Huggett, R. (1980) *Systems Analysis in Geography*. Oxford: Clarendon Press.

Huggett, R.J. (1995) *Geoecology: An Evolutionary Approach*. London: Routledge.

Hutton, J. (1788) 'Theory of the earth; or an investigation of the laws observable in the composition, dissolution and restoration of land upon the globe', *Transactions of the Royal Society of Edinburgh*, I(II): 209–304.

Kerr, R.A. (2001) 'A variable sun paces millennial climate', *Science*, 294: 1432–3.

Lawton, J.H. and May, R.M. (eds) (1995) *Extinction Rates*. Oxford: Oxford University Press.

Lovelock, J. (2000a) *Gaia: A New Look at Life on Earth*. Oxford: Oxford University Press.

Lovelock, J. (2000b) *The Ages of Gaia*. Oxford: Oxford University Press.

Norrdahl, K. (1995) 'Population cycles in northern small mammals', *Biological Reviews*, 70: 621–37.

Peñuelas, J. and Fillella, I. (2001) 'Responses to a warming world', *Science*, 294: 793–5.

Pugh, D.T. (2000) 'Tides', in D.S.G. Thomas and A. Goudie (eds) *The Dictionary of Physical Geography*. Oxford: Blackwell, pp. 489–90.

Roberts, N. (1998) *The Holocene: An Environmental History* (2nd edn). Oxford: Blackwell.

Schumm, S.A. (1980) 'Some applications of the concept of geomorphic thresholds', in D.R. Coates and J.D. Vitek (eds) *Thresholds in Geomorphology*. Boston, MA: George Allen & Unwin, pp. 473–85.

Sepkoski, J.J. and Raup, D.M. (1986) 'Periodicity of marine extinction events', in D.K. Elliott (ed.) *Dynamics of Extinction*. New York, NY: Wiley, pp. 3–36.

Smith, D.E., Zuber, M.T. and Neumann, G.A. (2001) 'Seasonal variations of snow depth on Mars', *Science*, 294: 2141–6.

Strahler, A.N. (1952) 'Dynamic basis of geomorphology', *Bulletin of the Geological Society of America*, 63: 923–38.

Van der Lee, S. (2001) 'Deep below North America', *Science*, 294: 1297–8.

Von Bertalanffy, L. (1950) 'The theory of open systems in physics and biology', *Science*, 3: 23–9.

Von Bertalanffy, L. (1956) 'General system theory', *General Systems Yearbook*, 1: 1–9.

White, G. (1789) *The Natural History and Antiquities of Selborne* (reprinted 1962). London: Folio Society.

Willis, K.J. and Whittaker, R.J. (2002) 'Species diversity – scale matters', *Science*, 295: 1245–8.

Woodcock, N. and Strachan, R. (2000) *Geological History of Britain and Ireland*. Oxford: Blackwell Science.

# 15 Landscape and Environment: Biophysical Processes, Biophysical Forms

## Nick Spedding

### Definition

Landscape can be defined as the assemblage of objects that make up a particular part of the earth's surface. Physical geography primarily studies the 'natural' elements of the landscape: landforms, soils, vegetation cover, animal life and the visible aspects of weather and climate. Traditional definitions of landscape gave priority to appearance and stressed its holistic outlook. However, developments in physical geography in the second half of the twentieth century enlarged the concept's scope to incorporate the various biophysical processes that shape the earth's surface. As a result, contemporary understandings of landscape tend to combine form with process, description with explanation and synthesis with analysis.

### INTRODUCTION: WHAT IS LANDSCAPE?

This chapter is about how landscapes take shape inside the minds of physical geographers (see Chapters 16 and 17 for alternative interpretations of landscape). Landscape is a simple word, familiar from everyday use, but it has various meanings and interpretations attached to it. It is used as a generic term ('landscapes are what geographers study') and to identify something specific ('this geographer studied that particular landscape'). Landscape is simultaneously a label, a thing and a way of seeing. The word comes from the Dutch *landschap*, and was originally used by artists to denote rural scenery or its representation. Thus from the start landscape had strong pictorial associations – and these did not always find favour with those geographers seeking to establish geography as the scientific study of the earth's surface. Richard Hartshorne's (1939) treatise on 'The nature of geography', sought to purge landscape of its aesthetic and psychic connotations, preferring a firmly empiricist mapmaker's approach to landscape as the form of (a particular part of) the earth's surface (Hartshorne, 1939: 325–50; Stamp, 1961).

This core definition of landscape as the form of the earth's surface has the advantage of simplicity, but it fails to provide a sense of the disagreements between physical geographers over the use of landscape as a theoretical tool. There have been, and still are, two key points of argument. The first centres on the questions: what are the entities that make up landscape – what is 'in' and what is 'out' – and how do they relate to each other? It is common to think of landscape as an *assemblage* of individual forms – often to the extent that landscape is somehow more than the sum of its parts. This is a holistic perspective, in which relations between objects take on special importance. The opposite view to this rejects the significance of inter-object relationships and argues that it is preferable to study individual forms in isolation. This is analysis, not synthesis.

The second key issue is: how can we explain landscape? This is closely related to the first two questions. Different ideas about what landscape is are attached to different ideas about the appropriate form of explanation, involving sharp contrasts in their chosen treatments of time, space, scale and the type of relationships thought important. Two major schools of thought exist. The holistic school sees landscape as the complex product of a diverse series of events strung out across space and over time. The form – or *configuration* – of the landscape is *contingent*, or non-essential, in that it cannot be read off as the straightforward result of a single process or principle. Instead, to understand landscape from this point of view requires the reconstruction of the sequence of formative events that cumulatively shaped the landscape's form. Accordingly, Simpson (1963) described this as the 'historical' or 'configurational' approach to science. The opposite of this is 'non-historical' or 'immanent' (indwelling) science, which is concerned with the study of universal principles, such as the laws of physics, that are independent of specific instances of space and time (Simpson, 1963). The focus here is on single, abstract process relationships, such as the mechanics of river flow, winds or slope failure. This analysis sometimes extends to the study of associated individual forms, but – precisely because landscapes as a whole are not produced by a single process – rarely addresses larger, more complex phenomena.

The status of landscape as an intellectual concept has fluctuated considerably as fashion has variously favoured one or other of these two schools of thought. The tension between historical and non-historical science does much to explain the changing fortunes and meanings of landscape, and simultaneously provides a framework for understanding the development of physical geography since the late nineteenth century. In the rest of this chapter I try to trace some of the most significant debates. However, as a prelude I wish to introduce my 'home' landscape of the Cairngorms, Scotland, in order to 'set the scene ' for what follows.

## THE CAIRNGORMS

As this heading confirms, we like to pick out and name particular portions of the earth's surface that are set apart from other areas by the distinctive features of their landscape. Although the exact limits of the region are debatable, as the controversy over where to draw the boundaries for a proposed national park demonstrated, locals, tourists and academics alike tend to agree that the Cairngorms (a mountain group in the northeastern corner of the Scottish Highlands) represent a distinct landscape of outstanding natural heritage (Murray, 1962; Gordon, 1993; Brazier et al., 1996; Glasser and Bennett, 1996; Watson, 1996; Scottish Natural Heritage, 2001).

The geographic and geological core of the Cairngorms is a mass of granite, emplaced deep in the earth's crust at the time of the Caledonian orogeny some 400 million years ago (Hall, 1991). Subsequent erosion has unroofed this batholith to create the current surface topography of rolling plateaux dissected by steep-sided valleys. The plateau top has an irregular cover of low, sparse arctic or alpine vegetation, interspersed with patches of sands and gravels, boulder fields or the spectacular tors that interrupt the otherwise smooth lines of the surface. Heather and blanket bogs are common at lower altitudes, and remnants of the ancient Caledonian Pine Forest are found on the valley floors. Impressive cliffs form large stretches of the valley walls, but slopes elsewhere are strewn with boulders, cut by deep linear scars or plastered with a smooth cover of pulverized rock debris. Except for those areas occupied by deep, narrow lakes, rivers wind between the rubble hummocks and the terraces that partly fill up the wide, flat valley floors.

The task of physical geography is not just to describe but also to explain the natural features of the earth's surface. What, then, are we to make of the Cairngorm landscape, whether summarized in prose or in some form of image (Figure 15.1(a), (b), (c))? It was difficult to write the above paragraph without giving too much away but, even without the help of words such as 'trough', 'corrie' or 'moraine' (or the pictures of Figure 15.1), I guess that most readers are now thinking glaciation. Readers with some knowledge of biogeography or ecology will have made links among granitic rock, high altitude, wind, cold, snow, wet and thin, acidic soils, so accounting for the characteristic alpine and tundra types of vegetation, with forest lower down where soils are

FIGURE 15.1  Aspects of the Cairngorms landscape: (a) typical Cairngorms scenery: Loch Avon and Ben Macdui from the north east (author's photograph)

thicker and climate is less harsh. These are just two aspects of the landscape. There are many other 'landscape questions' we can ask, such as the following:

- Was the plateau surface covered by ice in the past and, if so, why does it not show more widespread evidence of glacial erosion?
- How did the tors form, and how old are they?
- How was the plateau formed in the first instance?
- Has the distribution of vegetation always been the same?
- Has the landscape changed much since the ice disappeared and, if so, how and why?
- What will happen if the climate continues to warm up?

It is not my purpose to answer these questions here – try following up some of the references! Instead, I wish to stress the multitudinous aspects of landscape – lots of different bits, formed by lots of different processes and with a long history – and the diverse array of tools available to geographers seeking to answer some of these questions. A comprehensive answer to the question 'why is the Cairngorms landscape like it is?' requires a huge amount of knowledge, to the extent that it is scarcely possible. We would need to know the full history of tectonic and climatic change; the chemistry that controls the weathering of the minerals in the granite; the physics of past glacier

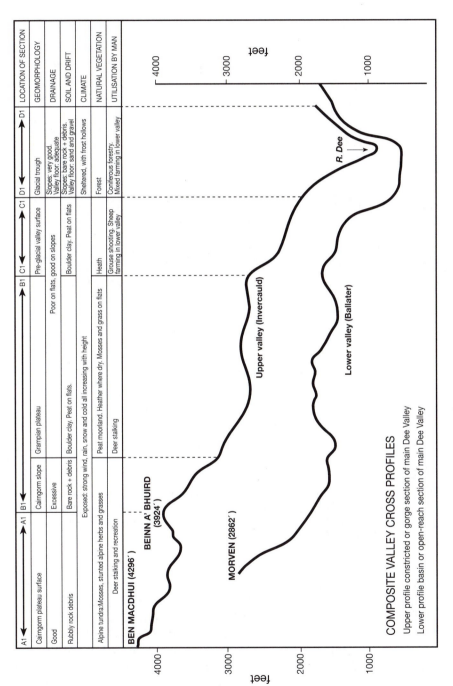

| A1 | | B1 | C1 | D1 | LOCATION OF SECTION |
|---|---|---|---|---|---|
| Cairngorm plateau surface | Cairngorm slope | Grampian plateau | Pre-glacial valley surface | Glacial trough | GEOMORPHOLOGY |
| Good | Excessive | | Poor on flats, good on slopes | Slopes: very good. Valley floor: adequate | DRAINAGE |
| Rubbly rock debris | Bare rock + debris | Boulder clay. Peat on flats. | Boulder clay. Peat on flats | Slopes: bare rock + debris. Valley floor: sand and gravel | SOIL AND DRIFT |
| | Exposed: strong wind, rain, snow and cold all increasing with height | | | Sheltered, with frost hollows | CLIMATE |
| Alpine tundra:Mosses, stunted alpine herbs and grasses | Peat moorland. Heather where dry. Mosses and grass on flats | | Heath | Forest | NATURAL VEGETATION |
| Deer stalking and recreation | Deer stalking | | Grouse shooting. Sheep farming in lower valley | Coniferous forestry. Mixed farming in lower valley | UTILISATION BY MAN |

BEN MACDHUI (4296´)

BEINN A´ BHUIRD (3924´)

MORVEN (2862´)

Upper valley (Invercauld)

Lower valley (Ballater)

R. Dee

feet
— 4000
— 3000
— 2000
— 1000

feet
4000 —
3000 —
2000 —
1000 —

COMPOSITE VALLEY CROSS PROFILES

Upper profile constricted or gorge section of main Dee Valley
Lower profile basin or open-reach section of main Dee Valley

*Source:* After Murray (1950) – is an excellent example of the type of regional monograph that sets out to define the landscape as the synthetic product of physical, biological and human forms

**FIGURE 15.1** Aspects of the Cairngorms landscape: (b) annotated cross-section of part of the Cairngorms to illustrate the relationships among topography, aspects of the physical environment and the biological response

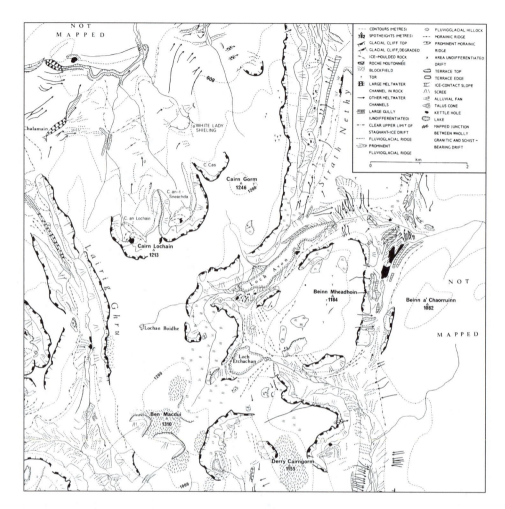

*Source*: Sugden (1970)

**FIGURE 15.1**  Aspects of the Cairngorms landscape: (c) geomorphological map of the central Cairngorms. Different features – such as the tors of the remnant Tertiary surface, the glacial troughs and corries, meltwater channels, terraces and solifluction forms – indicate different episodes of landscape change

flow, from which we can reconstruct processes, patterns and rates of glacial erosion; the tolerance of different plant species to different climatic factors, as well as the dynamics and impact of interspecies competition . . . plus lots more besides. The task of writing it all down, or building it into a suitable computer model, would be mind-boggling, even *if* we did have perfect knowledge and understanding – which, of course, we do not. Thus it is inevitable that geographers have sought to render the study of landscape tractable by selecting only parts of it at once. Some have concentrated on the overall form, others on the underlying processes, others on the history of change. The work of W.M. Davis provides the best example of this last theme.

## WILLIAM MORRIS DAVIS AND PHYSICAL GEOGRAPHY

William Morris Davis (1850–1934) was the most prominent Anglo-American physical geographer of the late nineteenth and early twentieth centuries. He is best remembered today as the author of the 'geographical cycle' (Davis, 1899), the first comprehensive theory from within physical geography of landscape as a coherent, regional entity (see, for example, Sparks, 1986, or, for a more recent treatment, Summerfield, 1991). Davis's research concentrated on landforms and landscapes – what we would now call geomorphology. This can give the impression that Davis had a narrow definition of physical geography, but this is not true. He was a champion of all aspects of the discipline, both physical and human. His 1906 essay on 'An inductive study of the content of geography' (Davis, 1954a) defines the discipline's purpose as the study of space-filling relationships between inorganic and organic (plants, animals and people) phenomena – a vision that, despite strong elements of environmental determinism, anticipated Sauer's later treatise on landscape (Sauer, 1925; see below).

Davis himself rarely made use of the specific term 'landscape'. At the time he was writing the word was still something of a neologism: it appears just twice in Davis's paper on 'The geographical cycle' (1899: 486), with its first use – '[T]o look upon a landscape of this kind' – demonstrating its aesthetic, rather than scientific, origins. Phrases such as 'an ideal system of topographical forms' (Davis, 1954b: 166) or 'system of terrestrial morphology' (p. 169) demonstrate awareness of spatial arrangements but, for the most part, Davis used 'land form' where we use 'landscape' today. The distinction between an individual form and the wider assemblage of the landscape was not important for Davis. Whereas subsequent work has emphasized diversity, using the motif of landscape as 'palimpsest' (a patchwork of superimposed forms of assorted origins and characteristics), Davis's cycle imposed uniformity upon the earth's surface. He did recognize geological contrasts, but the basic scheme assumed constancy of climate and processes so that the appearance of the land surface was read as a straightforward function of its age.

Writing in 2002 (or 1952) it is easy to find fault with Davis's concepts of landform/landscape and physical geography. To use more recent terminology, he gave undue emphasis to geomorphology, rather less to hydrology, and tended to skip over biogeography and climatology. The rigid character of the cycle left little room for diversity of materials, form and history: all aspects of landscape that we now incorporate using concepts such as scale and sensitivity (see below). Davis's treatment of landscape-forming processes was similarly restricted. Although the opening remarks of 'The geographical cycle' clearly state that the form of the land is a function of structure, process and time, processes were given little detailed treatment in his work. It is puzzling why this was so. The theoretical physics and experimental methods necessary to support process studies in hydrology and geomorphology were firmly established by Davis's time. Consider, for instance, Coulomb's (1776) equation for soil strength, the work of Forbes and Tyndall on glacier motion in the mid-nineteenth century (see Clarke, 1987), Darcy's law for pore-water flow, which dates from 1856, or du Boy's excess tractive force formula for river bedload transport, published in 1879. Process studies in ecology and meteorology did not take off until rather later (for example, see Bowler, 1992: 361–78 and 394–7, for further details). Chorley (1965) attributed Davis's strong preference for time as the decisive factor controlling landscape form – to the extent that time itself became a process – to the prevailing dogma of evolution (see also Osterkamp and Hupp, 1996). However, it is important to take into account Davis's self-image as a geographer and his view of geography as a relational discipline. From this point of view, it was the linkages between various earth surface phenomena associated with spatial coherence and historical change that were important. Abstract analysis of individual mechanisms was thus the preserve of other disciplines.

### The work of Davis's contemporaries

Davis's hold over physical geography in the time *c.* 1890 to *c.* 1920 was so great that histories tend to overlook other contributions. The work of Davis's contemporary, the geologist G.K. Gilbert, enjoys a fine reputation today, but this is because he (Gilbert) was adopted as an ancestral champion by geomorphologists of the 1950s and 1960s seeking to dismantle the Davisian edifice (for example, Chorley, 1962). In his own time Gilbert was something of an 'underappreciated fashion dude' (Sherman, 1996: 107).

J.E. Marr described his book on *The Scientific Study of Scenery* (1900) as an 'Introductory treatise on geomorphology', but sought to include 'all the existing features of earth, sky and sea . . . [that] cover a large proportion of the field of physical geography'. The coupling in the title of 'scientific' and 'scenery' is notable. Although much of the book consists of familiar material (if rather old-fashioned by today's standards) from the disciplines of geology and geomorphology, its introductory chapter emphasized the aesthetic aspects of landscape (or 'scene'), with its attributes of size, form, character, surface, colour and movement. If the reader was not convinced, Marr suggested that he or she consult two of the nineteenth-century's greatest authorities on the subject: the poet William Wordsworth and the art critic John Ruskin!

Although its influence did not last as long or reach as far as Davis's geographical cycle, T.H. Huxley's *Physiography* (first published in 1877) must also rank as a key document that helped to shape physical geography (Stoddart, 1986). Huxley shared Davis's impatience with the rote-learning of place facts characteristic of 'capes and bays' geography, and was determined to establish 'causality' at the root of earth surface studies. Whereas for Davis cause was primarily temporal, Huxley preferred to locate causality in the mechanical principles of the natural sciences, and proposed that these principles must be illustrated by means of the student's local area. For a short time this way of thinking was extremely important for geography. Mackinder (1887), for instance, in his famous 'Scope and methods' paper, set out to explain the geography of India starting with an understanding of 'the laws of Newton'. However, physiography's emphasis on scientific principles turned out to be its undoing. Natural science was difficult to apply to regional, human geographies, and was not itself geographical, said the geographers. Simultaneously the natural sciences grew in strength and stature as school and university subjects and had less need of the local landscape as an illustrative vehicle. Davis was one of the chief beneficiaries of the demise of physiography (Stoddart, 1986) but, because of his desire to synthesize process studies and an understanding of area, it was perhaps Huxley who was the more visionary prophet of landscape. However, it is Carl Sauer's (1925) essay on 'The morphology of landscape' that is credited with introducing the formal concept of 'landscape' into English language geography – and Sauer's ideas on what was important about landscape were vastly different from those of Huxley.

## CARL SAUER, LANDSCAPE AND PHYSICAL GEOGRAPHY

Along with Richard Hartshorne, Carl Sauer (1889–1975) dominates the general histories of American geography for the years *c.* 1920 to *c.* 1960 (see, for example, Livingstone, 1992, or Johnston, 1997; see also Chapter 17). Today, Sauer is best remembered as a champion of historical and cultural geography, and as a pioneer environmentalist (Leighly, 1963). Although climate change and the distribution of vegetation types formed important themes guiding his research on 'land and life', Sauer's work on, and enthusiasm for, physical geography as a whole tend to receive less attention these days (Stoddart, 1997). However, it helps to know something of Sauer's opinions if we are to start to comprehend how and why the discipline of geography developed between the two world wars, and with what consequences for the understanding of landscape from the perspective of the earth and environmental sciences.

### The morphology of landscape

Sauer wrote 'The morphology of landscape' soon after his appointment, in 1923, as Chair of the Department of Geography at the University of California, Berkeley. It was designed to establish the positions of both himself and his subject in the face of potential hostility from his powerful colleagues in the Department of Geology (Stoddart, 1997). In the essay he set out to define the identity, methods and study objects of academic geography (much as did Hartshorne in 'The nature of geography' 14 years later). Drawing heavily on French and especially German geographers, it was a manifesto for regional geography that had as its core a theory of landscape – although it is important to remember that Sauer nursed a deep suspicion of 'grand theory', as his dislike of the cycle of erosion and the more extreme forms of environmental determinism showed. The aspects of Sauer's landscape that merit special consideration here are its holistic character, its lack of particularism and its relationship to causal processes. These are important because Sauer was convinced that, taken together, they set up geography as a respectable scientific endeavour.

Sauer's landscape was holistic because it was a harmonious relational entity. The facts of geography (note the emphasis on the empirical) were 'place facts', and the *relationships* between these facts – or 'morphologic forms' to use Sauer's preferred term – made up the landscape. Sauer insisted that landscapes were not unique – 'landscape is not simply an actual scene viewed by an observer' (Sauer, 1925: 322) – but had 'generic meaning'. By this he meant that the task of geography was not to describe individual areas but to identify specific *types* of landscape. In this process of classification lay the skill and the science. If we follow Sauer, a view of somewhere in the Swiss Alps – say the Matterhorn towering over Zermatt – is not landscape, but we can recognize an *assemblage* of features that defines an alpine landscape (think block diagrams of arêtes, pyramidal peaks, corries, troughs, etc.). Although Sauer felt that geography was not complete unless it included humans, he did admit the distinction between *natural* and *cultural* landscapes, and 'The morphology of landscape' does contain a lengthy discussion of the former. This was not greatly original, but it did help to establish the natural landscape as a formal

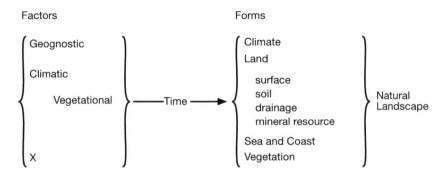

Source: After Sauer (1925)

**FIGURE 15.2**   Sauer's 'diagrammatic representation of the morphology of the natural landscape'. Sauer used 'geognosy' to indicate the non-historical part of geology concerned with describing the type, properties and surface distribution of rock materials

concept (Figure 15.2). It was the interaction of climate and earth surface materials over time that gave rise to the suite of surface features ('forms') that made up the natural landscape. This was a totality greater than the topography – and so, to some extent, Sauer's vision exceeds that of Davis. As well as the spatial arrangement of landforms, the weather and climate, the soils and the vegetation were all to be included (not least because these factors determined the potential for food to be gathered or farmed as the basis for human occupancy), and the whole had a history of development. The kind of geomorphic block diagram invoked in Figure 15.2 captures only part of this landscape. To trace the web of sophisticated relations in its spatial and temporal totality, prose – preferably rich and evocative – was the technique of choice.

What is perhaps most remarkable about Sauer's idea of landscape – at least in retrospect – is his blatant disregard for processes. In his view, geography's mission was to talk about the interrelationships of individual forms – or 'chorology'. It was a *phenomenologic* science (by which he meant 'to do with what was there to be seen') not a *genetic* science. He did recognize general 'factors' that shaped the landscape, but it was not necessary for geography to worry about how and why things came to be as they were: 'It is geographically much more important to establish the synthesis of natural landscape forms in terms of the individual climatic area than to follow through the mechanics of a single process, rarely expressing itself individually in a land form of any great extent' (Sauer, 1925, reprinted in Leighly, 1963: 336). Huxley's physiography was, for Sauer, the application of physics. It was certainly not geography, said Sauer, and scientists interested in processes were likely to be best employed in a department of dynamic geology (p. 347). Sauer's attitudes were soon picked up by others jumping on the regional geography bandwagon. The British geographer and Davisian geomorphologist, S.W. Wooldridge, for instance, in an essay strongly reminiscent of Sauer's 'Morphology', wrote that the geographical method was 'designedly and obstinately non-specialist' (1956 [1945]: 8–9), proceeding haughtily to confess at best a 'mild interest' in his 'younger friends returning to the field of geography from Service meteorology . . . assuming a rather preoccupied mien and promising to recast the whole field of meteorology in a synoptic or dynamical guise' (p. 16). It is important to stress that Sauer, Wooldridge, Hartshorne and lots of other geographers did firmly believe that their synthetic studies of landscape were scientific. However, by the early 1950s some geographers, the wider academic and scientific community and society at large did not agree. From this dissatisfaction we can trace the rise of a 'scientific' physical geography (although its practitioners were not always geographers by profession) that, with its emphasis on numbers and surface processes, did much to undermine previous understandings of landscape.

## THE REACTION AGAINST LANDSCAPE STUDIES

By the 1950s, much of academic geography, whether physical, regional or both, was historical and holistic in its outlook, with a strong preference for words and pictures. Sauer's colleague, John Leighly (1963: 6), wrote of 'The morphology of landscape' that '[its] positive effect . . . was, unfortunately, to stimulate a spate of detailed descriptions of small areas written in the ensuing twenty years, which had little value, either scholarly or practical'. In an earlier essay lamenting what he saw as the stagnation of physical geography, Leighly had charged Sauer with the rejection of the rational interpretation of physical forms, so lumping Sauer and Davis together as villains who, by way of their restrictive philosophical methodological statements, had cut physical geography off from its scientific roots (Leighly, 1955).

Leighly's criticism was rather tame in comparison with that coming from some geologists. Writing about some of his colleagues, Mackin (1963) identified an 'avant-garde', possessed of 'an almost evangelical zeal to quantify' who thought of Davis as 'an old duffer with a butterfly-catcher's sort of interest in scenery'. The head of this group (or its chief 'fashion dude': Sherman, 1996) was Arthur Strahler. His 1950 paper on Davis's work on slope development mixed an attack on Davis with an exhortation for geologists to take on the techniques of mathematics, statistics and physics necessary to do justice to the discipline of geomorphology, redefined as 'a geophysical science of almost terrifying complexity'. Strahler went on to note that 'it seems almost out of the question that this programme [of intensive scientific training] can be carried on by geographers' (Strahler, 1950: 210).

Given the contempt in which Strahler and others held traditional landscape studies, and its practitioners, it is not surprising that the debate was not courteous (see, for instance, Beckinsale, 1997: 6). However, the vitriolic language did spring from a genuine desire on the part of the reformers to make geomorphology into a truly 'scientific' discipline fit for the second half of the twentieth century, and there were enough of them able to handle the maths to bring about major changes to physical geography as a whole. The inward-looking practices of regional exceptionalism started to crumble as geography sought ideas and techniques outside its traditional disciplinary boundaries, and synthesis gave way to analysis. Landscape was the first victim of quantification. Reductionist theories derived from classical physics focus on a single, supposedly fundamental process, such as the turbulent flow of a river (Strahler, 1952). Regression equations, such as those of hydraulic geometry linking width, depth or flow speed to discharge, summarize empirical relationships for selected landscape features. The new quantitative methods, both rational and empirical, simply did not work for landscape's diverse assemblage of forms and attendant processes – so it was the landscape that was pushed aside. Horton's celebrated analysis of the drainage basin, which rivalled Davis's cycle of erosion in its scope, purported to capture the basin's essential features in a handful of equations and some neat semilogarithmic plots (see Chorley, 1995), but did little to convey the blend of size, form, character, surface, colour and movement that defined landscape according to Marr (1900).

## THE REVIVAL OF LANDSCAPE STUDIES

Short histories of physical geography can give the impression that in the 1960s, 1970s and 1980s work was done only on processes. This is not so; this was a time in which more sophisticated models were put forward to explain long-term landscape evolution, involving a wider understanding of what was to be included in the remit of physical geography. Walton (1968a), for instance, set out the case for a revision, rather than an outright rejection, of the Davisian framework that rested on an enhanced understanding of tectonics, climate change and ecology, such that climate, vegetation and soils – all of which were given little emphasis by Davis – were seen to control processes of erosion, transport and deposition, and so the form of the landscape. In some places, such as eastern Europe, quantification made little impact, and the standing of 'traditional' regional geomorphology at the core of all geography persisted unchallenged (see, for example, Simandan, 2002). However, this was certainly a time in which geographers increasingly chose smaller scales of analysis, gave preference to process rather than form and looked to physics and statistics for inspiration. This was also a period in which academic research became far more specialized so that work on many topics in hydrology, biogeography, meteorology and climatology in particular was lost to specialist departments of computer science, ecology, engineering, physics, etc. This trend was strongly related to the shift towards technical and mathematical methods in the environmental sciences in general, suggesting that Strahler's caustic comments on the limited abilities of geographers were not without truth. As this leakage continued, physical geography in the UK became increasingly dominated by geomorphology, and for the most part this was pursued as a pure, reductionist science (Newson, 1992). In North America, many departments of geography also lost much of their physical geography, here including geomorphology, to other disciplines (Johnston, 1997: 50–1). Teaching of physical geography was often restricted to descriptive, regional accounts of landforms, climate and vegetation as part of introductory courses – a tribute to the influence of Sauer – whereas research tended to adopt the insular, specialist qualities of pure science, as in the UK (see, for example, Graf, 1996: 444).

Since the 1990s, however, landscape studies have enjoyed something of a revival. Whereas previously science and process on the one hand, and landscape and environment on the other, tended to be seen as mutually exclusive, all four are now once more thought of as part of the same endeavour. That this is so can be attributed to three major developments:

1    A rethinking of the core philosophy and methods of the exemplar science, physics, resulting in a growing preparedness across many disciplines to work with more flexible definitions of what counts as good scientific practice.
2    New techniques for processing spatial information.
3    The rise to prominence of environmentalism and conservation as research agendas, and the enthusiasm for interdisciplinary work these have helped to promote.

All three of these themes have recently been put forward as the vehicle for reuniting the scattered subdisciplines of physical and

human geography (respectively, for example, Massey, 1999; Openshaw, 1991; Newson, 1992).

### Rethinking science

It is often argued that the environmental sciences suffer from a
'physics envy' that drives them to ape physics's core features, despite
the fact that these are not necessarily appropriate (Frodeman, 1995;
Massey, 1999). Physics in its classical, positivist guise is firmly
reductionist: it takes just a single, supposedly fundamental process at
a time. It is also anti-contextual: it relies on simple, heavily controlled
systems – in the lab, or on a sheet of paper, or on a computer – that
exclude the impact of time and space. The real world is far more
messy, and the techniques of physics cannot necessarily handle this.
To let in time and space, which happens as soon as the scientist steps
outside to confront the external environment, is to lose control. The
multitude of objects, processes and interactions does not readily
submit to isolation, observation or quantification. Thus, it is said, it is
necessary for the environmental sciences to adopt more flexible and
more realistic study standards (see also Chapter 2).

The calls for a return to interpretative and narrative methods in
geology and geomorphology represent one aspect of this new line of
thought (Frodeman, 1995; Harrison, 1999; Baker, 2000). The traditional aspects of rigour at the core of positivist physics – control,
replicability and quantification – give way to the discipline imposed
by the need to tell a convincing story of landscape change, free from
wild plot devices and loose ends. This very much represents a revival
of Simpson's (1963) ideas of historical science; to a certain extent it
also involves a return to Davis's style of landscape explanation.
However, unlike Davis, we now have a much more sophisticated
understanding of the full range of earth surface processes, mechanisms
and patterns of climate change (for example, Alley, 2000) and different
tectonic frameworks (for example, Summerfield, 2000). To these factors we can add powerful concepts such as landscape *sensitivity*,
which is a measure of how susceptible a feature – landform, soils or
vegetation – is to change (Brunsden and Thornes, 1979; Thomas,
2001). Innovations of this kind add strength to our reasoning and free
us from charges of spinning fairy tales. If we have knowledge of
processes, the sequence, lengths and intensities of past episodes of
environmental change and, if we can tie these in to variations in
sensitivity, we are a long way towards an understanding, admittedly
for the most part qualitative, of landscape.

A second important aspect of the rethinking of science has been
the rethinking of physics itself. Classical physics is firmly and proudly
reductionist and deterministic, but in the last 20 years the utility of

these tenets has been attacked by developing theories of emergence, non-linear behaviour, chaos and complexity (Phillips, 1999a; Harrison, 2001; Manson, 2001). These are concerned with systems, pattern-forming and the transmission of change across space and over time, so it is not surprising that some geographers have been quick to pick up on the vocabulary. However, it is far from clear as yet whether or not some of these ideas are as significant as their proponents make out. It is difficult, for instance, to imagine exactly what mechanism it is that links the properties of river networks to the distribution of earth-quakes to the fluctuations of the stock market to traffic jams (cf. Bak, 1997). Critics suggest that these novel properties and principles are artifacts of the models' specifications, not genuine real-world features. Others accept the validity and potential of these ideas but argue that careful research linking the ideas to known processes and empirical problems is necessary (see Phillips, 1999b, for this kind of thinking).

### New spatial technologies

Traditionally, the map has been at the heart of geography's relation-ship with landscape, although the importance of maps was diminished as the enthusiasm for process studies rose. In some corners of the subject the map was lumped alongside the regional monograph as an anachronistic repository for 'mere description'. This is still the view of many postpositivist geographers, who have watched the resurrection of mapping technologies with a mixture of dismay, disgust and incom-prehension. This new geographic information science incorporates a wide range of techniques, including remote sensing, global positioning systems (GPS), digital elevation modelling and geographical informa-tion systems (GIS). These have produced an enormous increase in our ability not just to map the features of the earth's surface but also to display that information in different ways, and to examine the rela-tionships between different phenomena making up the surface. For instance, some GIS systems can switch at the press of a key from the traditional map view of a landscape to a three-dimensional perspective view, further enhanced by realistic renditions of surface colours and textures and 'flyover' animation sequences (Figure 15.3). Spatial over-lay techniques can be used to explore correlations between different landscape features, which can help to identify the types of processes that operate.

The impact of all these new techniques is controversial. Critics argue that the proliferation of technology simply means that we can churn out more data faster, and like to stress the negative aspects – such as its military applications. However, there are many positive aspects, one of which is the massive boost given to the concept of landscape. The seductive combination of the precise mathematics of

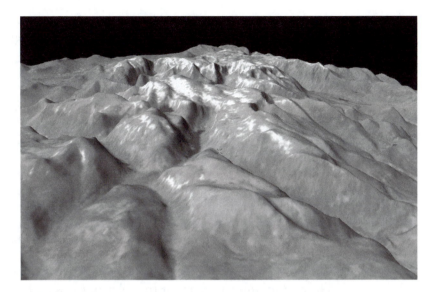

*Source*: Department of Geography and Environment, University of Aberdeen

**FIGURE 15.3**    Digital elevation model of part of the Cairngorms: extract from a 'flyover' sequence

topology with digital technology has replaced the fanciful conjectures of rambling, impressionistic prose, and has given regional studies the scientific credibility that was craved by the likes of Davis, Sauer and Hartshorne. Old concepts are now recast at the cutting edge of geography. For instance, Huggett and Cheesman's (2002) recent textbook on *Topography and the Environment* shows how digital elevation models can be used to revitalize the study of the relationships among terrain and climate, hydrology, soil, plants, animals and people. Their concept of the 'toposphere', which 'sits at the interfaces of the pedosphere, atmosphere and hydrosphere' and represents 'the totality of the Earth's physical land surface' (Huggett and Cheesman, 2002: 9), is much the same as that of Sauer's natural landscape (see also Chapter 10 on environmental spheres).

### *The stimulus from the environmental and conservation movements*

The legacy of Sauer's geography also figures strongly in calls for geography to return to 'the traditional strength of the field in establishing linkages between the natural and the social science traditions; its achievements in regional and global analysis and its preoccupation with the Earth as the home of people over Holocene and contemporary time scales' (Slaymaker and Spencer, 1998: ix; see also Stoddart, 1987). The threat from global environmental change is increasingly invoked as the *raison d'être* of geography, and this carries with it the necessity of a holistic approach (see, for example, Newson, 1992; Slaymaker and Spencer, 1998; Crofts, 1999; Gregory, 2000). The folklore has it that, had they not been led astray down the paths of reductionist process analysis, geographers would by now be well on the way to sorting out the current mess. The new respect that many geographers now have for landscape rests firmly with the environmental and conservation agendas. These first became important in UK geography in the 1960s; early work was strongly objective, or positivist, in its outlook, and retained landforms at the heart of landscape (see, for instance Linton, 1968; Walton, 1968b). More recent work tends to be more comfortable with the values it attaches to landscape. This new take on landscape relies less on landform for its definition. Instead it is much more forcefully ecological in its outlook, albeit an ecology that is strongly anthropocentric. Ecology and geographic information science make for a powerful partnership (see Huggett and Cheesman, 2002). This raises important questions about the relationships among the 'natural' world, the work of the scientific establishment and the wider community (for example, Frodeman, 2000). Key debates, both global (for example, biodiversity loss) and local (for example, the kind of land-use policy that is appropriate for the Cairngorms), are about power as well as fact. 'Natural' science is supposed to draw its authority from truth, but it is important not to overlook factors such as the size of one's computer or the people one knows. The study of landscape and environment, even for physical geography, is increasingly not just a problem of natural processes. It must grapple with cultural, ethical and political considerations too.

## CONCLUSION

Landscapes are large, complex systems that cannot be understood in their entirety. Material landscapes change over time as various bio-physical processes act in and on them. The mental representations that we have of these landscapes also change. As the historical survey shows, physical geographers striving to get at the essence of landscape have produced a wide range of different models that define and try to explain landscape in different ways. Three influences have been especially important:

1   *New discoveries.* For instance, climate change, the erosive action of glaciers, tectonic uplift and continental drift all figure strongly in the landscape history of the Cairngorms, yet the first two of these were only accepted in the mid-nineteenth century, and the last two in the mid-twentieth century.
2   *Academic fashions* (Sherman, 1996).

3   *Self-preservation*. Geographers tend to suffer from a perpetual anxiety that geography is about to be swallowed up by rival subjects and so feel it necessary to affirm the distinct identity of their discipline.

The impact of factors such as these makes landscape a flexible concept that is difficult to pin down. The interpretations that different physical geographers have given, and continue to give, to landscape are specific to particular times and places. However, this diversity does not stop notions of landscape persisting as a central feature of the discipline's thinking. To start to make sense of the historical geography of landscape as an idea is to take a significant step towards an understanding of the development of physical geography in its entirety.

## Summary

- The concept of 'landscape' is not fixed but has been subject to many different interpretations at different times and in different places. Traditional treatments stress the interaction of diverse objects at the earth's surface, and tend to be descriptive, historical and holistic in their outlook. Critics of this way of thinking have preferred to split landscapes up into their constituent parts, emphasizing the explanatory power associated with the analysis of individual processes.
- W.M. Davis's *geographical cycle*, put forward in the late nineteenth century, provided the first widespread theory of landscape within physical geography, although it concentrated on geomorphology and preferred to use time, not the action of earth surface processes, as its major causal agent.
- Writing in the 1920s, Carl Sauer popularized ideas of landscape as a synthetic entity produced by the spatial and temporal interaction of different earth surface forms, but his ideas also tended to overlook the action of earth surface processes, and so lacked explanatory power.
- After the Second World War, traditional understandings of landscape as a holistic entity attracted much criticism for not being sufficiently scientific. The new enthusiasm for mathematical methods, greater use of physics and analysis of individual processes undermined the standing of landscape studies as a core part of physical geography.
- In the last few years, the idea of landscape as a historically and spatially complex entity has enjoyed something of a revival. Important themes that have contributed to this include a rethinking of what kind of science is best suited to physical geography, the

development of new spatial technologies and the rise of the environmental and conservation movements.

## Further reading

The writings of Davis (1954c) in *Geographical Essays* and Sauer in *Land and Life* (Leighly, 1963) set out some of the most important thinking prior to *c.* 1950 on the meaning and importance of landscape. In his paper 'What has happened to physical geography?' Leighly (1955) provides a useful review of the mid-twentieth-century standing of landscape studies as science and anticipates the enthusiasm for process analysis that was soon to marginalize traditional understandings of landscape. To follow up some of the themes identified as important for the recent revival of landscape ideas, see Harrison's (1999) article, 'The problem with landscape', and two books: *Earth Surface Systems: Complexity, Order and Scale* (Phillips, 1999b) and *Topography and the Environment* (Huggett and Cheesman, 2002). Harrison examines the return to historical narratives of landscape change, Phillips looks at how some of the ideas drawn from the 'new physics' can be used to help us understand landscapes, and Huggett and Cheesman demonstrate the boost that new technologies – in particular, techniques of digital terrain modelling – give to traditional ideas of landscape as the product of interactions between different objects and processes. For more information on the diverse aspects of the Cairngorms landscape, both in terms of subject-matter and techniques of study, take a look at a special issue of *Botanical Journal of Scotland* guest edited by McConnell and Conroy (1996).

*Note*: Full details of the above can be found in the references list below.

## References

Alley, R.B. (2000) *The Two-Mile Time Machine*. Princeton, NJ: Princeton University Press.

Bak, P. (1997) *How Nature Works: The Science of Self-organised Criticality*. Oxford: Oxford University Press.

Baker, V.R. (2000) 'Conversing with the earth: the geological approach to understanding', in R. Frodeman (ed.) *Earth Matters: The Earth Sciences, Philosophy and the Claims of Community*. Englewood Cliffs, NJ: Prentice Hall, pp. 2–10.

Barry, R.G. and Chorley, R.J. (1998) *Atmosphere, Weather and Climate* (7th edn). London: Routledge.

Bauer, B.O. (1996) 'Geomorphology, geography and science', in B.L. Rhoads and C.E. Thorn (eds) *The Scientific Nature of Geomorphology*. Chichester: Wiley, pp. 381–413.

Beckinsale, R.P. (1997) 'Richard J. Chorley: a reformer with a cause', in D.R. Stoddart (ed.) *Process and Form in Geomorphology*. Routledge: London, pp. 3–12.

Bowler, P.J. (1992) *The Fontana History of the Environmental Sciences*. London: Fontana.

Brazier, V., Gordon, J.E., Hubbard, A. and Sugden, D.E. (1996) 'The geomorphological evolution of a dynamic landscape: the Cairngorm Mountains, Scotland', *Botanical Journal of Scotland*, 48: 13–30.

Brunsden, D. and Thornes, J.B. (1979) 'Landscape sensitivity and change', *Transactions, Institute of British Geographers*, 4: 463–84.

Chorley, R.J. (1962) *Geomorphology and General Systems Theory. US Geological Survey Professional Paper* 500B.

Chorley, R.J. (1965) 'A re-evaluation of the geomorphic system of W.M. Davis', in R.J. Chorley and P. Haggett (eds) *Frontiers in Geographical Teaching*. London: Methuen, pp. 21–38.

Chorley, R.J. (1995) 'Classics in physical geography revisited: Horton, R.E. 1945: Erosional development of streams and their drainage basins: hydrophysical approach to quantitative morphology. BGSA, 56, 275–370', *Progress in Physical Geography*, 19: 533–54.

Clarke, G.K.C. (1987) 'A short history of scientific investigations on glaciers', *Journal of Glaciology* (special issue), 4–24.

Crofts, R. (1999) 'Geography matters for the environment of the twenty-first century', *Geography*, 84: 345–53.

Davis, W.M. (1899) 'The geographical cycle', *The Geographical Journal*, 14: 481–504.

Davis, W.M. (1954a) [1906] 'An inductive study of the content of geography', in *Geographical Essays*. New York, NY: Dover, pp. 3–22.

Davis, W.M. (1954b) [1894] 'Physical geography as a university study', in *Geographical Essays*. New York, NY: Dover, pp. 165–92.

Davis, W.M. (1954c) *Geographical Essays*. New York, NY: Dover.

Frodeman, R. (1995) 'Geological reasoning: Geology as a historical and interpretive science', *Bulletin of the Geological Society of America*, 107: 960–8.

Frodeman, R. (ed.) (2000) *Earth Matters: The Earth Sciences, Philosophy and the Claims of Community*. Englewood Cliffs, NJ: Prentice Hall.

Glasser, N.F. and Bennett, M.R. (eds) (1996) *The Quaternary of the Cairngorms: Field Guide*. London: Quaternary Research Association.

Graf, W.L. (1996) 'Geomorphology and policy for restoration of impounded American rivers: what is "natural"?' in B.L. Rhoads and C.E. Thorn (eds) *The Scientific Nature of Geomorphology*. Chichester: Wiley, pp. 443–73.

Gregory, K.J. (2000) *The Changing Nature of Physical Geography*. London: Arnold.

Hall, A.M. (1991) 'Pre-Quaternary landscape evolution in the Scottish Highlands', *Transactions of the Royal Society of Edinburgh: Earth Sciences*, 82: 1–26.

Harrison, S. (1999) 'The problem with landscape', *Geography*, 84: 355–63.

Harrison, S. (2001) 'On reductionism and emergence in geomorphology', *Transactions, Institute of British Geographers*, 26: 327–39.

Hartshorne, R. (1939) 'The nature of geography', *Annals of the Association of American Geographers*, 39: 173–658.

Huggett, R. and Cheesman, J. (2002) *Topography and the Environment*. Harlow: Prentice Hall.

Johnston, R.J. (1997) *Geography and Geographers: Anglo-American Human Geography since 1945* (5th edn). London: Arnold.

Leighly, J. (1955) 'What has happened to physical geography?' *Annals of the Association of American Geographers*, 45: 309–18.

Leighly, J. (1963) 'Introduction', in *Land and Life*. Berkeley and Los Angeles, CA: University of California Press, pp. 1–8.

Linton, D.L. (1968) 'The assessment of scenery as a natural resource', *Scottish Geographical Magazine*, 84: 219–38.

Livingstone, D.N. (1992) *The Geographical Tradition*. Oxford: Blackwell.

Mackin, J.H. (1963) 'Rational and empirical methods of investigation in geology', in C.C. Albritton (ed.) *The Fabric of Geology*. Reading, MA: Addison-Wesley, pp. 135–63.

Mackinder, H.J. (1887) 'On the scope and methods of geography', *Proceedings of the Royal Geographical Society*, 9: 141–60.

Manson, S.M. (2001) 'Simplifying complexity: a review of complexity theory', *Geoforum*, 32: 405–14.

Marr, J.E. (1900) *The Scientific Study of Scenery*. London: Methuen.

Massey, D. (1999) 'Space-time, "science" and the relationship between physical geography and human geography', *Transactions, Institute of British Geographers*, 24: 261–76.

McConnell, J. and Conroy, J.W.H. (eds) (1996) 'Environmental history of the Cairngorms', *Botanical Journal of Scotland*, 48.

Murray, R.M. (1950) 'Upper Deeside: A Regional Study'. Unpublished Honours thesis, Department of Geography, University of Aberdeen.

Murray, W.H. (1962) *Highland Landscape: A Survey*. Aberdeen: Aberdeen University Press.

Newson, M. (1992) 'Twenty years of systematic physical geography: issues for a "new environmental age" ', *Progress in Physical Geography*, 16: 209–21.

Openshaw, S. (1991) 'A view on the GIS crisis in geography: or, using GIS to put Humpty Dumpty together again', *Environment and Planning A*, 23: 621–8.

Osterkamp, W.R. and Hupp, C.R. (1996) 'The evolution of geomorphology, ecology and other composite sciences', in B.L. Rhoads and C.E. Thorn (eds) *The Scientific Nature of Geomorphology*. Chichester: Wiley, pp. 415–41.

Phillips, J.D. (1999a) 'Divergence, convergence and self-organisation in landscapes', *Annals of the Association of American Geographers*, 89: 466–88.

Phillips, J.D. (1999b) *Earth Surface Systems: Complexity, Order and Scale*. Oxford: Blackwell.

Sauer, C.O. (1925) 'The morphology of landscape', *University of California Publications in Geography*, 2: 19–54 (reprinted in Leighly, J. (1963) *Land and Life*. Berkeley and Los Angeles, CA: University of California Press).

Scottish Natural Heritage (2001) *Report on the Proposal for a National Park in the Cairngorms*. Edinburgh: Scottish Natural Heritage.

Sherman, D.J. (1996) 'Fashion in geomorphology', in B.L. Rhoads and C.E. Thorn (eds) *The Scientific Nature of Geomorphology*. Chichester: Wiley, pp. 87–114.

Simandan, D. (2002) 'On what it takes to be a good geographer', *Area* 34, 3: 284–93.

Simpson, G.G. (1963) 'Historical science', in C.C. Albritton (ed.) *The Fabric of Geology*. Reading, MA: Addison-Wesley, pp. 24–8.

Slaymaker, O. and Spencer, T. (1998) *Physical Geography and Global Environmental Change*. Harlow: Longman.

Sparks, B.W. (1986) *Geomorphology* (3rd edn). London: Longman.

Stamp, L.D. (1961) 'Landscape', in *A Glossary of Geographical Terms*. London: Longman, pp. 287–8.

Stoddart, D.R. (1986) 'That Victorian science', in *On Geography*. Oxford: Blackwell, pp. 180–218.

Stoddart, D.R. (1987) 'To claim the high ground: geography for the end of the century', *Transactions, Institute of British Geographers*, 12: 327–36.

Stoddart, D.R. (1997) 'Carl Sauer: geomorphologist', in D.R. Stoddart (ed.) *Process and Form in Geomorphology*. London: Routledge, pp. 340–79.

Strahler, A.N. (1950) 'Davis' concepts of slope development viewed in the light of recent quantitative investigations', *Annals of the Association of American Geographers*, 40: 209–13.

Strahler, A.N. (1952) 'The dynamic basis of geomorphology', *Bulletin of the Geological Society of America*, 63: 923–38.

Sugden, D.E. (1970) 'Landforms of deglaciation in the Cairngorm Mountains, Scotland', *Transactions, Institute of British Geographers*, 201–19.

Summerfield, M.A. (1991) *Global Geomorphology*. Harlow: Longman.

Summerfield, M.A. (ed.) (2000) *Geomorphology and Global Tectonics*. Chichester: Wiley.

Thomas, M.F. (2001) 'Landscape sensitivity in time and space – an introduction', *Catena*, 42: 83–98.

Walton, K. (1968a) 'The unity of the physical environment', *Scottish Geographical Magazine*, 84: 5–14.

Walton, K. (1968b) 'The approach of the physical geographer to the countryside', *Scottish Geographical Magazine*, 84: 212–18.

Watson, A. (1996) 'Internationally important environmental features of the Cairngorms: research, and main research needs', *Botanical Journal of Scotland*, 48: 1–12.

Wooldridge, S.W. (1956) [1945] 'The geographer as scientist', in *The Geographer as Scientist: Essays on the Scope and Nature of Geography*. Edinburgh: Thomas Nelson, pp. 7–25.

# 16 Landscape and Environment: Natural Resources and Social Development

## Ian G. Simmons

### Definition

Humans have fundamentally altered the biological and physical components of the earth. Periods of human-induced change can be classified by their access to extra-somatic energy sources which are different in each of the hunter-gatherer, solar-powered agricultural, fossil-fuel industrial and 'postindustrial' phases. Crucial to these changes is the role of technology.

### INTRODUCTION: DIFFERENT TIMES, DIFFERENT OUTLOOKS

Definitions of the chapter's title words are plentiful and not necessarily agreed. 'Landscape' can be taken as being inevitably human: the result of the way in which we 'see' the world about us, evaluate its meaning and worth and then act. In some contrast, 'environment' is sometimes taken to denote all the physical and biological components of the planet which are not human. Already there is the difficulty of hybrid forms: a Border terrier is demonstrably not human but its very shape and existence as a breed are due to human actions during the nineteenth century. So really both terms include the biophysical and the social and any discussion under this chapter's heading must address itself to both the material world and the sphere of ideas.

Every human society of the last 10 000 years has left traces that include information about its relations with the non-human components of its world (Glacken, 1967; Simmons, 1996). Some are in the form of myths suitable for oral transmission, others are in the form of physical or biological remains, yet others are embedded in fragmentary but written records of quantities of grain or wood, and further examples are stored as scientific data in print or electronic media. Most traces derive from our use of the non-human materials of the earth as a resource or as yielding non-material benefits such as pleasure in scenic beauty or the sight of wild animals. Once a society has become agricultural in its subsistence pattern, its relationship with nature becomes uneven, for human societies then express the need to dominate their surroundings to create new genetic patterns in those species that become domesticated. On a larger spatial scale, societies create new ecosystems, sometimes deliberately as when a forest is cleared to make way for crops; at other times accidentally as when the cleared slopes shed water quickly and produce floods in the local rivers. Only in some hunter-gatherer societies is there any evidence for a kind of equality between humans and animals – for example, with people believing that they and the animals were of equal value and not to be dominated.

The ways in which humans have expressed this desire to exercise control over nature have been many and various: it can be seen in art of all periods from the cave paintings of Palaeolithic southern Europe to the transient 'sculptures' of Andy Goldsworthy, for example. Since the nineteenth century, a scientific paradigm[1] has been sought and this finds expression in the field of ecology: practitioners talk of 'man the ecological dominant', as they would describe the oak trees in a deciduous woodland where that species formed the canopy. Though many societies recorded in one form or another their transactions with, and alterations of, their environments, this was for centuries local knowledge. Not until the nineteenth century could global collections of information be made and the knowledge thus gained be made available, whether through cheaper printing technologies or easier travel to the new storehouses like the natural history museums in, say, Berlin or London. Compilers of worldwide knowledge emerged, such as Alexander von Humboldt (1769–1859), whose multi-volume work of 1845–62 is simply entitled *Kosmos*, or the American diplomat and scholar, George Perkins Marsh (1801–82), who collected many examples of human-created ecosystems and their consequences (see Glacken, 1967; Goudie and Viles, 1997). He began to formulate the kind of generalizations that underlie human impact studies today, including necessary admissions of ignorance.

Even though such travellers and scholars began to implant the idea that there was widespread evidence of human involvement in shaping the landscapes that were visible in any everyday sense, there was always a feeling that there were places and environments that were secure from human reach. The two obvious examples were the atmosphere and the oceans, and to them were often added great mountain ranges like the Himalayas and the Andes, and deserts such as the Sahara and the Atacama. Byron (1788–1824) could write of the oceans that human 'control stops with the shore' and the President of the Royal Society proclaimed in 1883 that 'nothing we do seriously affects

the numbers of the fish', when talking about the North and Norwegian Seas. Yet in 2000, cod fishing was closed in many of those
areas because of shrinking stocks. Since the pioneering work of
nineteenth-century scholars and writers, scientific effort and through
everyday experience, the impact of humans in almost all places has
been experienced, and now that upon the atmosphere is widely
perceived as one of the most threatening shapings that has been
brought about. That realization of the 'facts' of the changes has
been accompanied by considerable conceptual changes as well, which
are the foci here.

## THE DEPTH AND INTENSITY OF HUMAN-INDUCED CHANGE AND THEIR ACCOMPANYING CONCEPTS

Because some environmental changes are a source of anxiety, there
has been a lot of research into the history of such transformations,
especially where quantitative data on, for example, forest clearance or
desert expansion can be gathered or reconstructed. There are also
works of scholarship on the history of human ideas about nature and
environment (Glacken, 1967). The two are less often brought together
since there is a temptation – to be resisted if necessary – to link them
causally. Here, the major changes in the material relations between
humans and their surroundings are amplified by consideration of
some important conceptual notions that arise.

### Hunters and gatherers

Until about 8000 BC, hunter-gathers comprised 100% of the human
species and, indeed, some 90-plus per cent of our evolutionary time
has been spent as food collectors rather than food producers, as
foragers rather than farmers. The low population densities of such
societies, coupled with examples of 'environmental tenderness' in the
belief patterns of some groups, allowed the development of ideas
about hunter-gatherers which saw them as inhabitants of Eden or
children of nature, who lived only off the usufruct[2] of the earth and
who moved through ecosystems like corpuscles in a blood vessel, not
disturbing the equilibrium produced by organic evolution. Research
has shown, however, that the situation was often different. Examples
abound of hunting groups who have ruthlessly or carelessly exploited
their resource base and produced rapid and lasting environmental
change (Simmons, 1979; Roberts 1998). For every group who 'rested' a
deer or beaver-hunting area, there was one who simply killed all the
animals. In earlier millennia, hunters managed forests and savannas
with fire so as to maximize the numbers of their prey and so altered
the whole ecology of the region, since not all plants, for example, can

persist in the face of repeated burnings. So overall, the 1970s idea that we all needed to be more like them led to the revision of how any remaining societies should be treated by dominant political groups, but it was essentially irrelevant to the majority of us.

### Preindustrial agriculture

Between about 7000 BC and AD 1750 the numbers of hunters shrank markedly as they were replaced by agriculturalists (Simmons, 1996), some of whom farmed plant crops and others herded animals. The earth's surface was very noticeably affected by the production of food and fibre even before the coming of fossil fuels and machinery that are now almost ubiquitous. Preindustrial farming, however, is solar powered: it rests entirely on capturing the energy of the sun via photosynthesis. This does not mean that it is ineffective for, in many regions, surpluses can be produced, especially where the growing season can be lengthened or made more reliable by water management. Hence in addition to the make-over of the soils, vegetation cover and animal communities brought about by arable farmers and pastoralists, the surpluses allow non-food-producing uses of land and water. These uses are often commandeered by the politically powerful elements of the society. One example is the setting aside of areas as hunting grounds for an aristocracy for whom this sport was an essential part of their identity: examples abound from ancient Mesopotamia (c. 4000 BC) to medieval and early modern Britain. Another might be the creation of gardens which are usually combinations of pleasure together with the production of food, herbs and medicines.

The creation of surpluses allows the realization of immense projects provided that control of food and labour is present. The pyramids of Egypt could not have been built without the surpluses of wheat made possible by the control of the Nile; the lay brethren of the Cistercian abbeys of Yorkshire knew how to grow enough cereals and meat to feed the labourers who built Fountains and Rievalux. Food sold well beyond the farm gates fed the miners who dug out the salt that made the Archbishop of Salzburg rich enough to employ Mozart or who worked in the Thuringian coalmine in which J.S. Bach had a share. Surpluses need management and so writing was invented. Philosophy may bake no bread but the baking of bread allows philosophers to have time to think and write. Further, without rural surpluses, there could have been no cities.

### Industrialization

Once it was realized that coal was a concentrated source of energy that could be released in controlled conditions, the stage was set for an unprecedented set of leaps in environmental manipulation and, via increased food production, greater surpluses of energy. In effect, the exploitation of coal (followed by oil and natural gas) has provided immense subsidies to the gathering of solar energy and the capital to invent and construct machines which act as positive feedback loops for myriad activities.

Thus the non-essential activities described in the previous section have multiplied beyond most measures. The well-off of the world can eat meat and can use motors. A wide range of material possessions can be owned and discarded when outmoded. Environments can be reached which were formerly inaccessible: deep mines for minerals, remote places for trekking holidays or even sunny beaches at the end of a mere three hours in a jet aircraft. So a map of the earth as shaped by humans in say AD 1750 would look very different if redrawn for AD 2000. Not only has the human population grown immensely but also the range of manipulative activities has grown, as has the intensity of some of them. Beyond this, the actions of scientists have made it possible not merely to redirect the flows of energy and matter on the planet but (1) to escape it altogether as in space travel; (2) to introduce materials not present in nature at all, such as some chemical compounds found in pharmaceuticals and biocides; and (3) to intervene directly in the genetic composition of living organisms. All these activities have unexpected consequences in ecological and social terms.

Like agriculture, the non-material consequences have expanded as well. The intensification[3] of food production has meant that few people are now needed to grow the plants and tend the animals so that more individuals can be engaged in various types of manufacturing (itself shedding labour as automation increases) and providing services. The modern university would not be possible if its staff had to produce their own food, mine their own fuel and transcribe the books they ask their students to read. Social life becomes specialized ('For every man alone thinks he hath got/To be a phoenix, and that then can be/None of that kind, of which he is, but he', forecast John Donne in 1611) and it becomes apparently impossible to transform information into organized knowledge, let alone into wisdom.

### A post-industrial world

In 1945, the Russian biologist V.I. Vernadsky coined the term 'noosphere' for the web of human thought that would one day enmesh the world as did the atmosphere, lithosphere and biosphere (see Chapter 10 and Pepper, 1996). We might think that this has come about quite recently with the advent of instantaneous electronic communication, especially via the Internet. This technology is of itself not materially very substantial (compared with a rail or road system, for example) but its effects are great. In previous eras, capital formation tended to follow trade, i.e., the movement of materials and energy. Now a belt of digital data whizzes round the earth and spins off gobs of capital to finance a mine, a plane, a resort, an army, wherever there is confidence of an eventual return even though that repayment is itself more electronic digits. There is one thing in common with the previous stage: it creates considerable inequalities. About 20% of the world has given itself all kinds of choices under industrialism while the remaining 80% are unlikely to be able to access anything but the crumbs from the tables of those with the choice.

Environmentally, this is seen as the manipulation of poor places under pressures from outside. The growth of cash crops such as flowers in countries like Colombia for the markets of Europe causes social upheaval and environmental degradation. So do huge mines (see any map of Bougainville) or the receipt of toxic wastes not acceptable in the places where they were generated. Routed via the World Trade Organization (WTO), the developed world can generate enough cash to impose its manipulation anywhere. It may even have enough to prevent unpredictable climatic change due mostly to its consumption of fossil hydrocarbons, but don't bet on it short of a 'flip' from one climatic state to another rather than a smooth transition.

### Measures of outcomes

Since the 1960s there has been worldwide concern over the ways and extents to which the world has been changed materially by human actions. These have resulted in the formation of global bodies such as the UN Environmental Programme and the international negotiations surrounding climatic change, associated with the Rio Summit of 1992, for example (see UNEP, 1999, for a recent assessment). Many of these activities have been pushed along by measures of human impact on the earth's surface. Some are simple, like the increase in $CO_2$ concentration in the atmosphere, whereas others are complex and attempt to measure the amount of change, or the weight of the 'ecological footprint' (Chambers et al., 2000), or the extent to which lifestyle relies on unrenewable resources.

Two concepts materialize from these endeavours. The first is that they are possible. The heirs of Alexander Humboldt have access to billions of bits of data from many sources: official statistics, sample surveys, academic investigations, travellers' tales and satellite images do not exhaust the list. But we have the potential to picture the state of the planet in ways never before possible. The second is that the findings provoke concern: they do not fall into a total information void but often into policy-making. The response is rarely worldwide but often extends beyond national boundaries: satellite images of Amazonia have been instrumental in provoking a whole range of concerns from many countries beyond those of Brazil in particular. The web of Vernadsky's noosphere may be stronger than at first appears.

## EMERGING CONCEPTS WHICH LINK ACROSS TIME AND SPACE

The preceding material has a roughly chronological sequence; we now move to think about concepts which emerge from this succession which are not free of time and space but which link from one to the next. Since the shaping of the earth links both society and nature, ideas which encompass both will be emphasized, ending with the today's questions of whether and for whom and for what there is any significance in the ideas discussed.

### *The role of technology*

The use of technology to gain access to the materials of the environment seems to be as old as human societies, if not older (Smith and Marx, 1994). Its role in human affairs is debated for there is a school of thought that talks of society 'controlling technology for human welfare' as if it could simply be bought and controlled like a domestic vacuum cleaner. At its opposite we find those who believe in technological determinism, that society is made by its technologies and is simply a response – 'slave' might be the apposite word – to them. The many forms of technology from the control of fire in the Palaeolithic to the marvels of today's electronics do not need cataloguing here. What they have in common is high quantities of embedded energy. My computer does not need a lot of electricity to operate, but its manufacture, transport and packaging were all consumptive of energy. A room in the university devoted to computers may need 24-hour lighting and air-conditioning; a chemistry department's extractor fans may use as much as the rest of the institution put together. So the practices of society and the use of the earth are linked by energy access and, in particular, the flow of carbon tracks these connections.

The previous paragraph is replete with concepts. It seems likely, for instance, that Western societies are socially preprogrammed to accept technological innovation: there is an inbuilt acceptance that new technology is synonymous with progress and that resistance to it is simply reactionary. Material wealth can be measured not simply by GNP per capita but by the commercial energy consumption per capita, which acts as a surrogate for the environmental impact of that society's wealth. The measures usually exclude biomass energies such as fuelwood and dung (since their consumption is so difficult to measure) and so are to some extent incomplete. Yet since we can calculate energy in nature, from the volcanoes and earthquakes through to the heat loss from an anthill, *and* the human consumption of energy, there is a language which links both and which tells us something about each. Yet since it omits biomass in poorer countries, who is to say that it delimits the quality of life (a difficult concept in its own terms) rather than material throughput and waste production? But again, there seems to be a correlation between energy access and decreasing infant mortality. There is no doubt however that energy access is the facilitator of the Western view of the world and it should not surprise us when energy (e.g., as oil) is at the root of conflict of all kinds, from local blockades to regional wars (Klare, 2001).

### *Fragmentation and coalescence*

What we know of the thought-world of hunter-gatherers leads us to suggest that they often lived under the aegis of a unitary mythology[4] which linked people, the animals, the weather and the cosmos. Once agriculture became successful, a different world existed in which first stratification and then fragmentation have become commonplace. Viewed over a long period, we can see that social fragmentation into classes happened when the lord and lady withdrew from the common hall into the solar behind the dais, when the corridor allowed one class to sequester themselves, and when judicial death in public finally went behind the prison walls in most Western cultures. The fragmentation of the world seemed a natural thing in the late nineteenth century, when the atom was plotted, when painting became a series of points of light, moving pictures were conjured from many frames per second and even concentration camps (started in the Boer War and in Spanish Cuba) put a fence around undesired people. Examples from the twentieth century abound, with the personal stereo being the most obvious. Identity is bound up with individuality: ownership becomes a goal. In the social and political field, there is an accompanying tension between individual rights and collective responsibility.

At the same time, the world and its people are brought closer together. Overlapping networks of trade make it possible to imagine that a precious object from Wales could have been taken to Japan in about AD 400. Implosion of this kind followed the expansion of exploration and empire and was intensified by the steam ship and the railway. Printing allowed knowledge to be disseminated over wide areas and the current electronics boom can be seen as an intensi-fication of such a trend. Identity is bound up with a branch of McDonald's: having the same as 'them' is cool.

The alteration of nature has undergone parallel processes. Agri-culture segregates one class of land in particular but hunting grounds and gardens have ditches, fences and walls. Boundaries are set up to define resource-rich areas; nations go to war for more living space. Even large portions of the oceans are zoned to try to encourage the kind of ownership which does not deplete living resources. The apparently wild is also sequestered as national parks, nature reserves and the like, some of which are quite literally fenced. On the other hand, coalescence affects the non-human world. New technologies are sold or stolen abroad. Trade brings new economic species: the horse to the Americas, the tomato to Europe. It also brings unwanted immi-grants which may explode into new environments as have algae in the Mediterranean and American crayfish into English rivers. Perhaps most noticeable has been the spread of diseases, as in the Black Death or the post-First World War epidemic of influenza. Top of the list of current concerns is the way in which the whole globe will be affected by increased 'greenhouse gas' concentrations in the upper atmosphere, which are homogeneous and cannot be tied to any particular emitter: they cannot be owned. In a world of shifting species distributions and kaleidoscopic reshuffling of ecosystems, what does 'sustainability' mean, we might wonder.

The material outcome of the tensions between these two major processes is different at different spatial scales; in addition, the inter-action is so complex and so subject to contingency and synergy that prediction is virtually impossible. A GM crop may be tailored with great care in the laboratory but in the field is subject to the possibil-ities of organic evolution.

### Does it matter?

We all tend to be conscious of our personal space: in a crowded lift we stand stretched and the lecturer stays well out of spitting distance of the students, if only for the practical reason that most of them have colds. Writ large, it is easy to suffer from information overload and earth-concern fatigue: 'there's nothing I can do.' Indeed, to read the scientific semi-popular weeklies may be to live in a world of constant pain.

Put together, these observations add up to the neither new nor radical idea that the earth has meaning for many of us: after all, it is our home. It seems clear that human consciousness changed with the adoption of agriculture, for in both society and nature control became important especially when coupled with the means to make over at least part of the natural world in a cultural image. We ought also to think carefully about whether something similar happened in the late nineteenth century, though the case is perhaps less obvious. We cannot avoid confrontation with the concept that in our behaviour towards the environment, there is usually an 'ought': there are proper ways to treat the other components of the planet. That is, the question of ethics is involved (Elliot, 1991).

This debate about the rightness or wrongness of changes to the world can be carried on at a surface level of aesthetics, at the deeper level of the dignity of people who are malnourished or exploited to provide us with instant coffee, or at the rather fundamental plane of whether the conversions of the past and near future will produce unpredictable instabilities which will affect all humanity and quite likely a great deal of nature as well. There are few certainties in this debate: one is that some insects and rodents seem likely to survive almost everything; another seems to be the much more curious proposition that our species is destined to live forever rather than follow the usual evolutionary trajectory of a firework. And are we living in an interglacial? The idea of a human population of 10 billion pushed into the space left by, for example, the last Devensian glacial maximum, seems to lack attractiveness.

So we arrive at the question that has underlain every paragraph of this chapter like a watermark. Its key word, trailed above, is what Darwin described as 'that imperious little word, "ought" '. Ought we to be concerned at the loss of biodiversity in rift valley Africa, at the removal of forests in Indonesia, at the concreting of a valley in southern England, at the impact of earthquakes on settlements in El Salvador? If the answer is like that of Jim Royle,[5] 'I can't be arsed', then change course immediately: geography will give you a serious pain in the brain. If it is 'yes . . . but where can I possibly start?', take heart from the discussion of individuality above: there is no compulsion to conform. We can all get information from sources where the information has been subject to some kind of peer evaluation, we can all discuss it with friends and we can all act; consider the effect if every first-degree student in the English-speaking world drank only fair-traded coffee. Above all, remember coalescence. Our frontier as an individual is no longer a few centimetres from our nose: it is global and with it goes concern and responsibility.

## META-CONCEPTS?

Discussion always means that ideas get elaborated, often beyond recognition, into a morass of examples and detail. Let us end with some simplified versions of a few big ideas, conscious of the degree of generalization that is inherent in such words. First, we note that access to energy is a key to shaping the earth: first through fire and then via fossil-fuelled (and atomic) fire, channelled through technology, and producing atmospheric carbon as a significant waste product (Smil, 1994). Secondly, while technology is the means of change, one interpretation says that the end is control over the non-human. Thirdly, in the West at least, the growth of individualism confers the ability to work towards a diversity of changes in the evaluation and meanings of the non-human world. Lastly, the complexity of natural and social systems means that their interaction is never truly predictable: probabilities are the best that can be produced, though the laws of physics provide an outer envelope. The Spanish poet, Antonio Machado, writing in about 1913, can sum it up for us:

*Caminante, son tus hellas*          Traveller: your footsteps
    *el camino, y nada más;*              are the road – nothing else;

## Summary

- This discussion has encompassed the biological and physical components of the earth as well as the social and philosophical aspects of human thought and action.
- Access to energy is a key to shaping the earth, both materially and through knowledge; energy access permits the existence of the Western material economy and its thought patterns. Energy flow is a language that links ecology and social welfare.
- Surpluses from preindustrial agriculture allow a wider range of manipulations of land and water but they also allow the development of worlds of ideas by freeing some people from producing food and other necessities with their own hands. The city is a key location for the non-material as well as the tangible expression of this relationship.
- The noosphere may be low in material use but its impacts are not, and it is too soon to say much about longer-term outcomes of the implosion of communications in the developed world.
- It is a wholly recent idea that comparable data on the environment can be gathered for the whole world and that their use may globalize responses to their findings.
- We cannot be separate from the process of the planet and this means examining our behaviour. The complexity of natural and social systems means that their interaction is never truly predictable: probabilities are the best that can be produced, though the laws of physics provide an outer envelope. Shaping the earth has an element of 'right' behaviour, which means an understanding of the notions of environmental ethics.

## Further reading

For the current facts about the human interaction with the materials of the earth, the many publications of the World Resources Institute (Washington, DC) are good sources. There are yearly assessments of *The State of the World*, *World Resources* and specialist topic booklets. A list of the latest and the content of some of them is available on www.wri.org. The United Nations Environment Programme (e.g. 1999) publishes volumes of data periodically, and its millennium assessment is *Global Environmental Outlook 2000*; the UNEP contributions can be found on www.unep.org. Historical-empirical interpretations of the human–environment relationship are plentiful, and Simmons' (1996) *Changing the Face of the Earth* is no worse than most. Interpretations dominated by ideas are much rarer: students should take down and browse Glacken's (1967) *Traces on the Rhodian Shore* just to see what can be achieved by great scholars. Energy as a dominant mediator is dealt with best by Smil in several books, of which *Energy in World History* (1994) is most relevant here. An interesting collection on the theme of technology and society is Smith and Marx's (1994) *Does Technology Drive History?*. Measurements of our impress on the planet are the main theme of Chambers et al.'s (2000) *Sharing Nature's Interest. Ecological Footprints as an Indicator of Sustainability*. There is no simple guide to environmental ethics for those without a knowledge of Western philosophy but an entry point might be Elliot's essay, 'Environmental ethics', in Singer's (1991) *A Companion to Ethics*.

*Note*: Full details of the above can be found in the references list below.

### NOTES

1    A paradigm is where the thinking and practice occur within a single theoretical framework. In science, the theory assumes that the cosmos has regularities which are explicable in material terms.
2    The enjoyment of a product without destroying its source.
3    Intensification means more product per unit area per unit time.
4    A poetic, often fictionalized story which usually embodies a coherent view of the world and human actions within it.
5    A leading character in a British television comedy early in the twenty-first century, whose inertia was of international standard, though not up to that of his daughter.

## References

Chambers, N., Simmons, C. and Wackernagel, M. (2000) *Sharing Nature's Interest. Ecological Footprints as an Indicator of Sustainability*. London: Earthscan.

Elliot, R. (1991) 'Environmental ethics', in P. Singer (ed.) *A Companion to Ethics*. Oxford: Blackwell, pp 284–94.

Glacken, C.J. (1967) *Traces on the Rhodian Shore*. Berkeley, CA: University of California Press.

Goudie, A. and Viles, H. (1997) *The Earth Transformed: An Introduction to Human Impacts on the Environment*. Oxford: Blackwell.

Klare, T. (2001) *Resource Wars: The New Landscape of Global Conflict*. New York: Metropolitan Books.

Pepper, D. (1996) *Modern Environmentalism*. London: Routledge.

Roberts, N. (1998) *The Holocene: An Environmental History* (2nd edn). Oxford: Blackwell.

Simmons, I.G. (1979) *Biogeography: Natural and Cultural*. London: Edward Arnold.

Simmons, I.G. (1996) *Changing the Face of the Earth* (2nd edn). Oxford: Blackwell.

Smil, V. (1994) *Energy in World History*. Boulder, CO: Westview Press.

Smith, M.R. and Marx, L. (eds) (1994) *Does Technology Drive History?* London and Cambridge, MA: MIT Press.

United Nations Environment Programme (1999) *Global Environmental Outlook 2000*. London: Earthscan.

# 17 Landscape and Environment: Representing and Interpreting the World

**Karen M. Morin**

## Definition

In Anglophone cultural geography landscape tends to refer to a physical area visible from a particular location, as well as an ideological or social process that helps (re)produce or challenge existing social practices, lived relationships and social identities. Textual representations of landscapes, in paintings, film, advertising and numerous other media, are key to understanding the processes by which social practices and landscape are mutually constituted.

## INTRODUCTION

'Landscape' is a basic organizing concept in Anglophone cultural geography, but is equally foundational in fields as diverse as art, architecture, environmentalism, planning and in numerous of the earth sciences. This chapter focuses only on a select number of ways in which the term has been used throughout the twentieth century by Anglophone cultural geographers; that is, those in North America, Britain and in other English-speaking places (see Chapters 15 and 16 for other uses of the concept landscape). Thus what appears here is as much a *history* as a *geography* of landscape studies. And while it is important to keep in mind that landscape traditions differ across disciplines and places, even within Anglophone cultural geography the concept of landscape carries much ambiguity and complexity, with probably hundreds of nuances to the term. In addition, debates about landscape do not simply rest on what landscape *is*, but also on what landscape *does* – how it is produced and how it works in social practice. One fundamental aspect of this, as the title to this chapter implies, is that landscape always carries with it a set of 'representational practices'. These refer to how people see, interpret and represent the world around them *as* landscape, and how that represented landscape reflects and actually helps produce a set of lived relationships taking place 'on the ground'.

This chapter begins with an historiography of the landscape concept in Anglophone cultural geography. Next it demonstrates the usefulness of the landscape concept to studies of how social and cultural conflict (or consensus) arises and prevails. Finally the chapter examines the key debates that have taken place in geography over landscape and landscape representation, around what is loosely categorized as Marxism and feminism. Overall the point of the chapter is to demonstrate the mutually constitutive role of landscape, landscape representation and social practice.

At the outset it should be noted that the term 'landscape' has been (confusingly) conflated with numerous other geographical categories such as region, area, nature, place, scenery (particularly the rural countryside), topography or landform, and environment. Nonetheless, cultural geographers have tended to emphasize the visual aspects of the physical world when they employ the term landscape. Therefore, a useful way to begin thinking about landscape, as distinct from these other geographical categories, is as a portion of the earth visible by an observer from a particular position or location – and especially as that which can be taken in a single view. (Of course, the 'position' or 'location' of the viewer – both physical location and social location – are never unmediated, as will be discussed below.) This persistent connotation of landscape as a particularly visual form of spatial knowledge that can be taken 'in a single view' derives from sixteenth-century Dutch landscape painting, with its emphasis on scenery.

Thus landscape may be thought of in the first instance as a 'thing' – an area or the appearance of an area, and the particular ways component parts of that area have been arranged to produce that appearance. From this vantage point we can talk about 'agricultural landscapes', 'urban landscapes', 'landscapes of consumption', 'modern' and 'postmodern' landscapes, 'symbolic landscapes', 'ordinary landscapes', 'heritage landscapes' and so forth.

But let it also be noted at the outset that landscapes have both material and ideological aspects. Landscapes have physical, material form or 'morphologies' that are literally produced through labour and other lived relationships (Mitchell, 1996). But landscapes are also represented in various media (film, painting, advertising), and they themselves are representations of lived relationships. Especially over the past 20 years, Anglophone cultural geographers have come to recognize how important representational practices are to the production of landscapes and, hence, to social relations and social structure (Cosgrove, 1984; Duncan, 1990; Rose, 1993).

Importantly, then, landscape is not only a 'thing' but must also be considered an ideological or symbolic *process* that has the power actively to (re)produce relationships among people, and between people and their material world. In this sense landscapes carry symbolic or ideological meanings that reflect back and help produce social practices, lived relationships and social identities, and also become sites of claiming or contesting authority over an area. Thus social practices and landscapes mutually constitute one other in an ongoing fashion. Monumental landscapes, for example, often carry laudatory messages about war heroes and military conquest. A nexus of social actors (historical societies, town planners, veterans' groups) and practices produces such landscapes but the messages deployed by the monuments – a particular version of the past that celebrates masculinist values, for example – actively reproduce those values in the present and thus can shape social practice (such as by reproducing a culture of war). Of course, such values never go unchallenged and can be undermined in numerous ways. The Vietnam War Memorial in Washington, DC, became a site for contesting the celebration of war, for example, as it highlights the suffering and loss of war rather than triumphal conquest.

## HISTORICAL TRAJECTORIES

Landscape studies were introduced into American geography in the 1920s by Carl Sauer, especially with his 'The morphology of landscape' (1925 [1963]). Sauer, influenced by German geographers such as Otto Schluter and the *Landschaft* school, reacted against the environmental determinism of his day by arguing that it was collective human transformation of natural landscapes that produced what he called 'cultural landscapes' (see Chapter 15 for an analysis of the way Sauer's work has been used in physical geography). Sauer's phenomenal influence on geographers' study of landscape over five decades (in the USA at least) cannot be understated. While Sauer himself was more concerned with physical and biological processes set in motion by humans that produced, for example, agricultural practices and patterns, his more enduring influence was on a whole generation of cultural geographers associated with the 'Berkeley School', who used his empirical observation method to study the morphological features of landscapes as evidence of cultural difference. His followers tended to study, for example, cultural artifacts such as house types and barn types to trace cultural hearth areas and diffusion of culture groups (e.g. Lewis, 1975).

Mid-twentieth century landscape studies in geography were also greatly influenced by the English historian W.G. Hoskins, who argued for detailed studies of landscape history (1955), and the American geographer, J.B. Jackson, who studied popular culture through vernacular landscapes such as trailer parks in the American Southwest (1990). Jackson was founder of the popular *Landscape* magazine, published for 17 years beginning in 1951. In 1979 Donald Meinig edited a collection of works, *The Interpretation of Ordinary Landscapes*, written by some of the most quotable landscape geographers working at the time – himself and J.B. Jackson, Peirce Lewis, David Lowenthal, Marwyn Samuels, David Sopher and Yi-Fu Tuan. This collection demonstrated both the continued interest in 'ordinary', everyday landscapes in Anglophone cultural geography (such as churches and houses), as well as how landscapes reveal social and personal tastes, aspirations and ideologies. To Meinig, landscapes themselves could be read as collective social ideologies and processes: 'symbols of the values, governing ideas, and underlying philosophies of a culture' (1979: 6).

*The Interpretation of Ordinary Landscapes* also demonstrated the use of one of landscape studies' most enduring and ultimately contentious metaphors, that of 'reading' and interpreting landscapes as 'texts'. Just as a book (text) is made up of words and sentences arranged in a particular order with meanings that we read, so landscape has elements arranged in a particular order that we can translate into language, grasp meaning and 'read'. Interpreting architectural forms and their arrangement, for example, *as* symbolic interactions among humans and their environment (e.g. the height of skyscrapers as symbols of power, modernity, public protection, etc.) would in many ways structure landscape debates in the 1980s and 1990s.

By the last two decades of the twentieth century, the textual metaphor helped usher in a number of new questions related not just to what landscape is but how landscape mediates social relations. Informed by critical social theory, geographers first challenged the assumption of their predecessors that cultural groups 'collectively' produced landscapes and 'read' them in the same way. Instead they insisted on acknowledging the patterns and processes of hierarchical social organization responsible for the morphological features observed. Thus landscape studies began to be focused on the unequal power relations – social, political and cultural – involved in producing landscapes and (in turn) social difference, by both historical and contemporary actors. Denis Cosgrove, for example, in his *Social Formation and Symbolic Landscapes*, defined landscape as a 'way of seeing' associated with the rise of capitalist property relations (1984: 13). He argued that the landscape concept enabled an erasure of class difference via media such as landscape paintings of landowners and their country property.

The works of James Duncan (1990; 1992) have been instrumental in clarifying the extent to which landscapes contain different meanings to different viewers, and how they act as 'intertextual' media through which often competing interpretations, discourses and knowledges intersect. Other geographers, such as Don Mitchell (1996), argue that landscape studies have relied too much on visuality and advocate, among other things, a focus on that which has been hidden from view, such as the histories of labourers whose work literally produces landscape.

Finally, geographers began rejecting the basic opposition that had persisted for so long in landscape analysis – that between subject and object, the viewer and that which is viewed. Representation in earlier landscape studies had assumed that some unmediated, transparent reality could be detected in empirically observed landscapes. More recent studies have emphasized that there is an inherent inseparability of the represent-*er* and the represent-*ed*. Thus the worlds we represent, whether as geographers, corporate executives or graffiti artists, reflect our own positionalities, values, interests, motivations and backgrounds.

The attacks on the World Trade Center in New York City in September 2001 highlighted the vastly different meanings that that corporate landscape represented for observers at numerous scales and locations, both 'before' and 'after' the attacks: as emblem of technological ingenuity, modernity, progress, success of global capitalism, patriotism and nationalism, to more decentred understandings – US political and economic vulnerability, anti-capitalism, a 'holy war' being waged against the USA, 'just desserts' or a 'wake-up call' for unjust American foreign policy and hegemony, mourning and loss of loved ones and livelihoods in the New York area and so on. The fact that these various meanings and interpretations all coexisted simultaneously forced a recognition that not only could the same landscape carry vastly different meanings to different observers but that the landscape itself was also a reference to a much larger set of social relationships, domestic and international, that required attention and contextualization.

Feminist landscape critics have been instrumental in exposing the problems associated with the former dualistic thinking (i.e. that an unmediated 'transparent' reality existed between observer and observed). Geographers' own embeddedness in the process of landscape interpretation and analysis became central to late twentieth-century geographical studies, with feminists such as Gillian Rose (1993) challenging the masculinist gaze of much landscape geography. Such late twentieth-century advancements in the study of landscape representation warrant a more detailed analysis, which follows in the next two sections.

## THE POLITICS OF LANDSCAPE

Representation of landscapes can take many forms – narrative descriptions, drawings, paintings, maps, planning documents, engravings, photographs and films, among others. Trevor Barnes and James Duncan's edited collection, *Writing Worlds: Discourse, Text and Metaphor in the Representation of Landscape* (1992), examined numerous such forms of landscape representation. These authors asserted that landscape representation and interpretation required contextualization of author and audience, an outline of the rhetorics and tropes (figures of speech) employed to convey meanings, and an analysis of the processes by which readers become convinced that meanings conveyed are the 'natural' order of things in the world.

What cultural geographers might generally refer to as the 'politics of place' is central to an analysis of landscape. Anglophone cultural geographers of the 1980s and 1990s have emphasized that landscapes are social products, the consequence of how people, particularly dominant groups of people, create, represent and interpret landscapes based on their view of themselves in the world and their relationships with others. While authority lies with those who can 'produce landscapes as property' (in Don Mitchell's words) as well as control their representation, there is always room for contestation of that authority. In this sense, more recent landscape studies include a decidedly political component as they highlight the social and cultural conflicts and relationships, especially unequal power relations based on race/ ethnicity, class, gender and sexuality, that are involved in the creation, representation and interpretation of landscapes.

Landscapes of graffiti, for example, highlight both hegemonic and subversive representations and interpretations of landscape. Dominant or hegemonic readings of graffiti, for instance those articulated by a mayor's office or transportation authorities, might interpret graffiti as simply destruction of property, a 'crime' against the city. But graffiti has also been variously understood as a means for those with no other power to mark and stake out territory, or a means to challenge the existing social order by drawing visual attention to the situation of those marginalized 'others' in the city. Alternative readings of landscapes always exist, and landscapes can always be read in ways not intended. Tim Cresswell (1996), for example, shows that many graffiti makers think of themselves as creating art – an intention behind graffiti landscapes rarely acknowledged by more powerful voices.

Much of the current social theory work in landscape analysis examines the relationships among a triad of phenomena: social structure and ideologies (power, ideas and values); the creation of certain types of landscapes to reflect those ideologies; and the discourses or system of language and written works that are involved in the production, representation and interpretation of those landscapes. In this type of analysis each part of the triad may be referred to as a text – the culture/society and the visible landscape as signifying systems and thus metaphorical texts, and the written or spoken word about that landscape as actual text. Duncan (1990) refers to the transformation of ideas from one medium to another as the 'intertextual' nature of landscapes. Thus the context for any text is other texts. This way of thinking provides a frame for conceptualizing relationships among an array of phenomena – social structure; social practices, especially the exercise of different forms of power; the physical landscape; and landscape representation – which all work to produce and re-produce one another in an ongoing fashion.

As previously discussed, critical social theorists have tended to highlight the extent to which multiple interpretations, or layers of meaning, are embedded within landscapes and their representations (e.g. skyscrapers, graffiti). This is important because landscape as a site of struggle for challenging the dominant social order often rests on effective interpretations – by geographers or anyone else. Therefore it seems essential to recognize that meanings are not inherent in concrete objects or the physical world but that they are socially ascribed to objects and that they change over time, and with the particular perspectives and social positioning of the viewer. Thus not only is every landscape capable of multiple readings but also every landscape has been produced by multiple actors for whom no single intention can be inferred. As Duncan (1990: 12–13) explains:

> Descriptions are not mirror reflections; they are of necessity constructed within the limits of the language and the intellectual frameworks of those who describe . . . descriptions can only have meaning in such a context-bound sense . . . [and] in addition, not everything that has causal power can be observed or experienced.

This possibility that landscapes contain multiple layers of meaning has been challenged (see below). Suffice to say at this juncture that if the sense that the number of plausible readings of a landscape is limited only by the number of potential readers, one should be cautioned that not all landscape interpretations are equal; there are good ones and bad ones. 'Good ones' connect contextualized understandings of social relations and practices – of both the observer and the observed – with the physical morphology on the ground, demonstrating how they produce and reproduce one another.

One cogent model of how social ideologies and social practices worked to create, represent, and reproduce landscape is Duncan's (1992) study of the redevelopment of the Shaughnessy Heights neighbourhood in Vancouver, British Columbia. Duncan discusses a successful attempt in the 1980s by a small group of élite, mansion-owing families to return the neighbourhood into one originally envisioned by the Canadian Pacific Railway; one that was a genteel, picturesque reproduction of the English country house and garden (as opposed to one of multi-family units and 'slip-ins'). These property owners managed to appropriate a nexus of interests to their own advantage – the City Council and planning commission's commitment to the preservation of green space and historic buildings, as well as to neighbourhood self-determination. As Duncan also found out, surprisingly, Vancouver's working-class people, whose best interests would not seemingly be served by the preservation of such an élite landscape, nevertheless supported it. To them it represented a beautiful space in which all Vancouverans could take pride, meanwhile promising the possibility of upward mobility. Duncan effectively shows in this work how representation of landscape became a way not just of seeing the world but also of 'world making'.

## KEY DEBATES

As one might guess, one of the most significant developments in Anglophone landscape studies was the movement in the 1980s towards approaches advocated by the 'new' cultural geographers, a shift that began first in Europe under the influence of an emerging cultural studies paradigm. The shape this discussion took was not so much a conversation between different approaches as much as it was a one-sided rejection of the old school (such as Sauer's Berkeley school) by adherents of the new. Little resistance seemed to follow, although many geographers continue to study landscapes in the earlier tradition(s). A much more vociferous debate has ensued, though, from two other quarters within cultural geography. The first can be labelled the 'image versus reality' debate, and the second has been put forward by feminists dissatisfied with the seemingly intractable masculinism of much landscape studies.

The first conversation has at its foundation a difference of opinion as to whether or in what sense landscape contains some sort of reality beyond its representation. Much landscape work in recent years originating from a neo-Marxian standpoint (following, for example, Cosgrove, 1984) has highlighted questions about what exactly is the relationship between the concrete, physical, material world – the morphological aspect of landscape – and its representation. Geographers such as Don Mitchell worry that landscape studies that are concerned only with representations (e.g. Barnes and Duncan, 1992) seem to leave the 'real world' of landscape modes of production and reproduction behind as objects of study. In this way of thinking, meanings produced in and through language, texts, discourses, iconography and symbolism neglect the 'brute reality' of landscapes and thus represent a 'dangerous politics' (Mitchell, 1996: 27).

Materialist approaches such as Mitchell's emphasize that linguistic or representational expressions are important aspects of landscapes, but that landscapes are not fundamentally linguistic entities; that there is a world outside the linguistic that is experienced (if not 'seen') and that performs a different function from representations. In a debate published in *The Professional Geographer*, Judy Walton, Don Mitchell and Richard Peet argue the point, and to paraphrase Mitchell, the 'morphology of landscape, *no matter how it is represented*' has a role in social life (1996: 99, emphasis in original). Elsewhere he argues that if landscape 'is indeed a relation of power' there cannot be multiple interpretations of it, since that would defeat its ideological function – i.e. one that depends on the imposition of a dominant social order (1996: 27). Mitchell's *The Lie of the Land* (1996) makes an important point about the role of labour within the expanding capitalist economy in California's San Joaquin Valley. Mitchell uses this example to illustrate how relations of production are involved in shaping any landscape; in effect, that we must pay attention to how landscapes 'get made' in addition to how they are then re-presented as landscape. In this case of southern California, that representation is an aesthetic, pastoral depiction of thriving agriculture that is the product of (otherwise invisible, exploited) labour (1996: 16).

Part of the problem with the 'representation versus reality' debate is that it is based on binary, dualistic thinking that in the end is counterproductive. In *The Professional Geographer* debate, Walton (1996: 99) poses the question: 'Where is the pure materiality or physicality of an object (or landscape) beyond our interpretations of it?' In other words, to her there is always a cultural filtering process that brings reality to us through language. We only know landscapes, therefore, through our readings of them. Rather than set up a false dualism, it seems more productive to focus on the necessarily discursive constitution of the material world. Representations, then, are not reflective or distortive images of some real, pre-interpreted reality, but they themselves materially constitute reality. Thus 'reality' is indistinguishable from its representation and, in this sense, the much more important question is how representations are produced and contested.

That brings up a final and key debate surfacing in discussions of landscape representation in geography. Discussions of 'gendered landscapes' – everything from homes to downtowns to suburbs to shopping malls to workplaces to national monuments to natural environments – have drawn the attention of Anglophone feminist geographers since the 1970s, though this work has for the most part remained at the sidelines of landscape studies in geography. Much of the feminist work focuses on how landscapes construct, legitimate, reproduce and contest gendered and sexualized identities, or how women's relationships to landscapes (in both their representation and interpretation) differ from men's. Other work has more explicitly concerned itself with the politics of representation per se, especially in attempting to confront the relationship between observer and observed (see also Chapter 4 on some of these issues).

Several analyses of gender differences in the representation and interpretation of western American landscapes have appeared. Janice Monk and Vera Norwood's edited volume, *The Desert is No Lady* (1987), provided one of the first attempts at counteracting the masculine landscape tradition in geography. This volume and others of the period and genre drew on the foundational work of Annette Kolodny (1984), who discussed American expansionism in North America and the differences between men's relationships with the untamed wilderness as one of domination and conquest, versus women's tendency to relate to it in terms of an 'idealized domesticity', of locating a home surrounded by landscape features such as cultivated gardens and wildflowers.

Until *The Desert is No Lady* appeared, much of the work on landscape interpretation of the American West had emphasized the perspectives of nineteenth-century male settlers, industrialists, politicians, military men and railroad boosters. Such men viewed western landscapes either as the setting for the 'great male adventure story' or as a platform for large-scale mastery and subduing of the land and accumulation of wealth. Monk and Norwood showed that women did not necessarily share this masculine (and masculinist) vision of the southwestern desert landscape, and questioned the appropriateness of these images for women. Their collection demonstrates how Hispanic, Native American and Anglo women imaged the American southwest in a way that was both different from men's and also quite unlike each other's. Women writers, photographers and artists envisioned the desert land not in terms of its material resources to be exploited, a land awaiting metaphorical rape, but as a strong woman, unable to be conquered. The women artists' imagery turns out to be sexual (like men's), though not in terms of domination or suppression but in terms of affinity and connection, of uniting with the productive and reproductive energy of the earth.

Feminist landscape studies in the American West more recently have integrated an analysis of gender constructs with numerous others axes of social identity in their assessments of women's landscape representation. Jeanne Kay Guelke (1997), for instance, demonstrates how women involved in recreating a Mormon 'Zion' on the Utah frontier were deeply embedded, as faithful religious women, in economic development of the region. To Kay Guelke (1997: 362), religious constructs pre-empted associations of nature with the female body, which had prevailed as the most common of landscape metaphors: land as Great Mother, enticing temptress, dangerous or uncontrollable hag or fury, and so on. Thus many of these Mormon women perceived themselves as willing and active participants in the subduing and conquering of nature, and transforming wild landscapes into productive agricultural ones. Karen Morin's (1998) study of British Victorian women travellers' representations of some of the same western American landscapes shows that attention to mode of transportation, type of engagement with the land, domestic and imperial social relations and Romantic literary conventions, all converged to produce largely negative representations of the region during their late nineteenth-century travels.

One feminist critique that has found its way into more mainstream human geography is Gillian Rose's (1993) study of geography's traditions in fieldwork and landscape analysis. Informed by a larger feminist corpus which highlights situated and partial knowledges and the positionality of the researcher (or observer) of objects, people and landscapes, Rose argues that geographical representation through fieldwork and landscape analysis reveals deeply embedded masculinist cultural values and knowledge.

To Rose, geography's traditions involve a masculinist way of seeing landscape that is not just one of a relation of mastery or domination, but one of (white, bourgeois, heterosexual) pleasure in looking at landscape that has been constructed as feminine. Part of her commentary revolves around the same painting that Cosgrove deconstructed in his (1984) study, Thomas Gainsborough's 'Mr and Mrs Robert Andrews' (c. 1748), which codified a particular way of seeing the land that helped naturalize and celebrate capitalist property and the rights of owners (see Figure 17.1). Rose, however, rightly claims that Cosgrove's interpretation misses the different relationships that men and women had to the surrounding landscape; the painting reminds us that only men were landowners, and women's role was principally reproductive (1993: 92–3). In this and other landscape paintings, women appear passive or prostrate, as commodities of the male gaze. Not only do such landscape images themselves associate women with a feminized landscape but, as Rose points out, geographers reinforce sexism and masculinism by their inattentiveness to

**FIGURE 17.1**   Thomas Gainsborough: 'Mr and Mrs Robert Andrews' (*c.* 1748)

gender roles and relations in landscape representation. As most critical theorists would agree, these are not 'innocent', detached representations but they refract and reinforce lived gender roles and relations.

Other feminist geographers have suggested possibilities for other types of homo-erotic and female heterosexual gazes on the landscape. For example, Catherine Nash (1996) examines Diane Baylis's photograph 'Abroad' as a representation of the male body as aesthetic nature. Still, Rose's larger observation holds; geographers have not generally 'problematized' themselves as authors of landscape representation or interpretation. For all their success in carefully contextualizing landscape representation, Barnes and Duncan (1992), for example, allow the geographer himself to remain unmarked and disembodied. To Rose (1993: 112), a 'feminine' resistance to such hegemonic ways of seeing is necessary. Such resistance promises to:

> 'dissolv[e] the illusion of an unmarked, unitary, distanced, masculine spectator, [while] permit[ting] the expression of different ways of seeing among women . . . Strategies of position, scale and fragmentation are all important for challenging the particular structure of the gaze in the discipline of geography'. (Rose, 1993: 112)

While much feminist work has demonstrated the mutual constitution of gendered landscapes and women's gendered identities (e.g. Morin, 1998), a recent turn in landscape studies has directed attention to relationships among men, masculinity and landscape. Rachel Woodward (2000), for example, examines the processes by which

**FIGURE 17.2**   Military masculinity and landscape

military masculinity and the landscape of Britain's rural countryside are mutually constituted. Woodward examines five sources of information – Army recruitment materials, general publicity, basic training information and videos, mass-market paperbacks about military adventures and television documentaries on military life. She shows how essential a particular construction of rurality itself is to the construction of 'warrior hero' – it is dangerous, rough and hazardous. The rural countryside in the Army documentation is not that of idyllic community and nature in harmony, but is rather a harsh, threatening landscape against which the new recruit is pitted, and out of which his requisite physical and mental attributes will arise through its 'conquest' (see Figure 17.2). Thus this representation of the rural serves the dual purpose of articulating and legitimating one hegemonic type of military masculinity, as well as constructing the rural itself as a legitimate place to bear arms.

## CONCLUSION

Students of landscape today might consider themselves fortunate to enjoy the fruits of their precursors' work. 'Traditional' landscape geographers' attention to ordinary and 'everyday' landscapes, especially in attention to their morphological aspects, has ultimately and justifiably endured. The materialist-poststructuralist debate in landscape studies that raged in the 1990s seems less worrisome today. Materialists seem better attuned to the importance of landscape representation in the construction of 'reality', and those who have been most interested in linguistic or discursive analysis of landscape representation also seem more engaged today with what is 'on the ground'. Theirs has been a productive debate. Focus on less visually orientated modes of representation has proven worthwhile (e.g. Mitchell, 1996), as has the entire corpus of work that sees contests over representational practices key to challenging the existing social order (e.g. Duncan, 1992; Cresswell, 1996). The situation on the feminist front is encouraging, but much work remains to be done. Attention to gendered landscapes within which various masculinities are produced seems a fruitful direction, as does attention to the myriad ways in which landscape helps produce and mediate national, ethnic and sexual difference.

## Summary

- Landscapes have both material and ideological aspects; they have physical, material form, are represented in various media and are themselves representations of lived relationships.
- Landscapes carry symbolic or ideological meanings that reflect back and help produce social practices, lived relationships and social identities, and also become sites of claiming or contesting authority over an area. Social practices and landscapes mutually constitute one other in an ongoing fashion.
- While early twentieth-century Anglophone landscape studies focused more on morphological features and cultural difference read through them, later studies argued that landscapes are not 'collectively' produced by culture groups but rather act as intertextual media through which competing authority, interpretations, discourses and knowledges intersect.
- Acknowledgement of the physical and social location of the observer of landscape is paramount. That which is viewed is inseparable from the viewer.
- Understanding the power of landscape to challenge or subvert the existing social order has been of primary concern to many cultural geographers.
- Two key critiques appear in geography's landscape studies: Marxian and feminist. The first takes a materialist orientation to argue that many landscape geographers focus too heavily on landscape representation to the expense of morphology – a false dichotomy that seems itself to ignore the fact that representations materially constitute 'reality'. The second, more enduring, challenge focuses on the masculinism of landscape studies. Much headway has been made in this area but much also remains to be accomplished.

## Further reading

Sauer's (1963 [1925]) paper, 'The morphology of landscape', is a foundational statement in American cultural geography. It argues for an empirical observation method to study the morphological features of the landscape as evidence of cultural difference. Subsequently, Marxist, poststructuralist and feminist approaches to geography have led to the emergence of a diverse range of ways of understanding landscape. One of the first statements in Anglophone cultural geography that brought a Marxist sensitivity to artistic representations of landscape is Cosgrove's (1984) *Social Formation and Symbolic Landscapes*. In this book he understands landscape as a 'way of seeing' associated with the rise of capitalist property relations. Duncan's (1990) *The City as Text* helped usher in poststructural linguistic theory into geography's landscape studies by analysing the creation of the urban landscape of the precolonial Kandyan kingdom in Sri Lanka. In this book, attention is paid to layers of landscape signification, rhetorical devices, power relations and intertextuality. Rose's (1993) *Feminism and Geography* challenges the masculinist foundation of geography's history and geographical knowledge, including a critique of the 'masculinist gaze' embedded in landscape studies. Some of the tensions between these different ways of viewing landscape are evident in a special issue of the journal *The Professional Geographer* (1996) in which Judy Walton, Don Mitchell and Richard Peet debate the tensions between materialist and poststructuralist interpretations of landscapes.

*Note*: Full details of the above can be found in the references list below.

## References

Barnes, T. and Duncan, J. (1992) *Writing Worlds: Discourse, Text and Metaphor in the Representation of Landscape*. London: Routledge.

Cosgrove, D. (1984) *Social Formation and Symbolic Landscapes*. London: Croom Helm.

Cresswell, T. (1996) *In Place/Out of Place: Geography, Ideology, and Transgression*. Minneapolis, MN: University of Minnesota Press.

Duncan, J. (1990) *The City as Text: The Politics of Landscape Interpretation in the Kandyan Kingdom*. Cambridge: Cambridge University Press.

Duncan, J. (1992) 'Élite landscapes as cultural (re)productions: the case of Shaughnessy Heights', in K. Anderson and F. Gale (eds) *Inventing Places*. Melbourne: Addison Wesley Longman, pp. 53–69.

Hoskins, W.G. (1955) *The Making of the English Landscape*. London: Hodder & Stoughton.

Jackson, J.B. (1990) 'The house in the vernacular landscape', in M. Conzen (ed.) *The Making of the American Landscape*. Boston, MA: Unwin Hyman, pp. 355–9.

Kay Guelke, J. (1997) 'Sweet surrender, but what's the gender? Nature and the body in the writings of nineteenth-century Mormon women', in J.P. Jones et al. (eds) *Thresholds in Feminist Geography: Difference, Methodology, and Representation*. Lanham, MD: Rowman & Littlefield, pp. 361–82.

Kolodny, A. (1984) *The Land before Her: Fantasy and Experience of the American Frontier, 1630–1860*. Chapel Hill, NC: University of North Carolina Press.

Lewis, P. (1975) 'Common houses, cultural spoor', *Landscape*, 19: 1–22.

Meinig, D.W. (ed) (1979) *The Interpretation of Ordinary Landscapes: Geographical Essays*. New York, NY: Oxford University Press.

Mitchell, D. (1996) *The Lie of the Land: Migrant Workers and the California Landscape*. Minneapolis, MN: University of Minnesota Press.

Monk, J. and Norwood, V. (eds) (1987) *The Desert is No Lady: Southwestern Landscapes in Women's Writing and Art*. New Haven, CT: Yale University Press.

Morin, K.M. (1998) 'Trains through the plains: the Great Plains landscape of Victorian women travelers', *Great Plains Quarterly*, 18: 235–56.

Nash, C. (1996) 'Reclaiming vision: looking at landscape and the body', *Gender, Place and Culture*, 3: 149–69.

*The Professional Geographer* (1996) Special issue on landscape, 48: 94–100.

Rose, G. (1993) *Feminism and Geography: The Limits of Geographical Knowledge*. Minneapolis, MN: University of Minnesota Press.

Sauer, C. (1963) 'The morphology of landscape [1925], in J. Leighly (ed.) *Land and Life: A Selection of the Writings of Carl Ortwin Sauer*. Berkeley, CA: University of California Press, pp. 315–50.

Woodward, R. (2000) 'Warrior heroes and little green men: soldiers, military training, and the construction of rural masculinities', *Rural Sociology*, 65: 640–57.

# Index